Advancements in Renewable Energy and Green Hydrogen

Maria Simona Raboaca
National Research and Development Institute for Cryogenic and Isotopic Technologies, Râmnicu Vâlcea, Romania

Saheb Djohra
Renewable Energy Development Centre, Algeria

Omar Moussaoui
University Mohammed First, Oujda, Morocco

Traian Candin Mihaltan
Technical University of Cluj, Romania

Mustapha Koussa
Renewable Energy Development Centre, Algeria

IGI Global
PUBLISHER of TIMELY KNOWLEDGE

A volume in the Advances in Chemical and Materials Engineering (ACME) Book Series

Published in the United States of America by
IGI Global
Engineering Science Reference (an imprint of IGI Global)
701 E. Chocolate Avenue
Hershey PA, USA 17033
Tel: 717-533-8845
Fax: 717-533-8661
E-mail: cust@igi-global.com
Web site: http://www.igi-global.com

Library of Congress Cataloging-in-Publication Data

Names: Raboaca, Maria, 1981- editor. | Saheb, Djohra, 1964- editor. | Moussaoui, Omar, 1978- editor. | Mihaltan, Traian, 1983- editor. | Koussa, Mustapha, 1959- editor.
Title: Advancements in renewable energy and green hydrogen / edited by: Maria Raboaca, Djohra Saheb, Omar Moussaoui, Traian Mihaltan, Mustapha Koussa.
Description: Hershey PA : Engineering Science Reference, [2024] | Includes bibliographical references. | Summary: "By establishing a decentralized and unchangeable ledger, it enables efficient and secure recording of transactions, certification of green hydrogen sources, and tracking of carbon emissions. Through blockchain-based platforms, the trading and certification process is streamlined, guaranteeing the authenticity and sustainability of green hydrogen"-- Provided by publisher.
Identifiers: LCCN 2023048653 (print) | LCCN 2023048654 (ebook) | ISBN 9798369310144 (hardcover) | ISBN 9798369310151 (ebook)
Subjects: LCSH: Hydrogen as fuel--Technological innovations. | Clean energy--Technological innovations.
Classification: LCC TP359.H8 I59 2024 (print) | LCC TP359.H8 (ebook) | DDC 665.8/1--dc23/eng/20240404
LC record available at https://lccn.loc.gov/2023048653
LC ebook record available at https://lccn.loc.gov/2023048654

This book is published in the IGI Global book series Advances in Chemical and Materials Engineering (ACME) (ISSN: 2327-5448; eISSN: 2327-5456)

British Cataloguing in Publication Data
A Cataloguing in Publication record for this book is available from the British Library.

All work contributed to this book is new, previously-unpublished material. The views expressed in this book are those of the authors, but not necessarily of the publisher.

For electronic access to this publication, please contact: eresources@igi-global.com.

Advances in Chemical and Materials Engineering (ACME) Book Series

J. Paulo Davim
University of Aveiro, Portugal

ISSN:2327-5448
EISSN:2327-5456

Mission

The cross disciplinary approach of chemical and materials engineering is rapidly growing as it applies to the study of educational, scientific and industrial research activities by solving complex chemical problems using computational techniques and statistical methods.

The **Advances in Chemical and Materials Engineering (ACME) Book Series** provides research on the recent advances throughout computational and statistical methods of analysis and modeling. This series brings together collaboration between chemists, engineers, statisticians, and computer scientists and offers a wealth of knowledge and useful tools to academics, practitioners, and professionals through high quality publications.

Coverage

- Thermo-Chemical Treatments
- Computational Methods
- Composites
- Polymers Engineering
- Heat Treatments
- Industrial Chemistry
- Nanomaterials
- Toxicity of Materials
- Materials to Renewable Energies
- Biomaterials

IGI Global is currently accepting manuscripts for publication within this series. To submit a proposal for a volume in this series, please contact our Acquisition Editors at Acquisitions@igi-global.com or visit: http://www.igi-global.com/publish/.

The Advances in Chemical and Materials Engineering (ACME) Book Series (ISSN 2327-5448) is published by IGI Global, 701 E. Chocolate Avenue, Hershey, PA 17033-1240, USA, www.igi-global.com. This series is composed of titles available for purchase individually; each title is edited to be contextually exclusive from any other title within the series. For pricing and ordering information please visit http://www.igi-global.com/book-series/advances-chemical-materials-engineering/73687. Postmaster: Send all address changes to above address.

Titles in this Series

For a list of additional titles in this series, please visit: http://www.igi-global.com/book-series/advances-chemical-materials-engineering/73687

FlexPDE and Finite Element Method Applications in Thermal Energy Storage and Cavities
Khashayar Hosseinzadeh (Sharif University of Technology, Iran) Mehdi Mahboobtosi (Babol Noshirvani University of Technology, Iran) and Davood Domairry Ganji (Babol Noshirvani University of Technology, Iran)
Engineering Science Reference • copyright 2024 • 380pp • H/C (ISBN: 9798369309322) • US $225.00 (our price)

Next Generation Materials for Sustainable Engineering
Shrikaant Kulkarni (Padmashri Dr. V.B. Kolte College of Engineering, India) Boris Kharisov (Autonomous University of Nuevo Leon, Mexico) A.K. Haghi (University of Coimbra, Portugal) and Vipul Srivastava (Lovely Professional University, India)
Engineering Science Reference • copyright 2024 • 442pp • H/C (ISBN: 9798369313060) • US $295.00 (our price)

Advancements in Powder Metallurgy Processing, Applications, and Properties
Shashanka Rajendrachari (Bartin University, Turkey) and Rayappa Shrinivas Mahale (Jain College of Engineering and Research, India)
Engineering Science Reference • copyright 2024 • 313pp • H/C (ISBN: 9781668493854) • US $275.00 (our price)

Sustainable Approach to Protective Nanocoatings
Rajni Garg (Galgotias College of Engineering and Technology, India) and Ansar Anjum (Galgotias College of Engineering and Technology, India)
Engineering Science Reference • copyright 2024 • 434pp • H/C (ISBN: 9798369331361) • US $345.00 (our price)

Innovations and Applications of Hybrid Nanomaterials
Virat Khanna (Maharaja Agrasen University, India) Prianka Sharma (Maharaja Agrasen University, India) and Priyanka Mahajan (Maharaja Agrasen University, India)
Engineering Science Reference • copyright 2024 • 336pp • H/C (ISBN: 9798369332689) • US $315.00 (our price)

Smart and Sustainable Applications of Nanocomposites
Rajni Garg (Galgotias College of Engineering and Technology, India) and Ansar Anjum (Galgotias College of Engineering and Technology, India)
Engineering Science Reference • copyright 2024 • 391pp • H/C (ISBN: 9798369310946) • US $300.00 (our price)

Emerging Applications of Plasma Science in Allied Technologies
Aamir Shahzad (Government College University, Faisalabad, Pakistan) and Maogang He (Xi'an Jiaotong University, China)

IGI Global
PUBLISHER of TIMELY KNOWLEDGE

701 East Chocolate Avenue, Hershey, PA 17033, USA
Tel: 717-533-8845 x100 • Fax: 717-533-8661
E-Mail: cust@igi-global.com • www.igi-global.com

Table of Contents

Detailed Table of Contents

Given the increasing global recognition of climate change as a pressing concern, the rising costs associated with petroleum products, and the declining expenses linked to renewable energy power systems, it is becoming more likely that renewable energy systems will have a substantial role in addressing electricity generation. The efficacy of the smart grid concept in the implementation of electric power networks has been demonstrated. The present work investigates the incorporation of renewable energy sources within smart grid systems. The introduction presents a comprehensive outline of the integration of distributed generation and renewable energy inside smart grid systems. This chapter provides an in-depth analysis of smart grids, focusing on their conceptual framework and delving into the advantages and disadvantages associated with the integration of renewable energy technologies inside smart grid systems. It is crucial to comprehend the pricing dynamics linked to renewable energy through an examination of the unique characteristics inherent in renewable energy sources.

With the rise of grid connected independent power producers (IPP), performance analysis of energy generation plants becomes an essential task. This chapter presents design and development of a performance analysis tool using MS Excel for typical IPP with a grid connected solar PV plant. The tool utilizes historical data to determine various energy-metrics, econometrics, and financial metrics. Additionally, the tool can determine various key parameters such as the energy sales to the utility, energy that cannot be sold, and the load that cannot be served. The tool will have the capability of capturing data, doing calculations, and visualizing the data and output from calculations through plots and graphs.

The PVF-Chart method consists of a combination of correlations and fundamental expressions for the hourly calculations of solar radiation at a given location. It uses long-term monthly average solar radiation and ambient temperature to predict the annual performance of a photovoltaic array. The chapter aims to analyse the long-term performances of a PV system used to cover an individual household situated in different Algerian climatic zones. The obtained results are compared to those presented in the literature and from which it can be concluded that the developed software program permits to calculate the PV system performances with a good exactitude which confirm its uses as a tool to analyse the long-term performance of PV system at any climate where only the monthly mean daily weather data is available. From these later, the hourly, the daily, and the annual performance of a photovoltaic array are predicted. For the preliminary assessment of the energy-producing of the studied PV project, the LCOE was calculated for three interest rates and two lifetimes.

Hydrogen can be produced in a clean, emission-free manner using solar technologies. Solar energy technology, specifically photovoltaic energy paired with PEM (proton exchange membrane), is required to complete the electrochemical water splitting process. The actual challenge of integrating renewable energy sources into PEM is developing the best possible set of operating parameters. Hydrogen generation flow in this study depends on a number of variables. Using two distinct solar energy methods with equivalent energy production, the authors discovered that the flow of hydrogen production may be altered. The best results were recorded in Illizi (Afra), where the annual power consumption was measured at 896.65 kWh and the flow rate was calculated to be around 4.59 L/h. The global annual loss under this setup is rather small at 10.6%.

Hydrogen (H2) is an energy carrier that offers both some of the benefits of fossil fuels and is clean with a low carbon footprint. Production of H2 from the water electrolysis using renewable energy is called green hydrogen and is the preferred pathway for future H2 production. In order to decarbonize the energy

Innovative technologies including geothermal desalination, anaerobic digestion, microbial fuel cells, and internet of things integration may be able to aid with the combined challenges of wastewater treatment and energy sustainability. Anaerobic digestion converts organic waste into biogas, a renewable energy source that may be used to produce heat and electricity. Microbial fuel cells (MFCs) use the metabolic processes of microbes to convert organic materials into electricity, allowing for the simultaneous treatment of wastewater and the generation of renewable energy. Geothermal desalination reduces the need for fossil fuels and uses heat from geothermal sources to turn brackish or salty water into freshwater, which is less harmful to the environment. IoT connection allows geothermal desalination systems and bioenergy from wastewater processes to collect, monitor, control, and optimise data in real-time.

This chapter examines electric vehicles (EVs) and their charging systems, focusing on contemporary infrastructure effectiveness. It explores hybrid vehicle architectures and recharging technologies, categorizing them into traditional and autonomous chargers. The following research examines photovoltaic and wireless charging systems, electronic components, and a simple management cycle for uninterrupted operation. This chapter proposes that extensive modelling and simulation studies utilising the MATLAB/Simulink programme are conducted to evaluate the energetic performance of an electric vehicle fitted with hybrid recharge equipment.

The clean energy curve is anticipated to avoid flattening with the help of investment stimulation, COVID-19 alleviation conditions, innovation, and infrastructure opportunities in low-carbon and digital technology. Over the coming years, the global renewable energy picture may ultimately resume its long-term trend towards the production of green electricity. The experience of the COVID-19 pandemic period may persuade countries to speed up their attempts to implement renewable energy policies before another global economic shock is caused by a different illness or an unanticipated weather calamity. Overall, there is optimism that the development of renewable energy will proceed.

Preface

In recent years, the world has witnessed a pivotal shift towards sustainable energy sources and the urgent need to mitigate climate change. As nations grapple with the challenges posed by fossil fuel dependency and environmental degradation, the quest for renewable energy solutions has become paramount. Among these solutions, green hydrogen stands out as a promising avenue towards a cleaner, more sustainable future.

Advancements in Renewable Energy and Green Hydrogen, crafted by the collaborative efforts of Maria Simona Raboaca, Saheb Djohra, Omar Moussaoui, Traian Candin Mihaltan, and Mustapha Koussa, emerges as a comprehensive guide for researchers, policymakers, industry professionals, and students navigating the complex landscape of renewable energy and hydrogen technology.

Green Hydrogen Production: A Sustainable Energy Carrier

Rooted in renewable energy sources such as solar, wind, hydro, and biomass, green hydrogen production emerges as a beacon of sustainable and clean energy. This handbook elucidates the intricate processes involved in electrolysis, where electricity generated from these sources drives the production of green hydrogen, offering a viable alternative to traditional fossil fuel-dependent methods.

Storage, Transport, and Safety Considerations

Understanding the nuances of green hydrogen storage and transport is paramount. Whether stored as compressed gas, liquefied at extremely low temperatures, or chemically converted materials like ammonia or methanol, this handbook navigates through the varied storage methods. The complexities of transporting green hydrogen, including safety measures, specialized vehicles, and alternative approaches like ammonia conversion, are thoroughly explored, emphasizing the significance of safety in handling this high-flammability energy carrier.

Standards, Security Regulations, and Power-to-Gas Concepts

Ensuring safe and reliable practices in green hydrogen production involves adherence to standards and security regulations. This handbook delineates key areas covered by these standards, encompassing production, quality, storage, handling, transport, distribution, and international harmonization. Additionally, the concept of Power-to-Gas-to-Power (P2G2P) is introduced, offering insights into the conversion of electrical energy into hydrogen, its storage, and subsequent reconversion into electricity—a pivotal concept in renewable energy integration and grid balancing.

Applications Across Industries and Power-to-X Technology

The versatility of green hydrogen finds applications across diverse industrial sectors. From serving as a clean fuel and feedstock in high-emission industries to fueling fuel cells for power generation, the applications are vast. This handbook sheds light on how green hydrogen can revolutionize industries traditionally reliant on fossil fuels, offering cleaner alternatives for high-temperature processes and serving as an energy storage solution for grid balancing. The Power-to-X concept, specifically Power-to-Hydrogen (PtH), is explored, emphasizing its potential role in creating a sustainable and low-carbon economy.

Digital Technologies Shaping Green Hydrogen

Digital technologies are at the forefront of revolutionizing the green hydrogen sector. This handbook delves into the role of digital platforms, blockchain technology, and the Internet of Things (IoT) in ensuring transparency, traceability, and efficiency in green hydrogen transactions. From real-time monitoring and management to the certification of green hydrogen sources, these technologies play a transformative role in shaping the future of the green hydrogen market.

A Handbook for Researchers, Students, and Industry Professionals

For anyone with concerns and scientific studies in renewable energy, solar energy resources, wind turbines, electrolyzers, fuel cells, hydrogen, real-time data acquisition, blockchain, IoT, Power-to-Gas-to-Power, and smart power-to-X, this handbook emerges as a comprehensive guide. Students, researchers, and industry professionals alike will find a wealth of knowledge within its pages, offering a roadmap to navigate the intricate landscape of digitalization and green hydrogen production.

As editors, we extend our gratitude to the contributors who have poured their expertise into this handbook. We envision this compilation as a valuable resource fostering understanding, innovation, and sustainable practices in the ever-evolving domain of green hydrogen production.

ORGANIZATION OF THE BOOK

As the editors of this comprehensive reference book, we are delighted to present a diverse collection of chapters that delve into the multifaceted realm of renewable energy, particularly in the context of smart grids, solar energy, green hydrogen, and sustainable practices. These chapters, contributed by esteemed authors and experts in their respective fields, offer a thorough exploration of opportunities, challenges, and advancements in the renewable energy landscape. From the integration of renewable sources in smart grids to the potential of green hydrogen in various industries, each chapter provides valuable insights and analyses. Let us provide you with a brief overview of each chapter.

Chapter 1: "Potential Sources of Renewable Energy in Smart Grids: Opportunities and Challenges" by Gamze Nalcaci, Nurcan Yarar, Gerhard Weber

This opening chapter lays the foundation for the book by addressing the global concerns of climate change and rising petroleum costs. The authors explore the role of renewable energy systems in electricity generation and investigate their integration into smart grid systems. The chapter offers an in-depth analysis of smart grids, highlighting their conceptual framework and examining the advantages and disadvantages associated with incorporating renewable energy technologies.

Chapter 2: "Performance Metrics Analysis Tool for Solar Energy Plants" by Unomasa Kasenina, KS Sastry Musti

Focusing on grid-connected Independent Power Producers (IPP), this chapter introduces a performance analysis tool developed using MS Excel for solar PV plants. The tool utilizes historical data to determine various energy, econometric, and financial metrics, providing a valuable resource for assessing the performance of solar energy plants. The authors emphasize the tool's capability to capture, calculate, and visualize critical parameters related to energy generation.

Chapter 3: "Long-Term Analysis of the Usability and Performance of Photovoltaic Systems" by Assia Koussa, Sabrina Boudra, Djohra Saheb, Mustapha Koussa, Abdelhak Razagui, Houdaifa Ounis

The authors present the PVF-Chart method, a unique approach combining correlations and fundamental expressions for hourly calculations of solar radiation. This method is employed to analyze the long-term performance of a photovoltaic system covering an individual household in different Algerian climatic zones. The chapter highlights the accuracy of the developed software program in predicting PV system performance, including the calculation of the Levelized Cost of Energy (LCOE).

Chapter 4: "Effect of Solar Energy Technology on Green Hydrogen Production" by Ilyès Nouicer, Bessem Kaghouche, Chaouki Ghenai, Meziane Fares, Kabouche Nourdine

This chapter explores the clean production of hydrogen using solar technologies, specifically through the electrochemical water splitting process. The authors investigate the challenges of integrating renewable energy sources into Proton Exchange Membrane (PEM) technology and provide insights into altering hydrogen production flow using different solar energy methods. The study identifies optimal conditions for hydrogen production, emphasizing the potential of solar energy in advancing green hydrogen technologies.

Chapter 5: "Green Hydrogen: A Clean Energy Solution for Electricity and Heat" by Ayşenur Öztürk Aydın, Gamze Bozkurt, Elif Daş

Examining hydrogen as a clean energy carrier, this chapter focuses on green hydrogen produced through water electrolysis using renewable energy. The authors explore various applications of hydrogen in power generation, such as fuel cells, gas turbines, and combined heat and power systems. Additionally, the chapter discusses the integration of heat pumps with fuel cells and the resource utilization improvement achieved by mixing hydrogen with natural gas in existing heating systems.

Chapter 6: "Energy Efficient Grid Interactive Building Optimization With Green Energy for Net-Zero Emission" by Dharmbir Prasad, Rudra Singh, Gaurav Gupta, Manish Kumar, Jatin Anand, Ranadip Roy

Addressing the increasing energy demands and the need for sustainability, this chapter explores grid-interactive building infrastructure. The authors present a case study in Kashinagar, Odisha, India, proposing a system that integrates solar and wind energy to reduce operating costs and achieve a net-zero emission outcome. Beyond economic and environmental benefits, the chapter highlights the potential for independent and resilient energy systems through the adoption of renewable sources.

Chapter 7: "Usage of Green Hydrogen in Steel Industry" by Prasad G

This chapter discusses the crucial role of hydrogen in the steel industry and its potential to reduce India's reliance on coal, natural gas, and petroleum. The author emphasizes the benefits of transitioning to hydrogen in terms of reducing emissions and promoting hydrogen adoption through government incentives. The chapter addresses the challenges faced by the steel sector in adopting hydrogen and calls for policy support to facilitate the shift towards emission-free steel production.

Chapter 8: "Safe Green Hydrogen: Production and Storage as Emerging Source of Energy for Sustainable Future" by Shuaibu Hassan, Sanjeev Kumar

Exploring the need for environmentally friendly and sustainable energy solutions, this chapter introduces green hydrogen as a promising option. The authors discuss the challenges posed by conventional energy sources and highlight the potential of green hydrogen in mitigating environmental impacts. The chapter sets the stage for further research into the production and storage of green hydrogen as a key contributor to a sustainable future.

Chapter 9: "Integrated Approach for Green Hydrogen Production and Sustainable Electrification of Railway Platform With Solar Energy Sources" by Dharmbir Prasad, Rudra Pratap Singh, Jatin Anand, Ranadip Roy, Sudhangshu Chakraborty

This chapter provides a detailed analysis of the feasibility of producing green hydrogen and electrifying railway platforms using solar energy and pumped hydro storage backup. The authors examine the potential benefits, including the reduction of greenhouse gas emissions, meeting climate objectives, and enhancing air quality. The study emphasizes the solar potential of the Hatia Railway Station area and its ability to generate sustainable energy through advanced technology.

Chapter 10: "Unveiling the Potential and Hurdles of Blockchain in Green Hydrogen Management: The Role of Blockchain in Green Hydrogen Supply Chain Management – A Literature Review" by Lynda Ait Bachir

Focusing on the role of blockchain in green hydrogen supply chain management, this chapter reviews existing literature to explore the potential of blockchain technology. The author highlights blockchain's intrinsic properties, including decentralization, immutability, privacy, and transparency, and discusses how these properties can revolutionize the green hydrogen supply chain. The chapter underscores the efficiency and sustainability improvements brought about by blockchain in the management of green hydrogen.

Chapter 11: "Utilizing Waste Water for Sustainable Energy Generation via IoT-Integrated Technologies" by Sabyasachi Pramanik

This chapter explores innovative technologies, including geothermal desalination, anaerobic digestion, microbial fuel cells, and Internet of Things (IoT) integration, to address wastewater treatment and energy sustainability challenges. The authors discuss how these technologies can convert organic waste into biogas, generate renewable energy, and desalinate water using geothermal heat. The integration of IoT allows real-time data monitoring, control, and optimization of these processes.

Chapter 12: "Intelligent Charging Methods for Electric Vehicles" by Aicha Degla

Examining electric vehicles and their charging systems, this chapter delves into contemporary infrastructure effectiveness. The author categorizes charging technologies into traditional and autonomous chargers and explores photovoltaic and wireless charging systems. The chapter proposes extensive modeling and simulation studies using MATLAB/Simulink to evaluate the energetic performance of electric vehicles equipped with hybrid recharge equipment.

Chapter 13: "Energy Pattern During Unforeseen Spells of COVID-19" by Ashok Asthana, Swati Srivastava

Concluding the book, this chapter discusses the impact of the COVID-19 pandemic on the clean energy sector. The authors anticipate the global renewable energy trend to continue, emphasizing the importance of investment stimulation, COVID-19 alleviation conditions, innovation, and infrastructure opportunities in low-carbon and digital technology. The chapter expresses optimism that the development of renewable energy will persist in the post-pandemic era.

In closing, each chapter in this edited reference book contributes valuable knowledge to the renewable energy landscape, offering a comprehensive exploration of opportunities, challenges, and innovations. We extend our gratitude to the contributing authors for their dedication and expertise, making this reference book a valuable resource for researchers, practitioners, and enthusiasts in the field of renewable energy.

CONCLUSION

As we draw the final curtain on the comprehensive exploration within the *Advancements in Renewable Energy and Green Hydrogen*, it is with great satisfaction and anticipation of the transformative impact this handbook holds in the realms of renewable energy, sustainability, and technological innovation.

The collaborative efforts of Maria Simona Raboaca, Saheb Djohra, Omar Moussaoui, Traian Candin Mihaltan, and Mustapha Koussa have birthed a compendium that not only dissects the intricate intersection of digitalization and green hydrogen production but also serves as a guiding beacon for researchers, students, and industry professionals navigating this dynamic landscape.

Navigating the Digital Frontier of Green Hydrogen

The handbook embarks on a journey through the digital frontier of green hydrogen, unraveling the symbiotic relationship between technology and sustainable energy. From digital platforms ensuring traceability and certification to the integration of green hydrogen into hybrid energy systems, each chapter contributes to a deeper understanding of how digitalization catalyzes and optimizes transactions critical to the green hydrogen market.

Sustainable Energy Carrier and Storage Innovations

The significance of green hydrogen as a sustainable energy carrier is underscored, highlighting its roots in renewable energy sources and its potential to revolutionize traditional industries. The nuanced exploration of storage methods, including compressed gas, liquefaction, and chemical conversion, lays the groundwork for sustainable and efficient solutions in the storage and transportation of green hydrogen.

Safety, Standards, and Future Concepts

Ensuring the safe production, storage, and transport of green hydrogen is given due diligence, emphasizing the importance of adherence to safety measures, standards, and security regulations. Concepts like Power-to-Gas-to-Power (P2G2P) and Power-to-Hydrogen (PtH) offer glimpses into the future of renewable energy integration and a low-carbon economy, presenting both opportunities and challenges on the path to a sustainable energy transition.

Digital Technologies: Catalysts for Change

In the digital age, technology is at the forefront of shaping the green hydrogen sector. The handbook explores the role of digital platforms, blockchain, and IoT, showcasing how these technologies revolutionize transactions, certification processes, and decision-making in the green hydrogen market. The promise of real-time monitoring, transparency, and traceability is a testament to the transformative power of digitalization.

A Handbook for the Future

For students delving into the complexities of renewable energy, researchers exploring the frontiers of technology, and industry professionals seeking sustainable solutions, this handbook serves as a compass. Its pages are imbued with insights, knowledge, and a vision for a future where green hydrogen production seamlessly coalesces to shape a cleaner, more sustainable energy landscape.

As editors, we extend our heartfelt appreciation to the authors who have contributed their expertise, creating a compendium that encapsulates the spirit of innovation and sustainability. We invite readers to embrace the knowledge within these pages, fostering a collective commitment to a greener and more technologically advanced future.

Maria Simona Raboaca
National Research and Development Institute for Cryogenic and Isotopic Technologies, Râmnicu Vâlcea, Romania

Saheb Djohra
Renewable Energy Development Centre, Algeria

Omar Moussaoui
University Mohammed First, Oujda, Morocco

Traian Candin Mihaltan
Technical University of Cluj, Romania

Mustapha Koussa
Renewable Energy Development Centre, Algeria

Chapter 1
Potential Sources of Renewable Energy in Smart Grids:
Opportunities and Challenges

Gamze Nalcaci
Necmettin Erbakan University, Turkey

Nurcan Yarar
Necmettin Erbakan University, Turkey

Gerhard Wilhelm Weber
Poznań University of Technology, Poland

ABSTRACT

Given the increasing global recognition of climate change as a pressing concern, the rising costs associated with petroleum products, and the declining expenses linked to renewable energy power systems, it is becoming more likely that renewable energy systems will have a substantial role in addressing electricity generation. The efficacy of the smart grid concept in the implementation of electric power networks has been demonstrated. The present work investigates the incorporation of renewable energy sources within smart grid systems. The introduction presents a comprehensive outline of the integration of distributed generation and renewable energy inside smart grid systems. This chapter provides an in-depth analysis of smart grids, focusing on their conceptual framework and delving into the advantages and disadvantages associated with the integration of renewable energy technologies inside smart grid systems. It is crucial to comprehend the pricing dynamics linked to renewable energy through an examination of the unique characteristics inherent in renewable energy sources.

1. INTRODUCTION

Biomass, hydropower, geothermal, solar, wind, and marine energy are examples of renewable energy sources (RES). RES are primary, domestic, and clean or limitless energy types (Dinçer, 2001; Bilgen,

DOI: 10.4018/979-8-3693-1014-4.ch001

2004). According to the International Renewable Energy Agency (IRENA), global renewable generation capacity reached 3,064 gigawatts at the end of 2021 (https://www.irena.org). In 2021, the renewable generation capacity will have expanded by 257 GW (+9.1%). Asia added sixty percent of the new capacity in 2021, for a total of 1.46 terawatts (TW) of renewable power by 2021. China was the largest provider of the continent's new capacity, contributing 121 GW. Europe and North America, led by the United States, placed second and third, with the former contributing 39 GW and the latter 38 GW. In Africa, renewable energy capacity increased by 3.9%, and Central America and the Caribbean by 3.3%. Despite consistent growth, the pace in both regions is significantly slower than the world average, emphasizing the need for deeper international cooperation to optimize power markets and stimulate substantial investment in these regions. Table 1 outlines the most important renewable energy sources and their potential applications.

To achieve sustainable development, it is imperative to possess methods and techniques that enable measuring and comparing the environmental impacts associated with diverse human activities (Farhad, 2008). The utilization of fossil fuels has experienced a significant rise due to improvements in living standards, industrialization in developing nations, and the growth of the world population. Due to the considerable consumption of fossil fuels, there is a notable acceleration in the loss of fossil fuel reserves, which substantially impacts the environment. Consequently, it increases health risks and significantly threatens global climatic stability. Environmental reforms are increasingly gaining political acceptance in wealthier nations. To mitigate the release of greenhouse gases, there is a progressive shift towards adopting sustainable production methods, minimizing waste, reducing air pollution caused by vehicles, and promoting using renewable energy sources for decentralized energy generation (Sims, 2003).

The growing concern surrounding the energy dilemma has sparked a renewed focus on advocating for renewable alternatives to fulfill the increasing energy requirements of developing nations (Youm,

Table 1. Principal renewable energy sources and their application

Energy Source	Energy Conversion and Usage Options
Hydropower	Hydropower power plants
Biomass energy	• Space heating, • Hot water heating, • Electricity generation
Geothermal energy	• Electrical energy generation, • Heating/cooling applications such as central heating, central cooling, greenhouse heating, etc., • The delivery of heat for industrial activities such as drying, • Production of chemical compounds and minerals, including carbon dioxide, fertilizer, lithium, heavy water, and hydrogen. • Use in thermal tourism for thermal purposes • Aquaculture at low temperatures (up to 30 ° C), • Utilized in application and evaluation sectors, such as mineral-containing drinking water production.
Solar energy	• Solar electricity, • Solar water heating, • Solar heating, • Solar ventilation, • Solar lighting
Wind energy	• Electrical energy production • Pumping water • Renewable hydrogen
Ocean energy	Electrical energy production

(Panwar, 2011)

2000; Horst, 2009). This concern arises from the escalating usage of fossil fuels to satisfy our current energy demands. The overconsumption of fossil fuels has led to global warming due to increased carbon dioxide emissions. Consequently, it is imperative to prioritize the widespread adoption and promotion of renewable and environmentally friendly energy sources (Hall, 1991). To effectively monitor the emissions of greenhouse gases, an agreement was developed with the overarching goal of preventing pollution, which aligns with the objectives outlined in the Kyoto Protocol agreement (Holm-Nielsen, 2009). This study aims to assess the potential of renewable energy devices in meeting energy demands and their potential for mitigating carbon dioxide emissions.

The world's future significantly relies on using renewable energy sources (RES). Three distinct categories of energy resources exist: fossil fuels, renewable resources, and nuclear resources (Demirbaş, 2000). A renewable energy source refers to any form of energy capable of being harnessed and utilized again without being depleted. Examples of such sources include solar power, wind power, biomass power, geothermal power, and various other forms of renewable energy (Rathore, 2007; Zakhidov, 2008). Energy services can be provided without the emission of air pollutants or greenhouse gases by utilizing renewable energy sources to fulfill residential energy requirements. The advancement of renewable energy systems presents an opportunity to effectively tackle contemporary challenges, including enhancing the reliability of energy and water provisions, elevating the quality of life and employment opportunities for local communities, and safeguarding the enduring viability of remote desert and mountainous areas.

Furthermore, it will enable us to fulfill our nations' responsibilities outlined in the Kyoto Protocol. The implementation and management of renewable energy initiatives have the potential to mitigate rural-urban migration by generating employment opportunities in non-urban regions (Bergmann, 2008). One potential solution for meeting the energy demands of rural and small-scale areas in a reliable, cost-effective, and ecologically conscious way is the decentralized harvesting of renewable energy (Reddy, 1980; Ravindranath, 1995).

2. POTENTIAL RENEWABLE ENERGY SOURCES

In this study, potential renewable energy sources are gathered into six main groups, and these sources can be listed as follows: hydropower, biomass, wind energy, solar energy, and geothermal energy as shown in Figure 1. In this figure, hydropower takes 40% place of all RES, solar energy 28%, wind energy 27%, biomass 5%, and geothermal and ocean energy can be listed, respectively.

2.1 Hydropower

Hydropower is an environmentally favorable, renewable, and environmentally benign energy resource. As seen in Figure 1, it is observed that this particular source is responsible for generating around 78% of the global renewable electricity and contributes to 16% of the overall electricity production (Kuriqi, 2019). Hydropower and climate change exhibit a reciprocal relationship characterized by a two-way interaction. Figure 2 illustrates the significant role of hydropower, a prominent renewable energy source, in effectively mitigating greenhouse gas (GHG) emissions and combating the issue of global warming (Kuriqi, 2021). On the contrary, hydropower contributes to alterations in land utilization, resulting in elevated levels of CO2 emissions and expediting the progression of climate change. For example, the study conducted by (Alsaleh, 2021; Berga, 2008; Berga 2016) investigated the role of hydropower in both

Figure 1. Renewable energy sources distribution till 2022

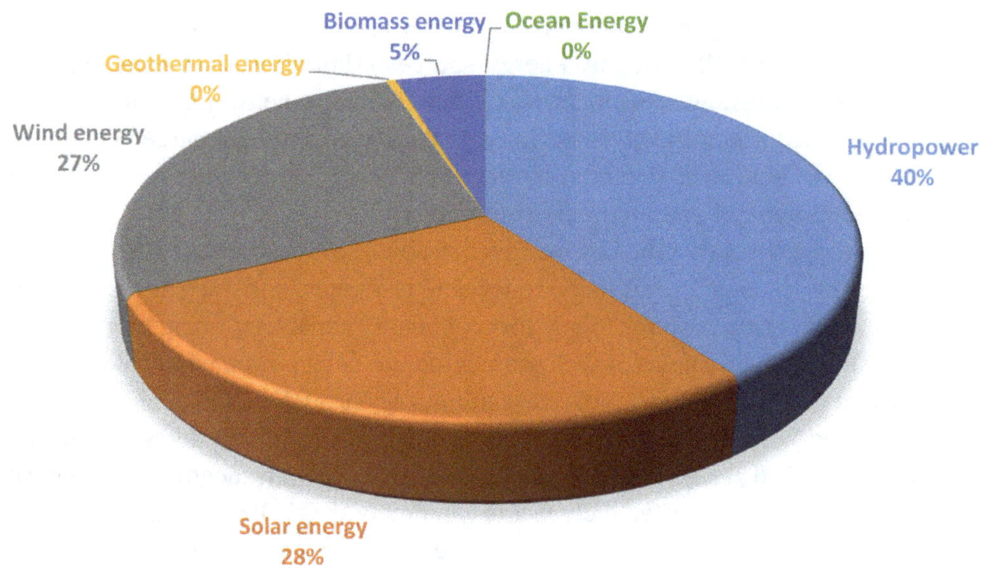

adaptation and mitigation strategies, suggesting that hydropower plays a significant role in mitigating the effects of global warming and climate change. A previous investigation by Alsaleh, (2022) examined the technical aspects of hydroelectric dams and their associated environmental consequences. The study highlighted the significance of cost-effective measures in mitigating the impacts of hydropower on safety, the economy, and the environment. Previous research (Alsaleh, 2021; Lehner, 2005) has examined the influence of global change on the hydropower potential of European nations. It has been determined that global warming and climate change can negatively affect the operational efficiency of around 25 percent of European hydroelectric facilities (Prakasam, 2020). Environmental flow and mitigation in hydropower projects were assessed using a geographic information system. The analysis reveals hydroelectric energy as a cost-effective and environmentally sustainable energy source. This finding aligns with a substantial body of prior research, exemplified by the studies referenced (Jiang, 2018; Prakash, 2012; Li, 2017; Guangliang, 2019).

The hydropower depicted in Figure 2 is derived from the harnessing of water to generate mechanical energy through the operation of turbines. Two primary categories of hydroelectric power plants exist those that incorporate dams and reservoirs and those that do not. Reservoirs associated with large hydroelectric dams can effectively store water for varying durations, whether short-term or long-term, to address periods of high electricity demand. Furthermore, the facilities can be separated into smaller dams, which can be utilized for various purposes, day or night use, seasonal storage, or pumped-storage reversible plants that serve the dual function of pumping and electricity production. Hydropower is generated on a smaller scale without dams and reservoirs, often through in-river facilities that maintain the river's natural flow. As a result of this, a significant number of individuals perceive small-scale hydroelectric power as a more environmentally favorable alternative. Figure 3 displays the installed capacity's temporal progression for renewable and blended hydropower sources. Figure 3 illustrates a consistent and progressive increase in the global capacity of hydropower over time.

Figure 2. Hydropower plant block diagram

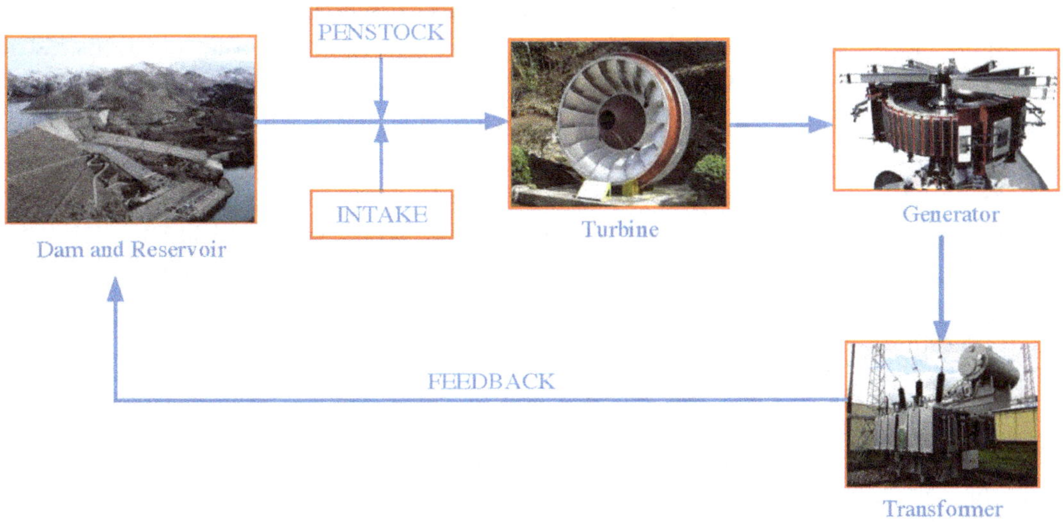

Figure 3. Installed capacity trends of hydropower (https://www.irena.org)

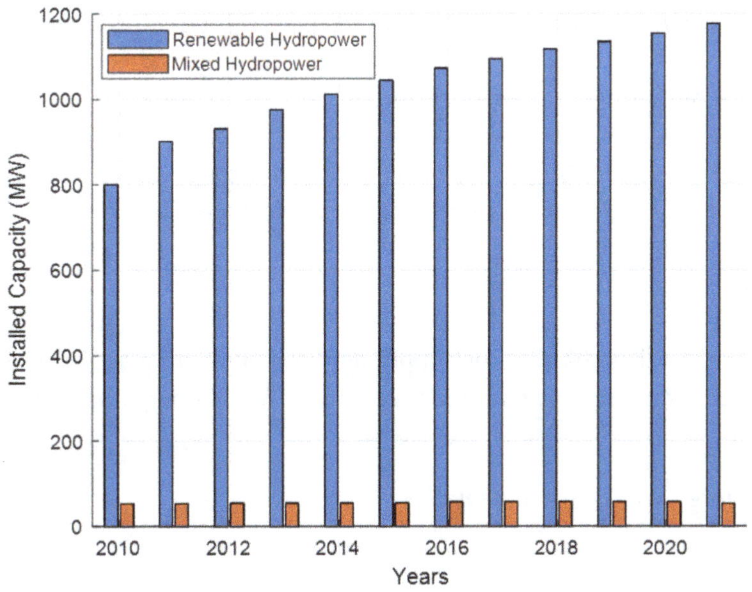

2.2 Biomass Energy

Biomass comprises various fundamental elements, predominantly sourced from agricultural, forest, and marine industries. The most prevalent elements in biomass are carbon, hydrogen, oxygen, and nitrogen. Biomass also consists of many components, such as cellulose, hemicelluloses, lignin, extractives, lipids, fats, proteins, simple sugars, starches, water, hydrocarbons, and other substances (Duca, 2022).

The features of vegetable raw materials and their elements, such as wood, branches, bark, shells, leaves, straws, and others, vary due to their varied types and origins (Toscana, 2013).

Using biomass leftovers and wastes as energy sources presents challenges due to various restrictions, such as material heterogeneity, high moisture content, poor biological stability, and low energy density (Toscana, 2015; Chia, 2020). Furthermore, it is essential to acknowledge that despite the abundance of potential, there exist notable challenges regarding the overall sustainability of biofuels and the composition of the biomass utilized in their production, which gives rise to various ethical and social considerations (Toscano, 2015; Oh, 2013). One approach to mitigate the conflict between food and non-food resources is the usage of lignocellulosic biomass, as depicted in Figure 4. This biomass includes agro-industrial wastes and agricultural residues (Aravani, 2022; Mai-Moulon,2021; Toscano, 2018). European production data from 2006 to 2015 indicates that these sources accounted for around 442 Mt/year (Prasad, 2020; Clarcq, 2017; Ilari, 2020).

Conventional biomass sources may prove inadequate in fulfilling forthcoming energy demands and sustainability criteria. There has been a suggestion that there is a need to make use of the broader range of agricultural wastes, by-products, and residues originating from the agro-industrial and agroforest sectors (Chia, 2020). However, the limited size and energy densities of this particular biomass material have implications for the expenses associated with harvesting and logistics and, to some extent, constrain its energy and environmental sustainability (Lautala, 2018, Duca, 2022). The potential impacts of physical features and chemical composition variations on the logistics, maintenance, and combustion efficiency of a biomass power plant could constrain its energy generation capacity and compromise its environmental sustainability. To attain its objectives, the bioenergy industry depicted in Figure 5 must enhance the efficiency of its supply chain (Toscano, 2022). One of the projects is the reconsideration of the structure of the supply chain, along with the improvement of its sustainability and logistics, as well as the enhancement of the quality of biofuel. It is achieved through the use of techniques such as the utilization of powder metallurgical agricultural residues (Manwell, 2010).

2.3 Wind Energy

Wind energy is a very suitable solution for supplying the increasing global energy demand due to its characteristics as a clean, renewable, and limitless resource. Wind energy has grown substantially glob-

Figure 4. Biomass energy production scheme

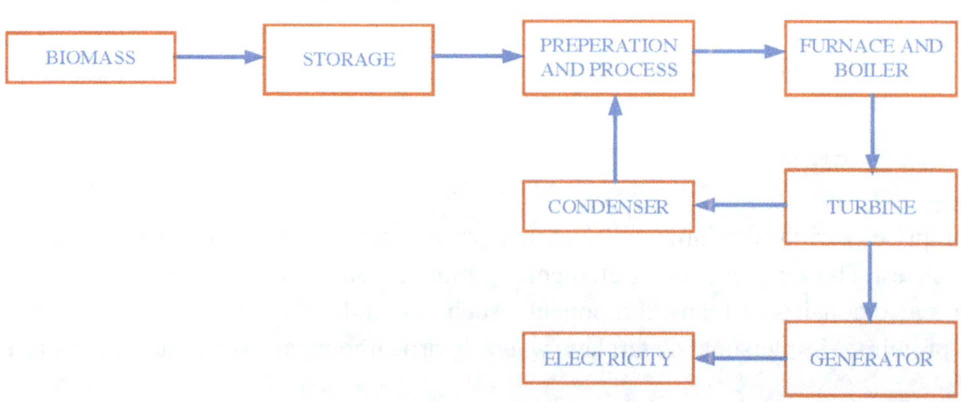

Figure 5. Installed capacity trends of biomass power (https://www.irena.org/)

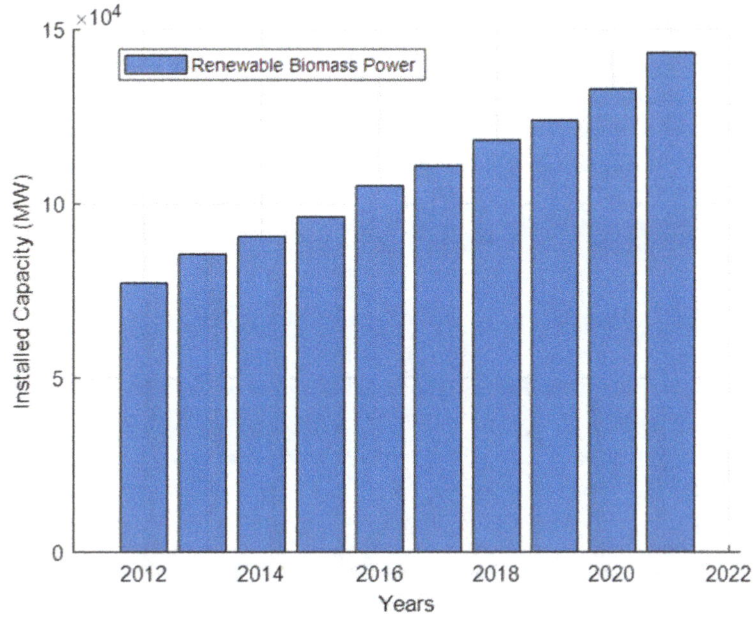

ally, particularly in the last twenty years. As depicted in Figure 7, the cumulative installed capacity of the wind industry had experienced significant growth, increasing from 1.29 GW in 1995 to 370 GW by 2015. The anticipated degradation of this scenario in the following years can be attributed to environmental considerations and the ongoing pursuit of energy stability, as seen by the expansion of installed capacity and the increasing size of wind turbines (WTs). Over time, these devices have transformed from rudimentary designs to intricate, multi-megawatt production units organized into expansive arrays, sometimes called "wind farms." The control systems play a crucial role in ensuring wind turbines' safe and effective operation, as depicted in Figure 6, due to their intricate nature (Lubosny, 2007).

Compared to alternative energy sources, wind power lacks controllability, giving rise to this phenomenon. The wind flux is a highly volatile phenomenon with temporal and spatial variability. Due to their inherent unpredictability, wind turbines (WTs) encounter challenges in an irregular and variable supply, diverse mechanical loads, and non-linear dynamics, all hindering efficient energy conversion. Due to

Figure 6. Wind energy block diagram

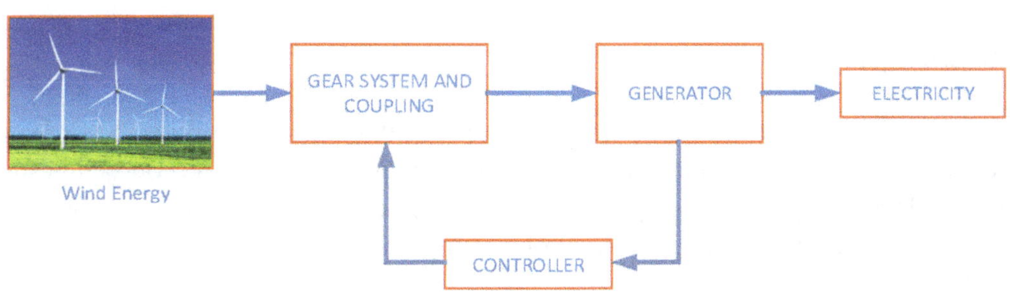

Figure 7. Global wind energy production (https://irena.org/wind)

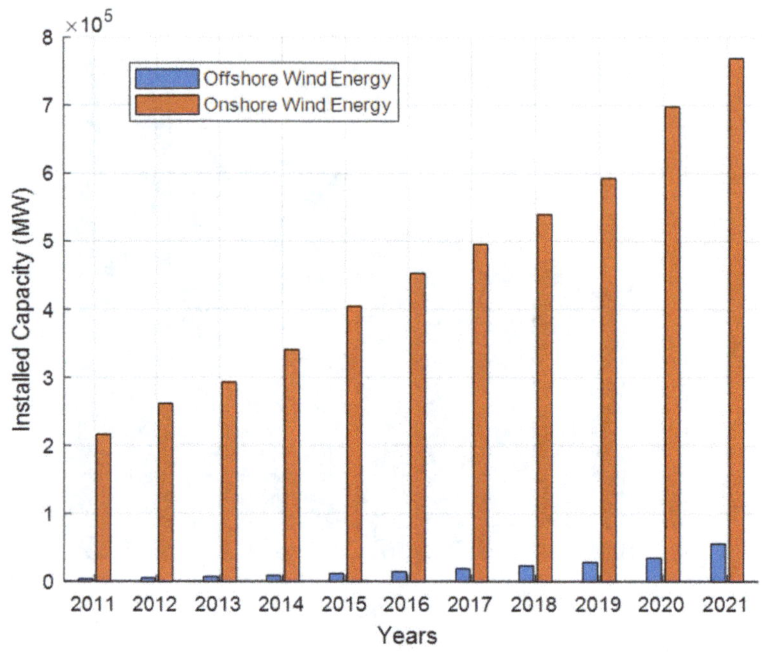

this rationale, water turbines necessitate control systems. Notwithstanding fluctuations in wind patterns, wind turbines provide reliable and efficient energy production. The control systems incorporated within wind turbines are engineered to optimize power generation, minimize dynamic and static mechanical strains, and ensure a reliable and uninterrupted provision of electricity to the electrical grid. The wind turbines (WTs) should be outfitted with operational control systems specifically intended to adjust the turbine characteristics to the appropriate set points to accomplish these objectives (Yingcheng, 2012). The most significant parts in governing wind turbines are the generator torque and blade pitch angle. The implementation of pitch angle control enables the regulation of the wind's input torque, resulting in consistent power generation and fewer mechanical strains. The WT rotor speed can be adjusted by using generator torque control, which enables the implementation of Maximum Power Point Tracking (MPPT) to optimize energy extraction from the wind flow.

To provide a reliable electrical power supply, wind turbines (WTs) must have grid integration management systems to regulate the power sent to the grid effectively. Integrating wind turbines into the power grid is challenging, rendering it a necessary activity. The frequency stability of the grid can be affected by the inherent unpredictability of wind.

2.4. Solar Energy

The sun is a significant and enduring source of energy for the Earth. The integration of sustainable development principles is crucial in the pursuit of viable energy solutions. The constant availability of solar energy renders it a highly appealing energy source. There is a widespread utilization and ongoing advancement of novel technologies to harness solar energy to generate power. The global distribution

Figure 8. Global solar energy map

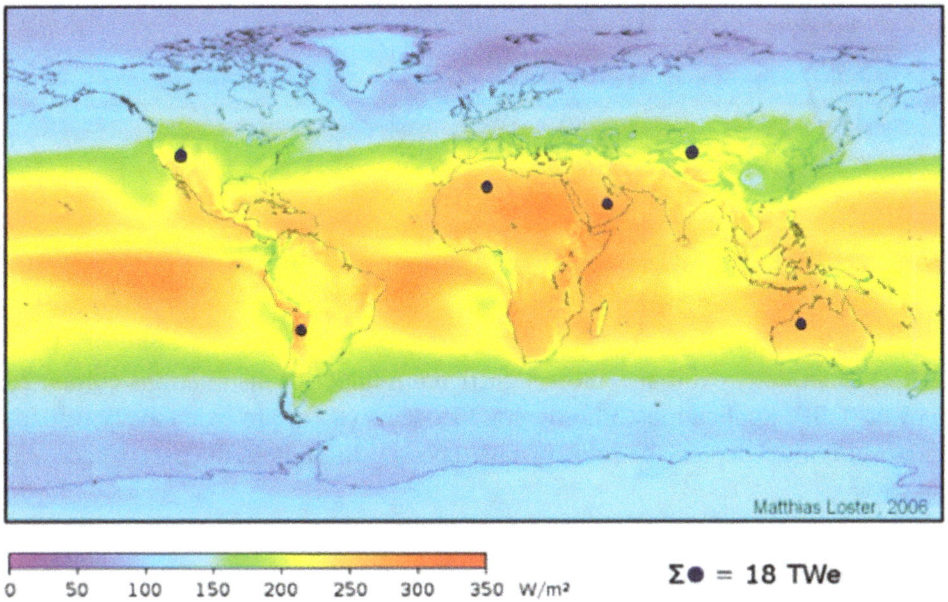

of solar radiation exhibits variability influenced by multiple factors, such as latitude, diurnal variation, climate, and regional variance (Al-Tameemi, 2016).

The mean solar irradiance received within the Earth's atmosphere is approximately 342 watts per square meter, with about 30% of this energy dispersed or reflected into space. According to Hart, (2015), 70% (equivalent to 239 W m2) is accessible for harvesting and capturing. According to a study by Luqman, (2015), the global yearly effective solar radiation exhibits a range of 60 to 250 W m2. Figure 8 illustrates the annual mean solar radiation intensity experienced on the Earth's surface.

The origins of solar energy may be traced back to the seventh century, during which mirrors were employed to harness and utilize solar light. The photovoltaic (PV) effect was first identified in 1839, and afterward, scientists dedicated several decades to developing this technology for power generation (Fraas, 2014). According to extensive research and development conducted by scientists globally over an extended period, solar technology has been categorized into two primary uses: solar heat and solar-powered photovoltaic (PV) systems. Recently, there has been a growing interest in photovoltaic technology, which utilizes semiconductors to convert sunlight into electrical energy directly (Mohanty, 2016). Photovoltaic (PV) devices have experienced a significant cost reduction and have gained widespread use due to their global applicability. As an illustration, from 2010 to 2018, there was a substantial reduction of 77% in the cost of photovoltaic (PV) devices. Additionally, the installed capacity of PV devices had a remarkable expansion of 100-fold between 2005 and 2018. Consequently, the photovoltaic (PV) systems depicted in Figure 9 have emerged as a significant means to attain cost-effective and environmentally friendly sustainable energy.

Globally, the installed capacity of solar energy systems, as depicted in Figure 10, has experienced significant and rapid growth. The installed capacity of photovoltaic (PV) technology changed considerably from 72,216 MW in 2011 to 843.06 MW in 2021. Similarly, the installed capacity of concentrated solar energy (CSP) experienced a notable increase from 1,705 MW in 2011 to 6,387 MW in 2021. Solar

Figure 9. Block diagram of PV systems

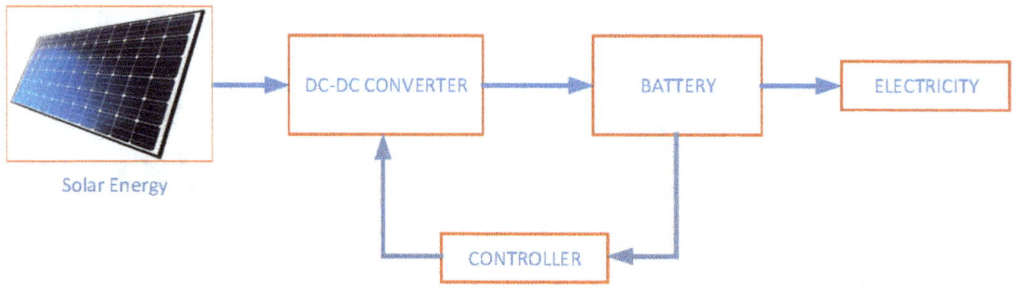

photovoltaic (PV) technology exhibits a more significant number of installations compared to concentrated solar power (CSP) applications. Photovoltaic systems (PVs) are extensively utilized in various space applications. Figure 2 depicts the solar-installed power throughout the past decade, encompassing a global perspective.

2.5 Ocean Energy

The Earth's surface is predominantly covered by oceans, which account for around 70% of its total area. Ocean energy, or marine energy, marine power, or hydroelectric power, encompasses several forms of renewable energy derived from the ocean. The oceans possess significant energy and are situated near most, if not the most densely populated areas. Ocean energy utilization has substantial potential to offer

Figure 10. Installed capacity trends of solar power

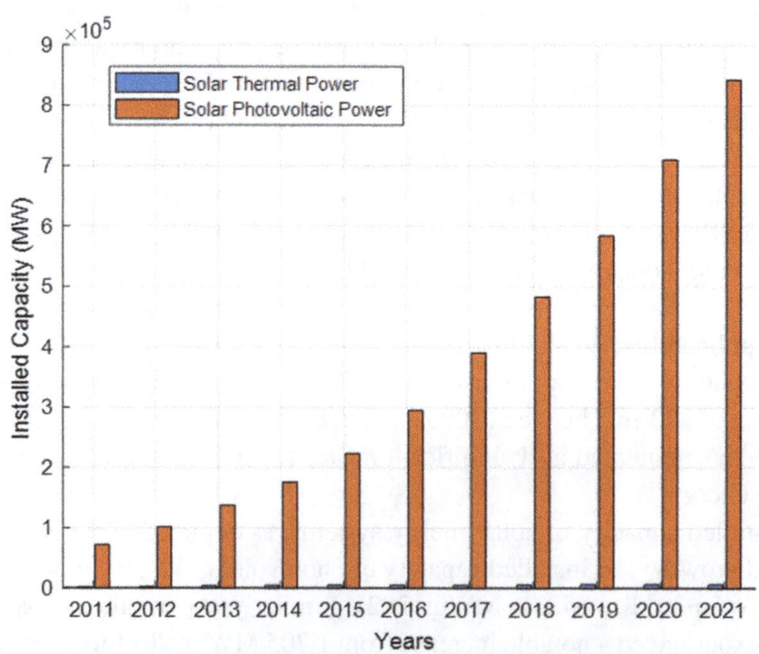

a considerable quantity of novel renewable energy globally. Ocean energy derived from the ocean can be categorized into three primary sources: Ocean Thermal Energy Conversion (OTEC), waves, and tides.

Ocean thermal energy conversion (OTEC) is a method of producing electricity by harnessing the ocean's natural thermal gradient. This process involves extracting heat from warm surface water to make steam, which is used to power a turbine. Simultaneously, cold water from the deep ocean is pumped to the top to condense the moisture into water. This technology has been documented in reference (Hoogwijk, 2008). Ocean Thermal Energy Conversion (OTEC) systems now dominate the ocean energy sector, as indicated by their significant market share. The global potential for OTEC is visually represented in Figure 11. In a typical day, around 60 million square kilometers (23 million square miles) of tropical oceanic regions assimilate solar energy, resulting in a heat content comparable to approximately 250 billion barrels of oil.

If a conversion efficiency of 1% can be achieved for converting stored solar energy into electrical energy, it has the potential to satisfy a significant portion of global electricity consumption. Compared to alternative forms of oceanic energy, the energy output derived through Ocean Thermal Energy Conversion (OTEC) is significantly greater, often ranging from one to two orders of magnitude higher.

Nevertheless, it is worth noting that thermal efficiency exhibits a significant deficiency, with the theoretical upper limit of efficiency estimated to be around 6% to 7%, according to reference (Meisen, 2009). Moreover, energy extraction is a challenging task, resulting in high costs. If this technology were able to achieve economic competitiveness with traditional power technologies, it would have the potential to generate gigawatts of electrical power.

Wave energy is a type of sustainable energy generated through the interaction of wind currents with unobstructed bodies of water. Previous studies have demonstrated the technological feasibility of harnessing the energy generated by ocean waves in offshore environments. Like numerous other environmentally friendly technologies, the development of wave energy systems commenced during the 1970s as a response to the prevailing oil crisis. In contrast to alternative offshore renewable energy sources like solar photovoltaic (PV), wind, or ocean current, wave energy exhibits a constant yet highly

Figure 11. OTEC energy worldwide potential map

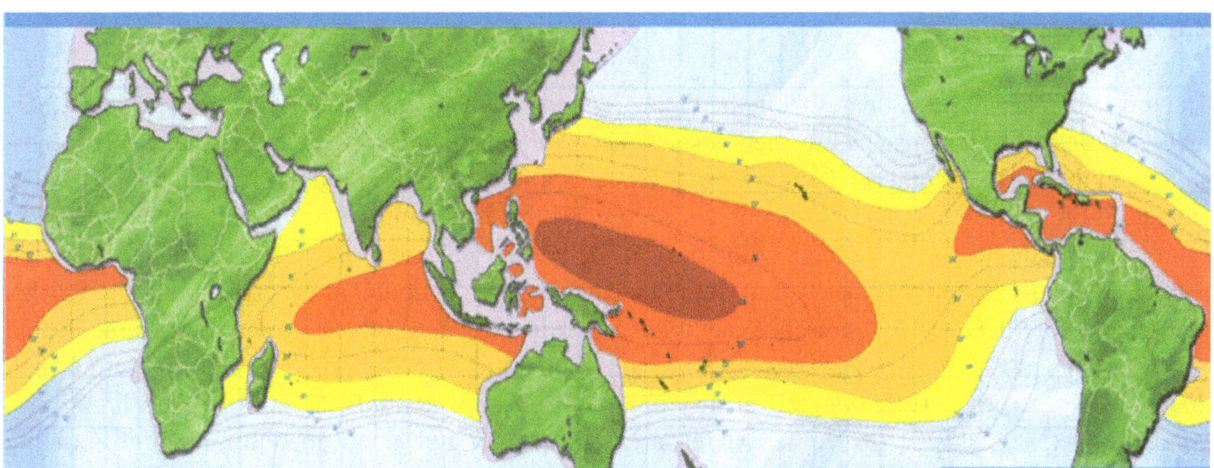

Figure 12. Ocean power electric generation of global

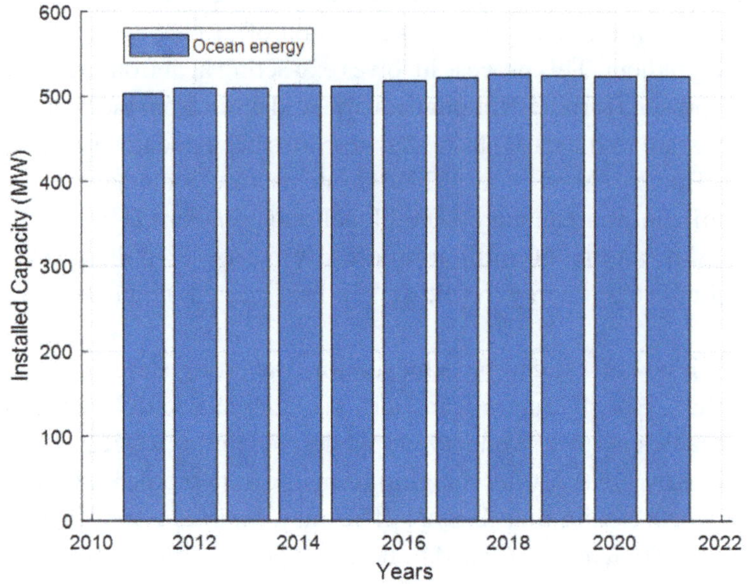

fluctuating nature. However, wave levels at a specific site may be reliably forecasted with reasonable accuracy several days ahead.

Tides refer to recurring fluctuations in the water levels of seas and oceans. Tides are a physical phenomenon that arises from the dynamic interplay of potential and kinetic energy within the Earth-Moon-Sun system. This phenomenon results in some areas of the world seeing considerably more significant local tidal fluctuations than others. There exist two distinct approaches for harnessing tidal energy (Hammar, 2012). One process involves harnessing the cyclical changes in sea level by implementing dams. At the same time, another method entails utilizing local tidal currents, akin to wind power, by deploying a device known as a "sea current turbine." Figure 12 depicts the temporal evolution of global ocean energy-derived electrical energy production during the past decade.

2.6 Geothermal

The term "geothermal" originates from the Greek language, where "geo" refers to the Earth and "thermal" pertains to heat. Consequently, geothermal energy or heat can be understood as the energy generated from the Earth. Geothermal energy is classified as a form of renewable energy, indicating that its availability remains unaffected by resource scarcity and the escalating cost of fossil oil (Faisal, 2020; Nakomcic-Smaragdakis, 2012). One of the benefits of geothermal energy is its environmentally benign nature. Geothermal energy is derived from the thermal energy generated and stored within the Earth's mantle and core. Deep faults' presence facilitates water ingress into these heat sources. As the heated water ascends, it forms a geothermal reservoir with temperatures above 350°C or even higher (Schiel, 2016). The geothermal fluid contains energy in vapour-phase water, which can exist as a liquid or a combination (Malafeh, 2015). Geothermal resources are categorized as low, medium, and high enthalpy resources based on the temperatures of the reservoir fluid (Idrus, 2016).

Geothermal energy utilization has historically been classified into two primary classifications—the utilization and production of electricity. Recently, the subject matter has been organized into three primary areas. The practical utilization of geothermal energy encompasses diverse applications, including but not limited to space heating and cooling, industrial processes, greenhouse cultivation, aquaculture, and therapeutic wellness facilities. Geothermal energy has widespread industrial applications, notably in food dehydration, milk pasteurization, and gold mining (Lee, 2001). The direct utilization of geothermal energy has demonstrated its technological prowess, reliability, economic viability, and environmental acceptability on a global scale. Direct applications can utilize geothermal resources at both high and low temperatures. Nevertheless, using geothermal energy by direct application is typically limited to specific locations due to the impracticality of transporting steam and hot water across vast distances from the geothermal field.

The electricity generation process with geothermal energy, as seen in Figure 13, involves using water and water vapour at elevated temperatures ranging from around 150 to 370°C. Hence, geothermal power facilities are typically situated at significant depths (about 1/2 mile) beneath the Earth's surface. These plants are constructed near-surface reservoirs that provide the requisite hot water resources (Melikoglu, 2017). In contrast, geothermal heat pumps utilize resources near the Earth's surface for heating and cooling buildings. This system's fundamental idea is the transfer of thermal energy from the soil to the structures during winter, while the process is reversed during summer.

3. PRODUCTION OF GREEN HYDROGEN

Hydrogen, as the most prevalent chemical element on Earth, is found in approximately 75% of all matter. Nevertheless, the presence of chemical elements such as oxygen, which combines with hydrogen

Figure 13. Geothermal energy distribution according to years

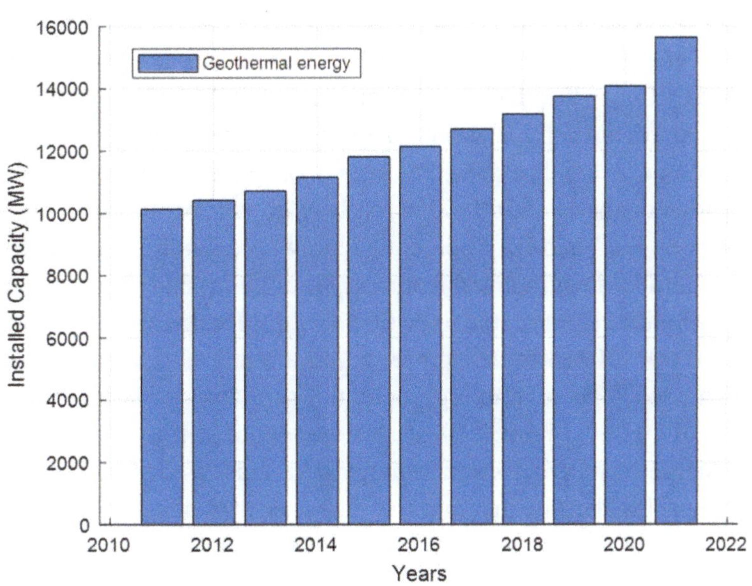

to make water, or carbon, which forms organic molecules, is always observed alongside it. Throughout history, humankind has extensively utilized a particular substance as a primary resource in the chemical industry, metallurgy, and as a source of energy. However, due to its unavailability in a pure form in nature, it necessitates a process of artificial production. The method employed for hydrogen production plays a crucial role in determining its viability as a clean and sustainable fuel source.

The term "green hydrogen" refers to hydrogen that is produced without the release of harmful emissions, thereby aligning with sustainability principles. One fuel that is now being touted as the primary energy carrier for attaining worldwide decarbonization and meeting the established objectives for 2050 in the battle against climate change. Hydrogen, being the primary chemical element, holds the distinction of being the most basic in nature and occupies the initial position in the periodic table, denoted by its atomic number of 1. This particular substance possesses a low weight, exhibits the ability to be conveniently stored, and does not inherently produce any emissions that contribute to environmental pollution. Given these particular attributes, it possesses the ideal qualities to be considered a suitable contender for use as a fuel source. Nevertheless, it is important to note that hydrogen does not function as a main energy source, but rather as an energy vector. This implies that the production of hydrogen necessitates a chemical process (Ishaq et al., 2022). It is commonly asserted that hydrogen is a renewable fuel; nevertheless, it is important to note that this assertion is not universally valid. The renewability of hydrogen is contingent upon the renewability of the extraction technology employed.

Green hydrogen refers to hydrogen that is produced using renewable energies, resulting in a fuel that is both environmentally friendly and sustainable. This type of hydrogen has a zero pollution index and can serve as an important energy source and raw material. The significant contribution of this technology in combating climate change stems from its capacity to substitute fossil fuels in sectors and applications that have traditionally posed challenges for decarbonization efforts. Furthermore, it has promise as an energy storage solution.

4. OPPORTUNITIES AND CHALLENGES

The utilization of solar energy as a renewable energy source is characterized by its nearly limitless potential. The average solar energy received by the Earth's surface is around 343 W/m2. When the Earth's surface area is multiplied by this value, the result is approximately 5x1014 square meters and 1715x1014 watts. Nevertheless, it is worth noting that about 30% of the incident sunlight is reflected, whereas only around 30% of the Earth's surface is comprised of land. Consequently, the amount of solar energy that may be effectively harnessed stands at 360x1014 W. It is imperative to allocate less consideration to the region that encompasses less than fifty percent of the whole landmass, including forested regions, areas covered by ice, and similar geographical features. Throughout one year, the cumulative quantity of solar energy that can be acquired is estimated to be 66x1022 Joules. In 2018, global energy consumption amounted to an estimated 600x10^18 Joules, representing a fraction of less than 0.1% of the total solar energy that may be harnessed. The solar energy potential available for harnessing exceeds our energy requirements by a factor of 10,000. Figure 1 illustrates the identification of regions possessing substantial energy potential capable of meeting global energy demands, as denoted by black dots measuring 200 kilometers in diameter. Practically, meeting the global energy demands only from solar sources is deemed unfeasible.

Figure 14. Green hydrogen with wind power

$$E = A \times H \times r \times PR \tag{1}$$

The relationship between the daily energy production (E) of a photovoltaic (PV) system and the total PV area (A), panel efficiency (r), average solar radiation amount (H), and performance ratio (PR) is one of direct proportionality. The equation presented in Khatib, (2012) demonstrates that the quantity of energy generated by solar energy is contingent upon numerous parameters. The efficiency of photovoltaic (PV) panels is subject to fluctuations due to factors such as panel pollution, dust accumulation, and cloud cover (Alsharif, 2018) Utilizing solar energy systems on a large scale yields advantageous outcomes for the environment, economy, and society. The primary advantage lies in mitigating carbon emissions since each kilowatt-hour of solar energy generated decreases by 0.7 kilograms of CO2 emissions. Additionally, it exhibits little operating and maintenance expenses.

The cost associated with the widespread adoption of solar energy poses a significant drawback. The International Energy Agency (IEA) uses the leveling cost of energy (LCOE) methodology to determine the financial feasibility of generating electricity using various technologies, as stated in their report (Ohunakin, 2014). The LCOE, or Levelized Cost of Electricity, is a metric that quantifies the cost per unit of electricity generated, considering all relevant expenses such as capital investment, operational costs, maintenance expenses, and fuel costs. Its calculation involves the utilization of the following equation.

$$LCOE = \frac{OC}{CF \times 8760} \times CRF + OMC + FC \tag{2}$$

$$CRF = \frac{r \times (1+r)^T}{(1+r)^T - 1} \tag{3}$$

In this context, we define OC as the daily construction cost, OMC as the annual operation and maintenance cost, FC as the annual fuel cost, CRF as the capital recovery factor, CF as the capacity factor, r as the discount rate, and T as the economic life of the plant. During the period spanning from 2010 to 2019, a significant reduction of 82% was noted in the levelized cost of electricity (LCOE) associated with photovoltaic (PV) technology. By the end of 2019, the LCOE for PV had reached a value of US$ 0.068 per kilowatt-hour (kWh). Notwithstanding these improvements, the cost of photovoltaic (PV) technology remains relatively high compared to fossil fuel technologies.

Solar energy is characterized by its stochastic nature, resulting in intermittent availability and limited energy production exclusively during daylight hours. The phenomenon is influenced by abrupt changes in cloud cover or precipitation patterns (Kusakana, 2013). The abovementioned issue can be mitigated by utilizing short-term and long-term weather forecasts coupled with effective planning strategies. It is essential to integrate storage mechanisms and diversify the energy sources to have a sustainable and uninterrupted energy supply. During the operational period of energy generation, the photovoltaic (PV) system provides a direct power supply to the load, while any surplus electrical energy is afterward stored. During periods of energy unavailability, the energy supply is sourced from the working battery.

Installing photovoltaic (PV) plants necessitates substantial land tracts, which are exclusive to this purpose and cannot be utilized for other activities. The land utilization in question is likewise a matter of apprehension. The longevity of solar panels typically ranges from approximately 20 to 30 years. There is anticipated to be a significant proliferation of photovoltaic (PV) power plants in the forthcoming years. However, the concentration of waste panels in a singular location presents a substantial environmental hazard. Hence, it is essential to strategize the management of waste photovoltaic (PV) panels in the forthcoming period.

Global wind speeds are seeing an upward trend, a development that bodes well for the prospects of wind energy. There has been a notable upward trend in wind speed over North America, Asia, and Europe since 2010. Over less than ten years, there has been an observed rise in the worldwide mean wind speed from 7 meters per hour to 7.4 meters per hour. The average wind turbine experiences a corresponding enhancement in efficiency of approximately 17%. In 2020, wind energy contributed around 1600 TWh of electrical power, representing a 5% share of global electricity generation and about 2% of overall energy consumption.

The worldwide recession in the 1970s was accompanied by a prolonged period of dropping global wind speeds. However, a subsequent reversal occurred in 2010, leading to an upward trend in wind speeds. During this particular time frame, scientists posited that the escalation of urban expansion and alterations in land use patterns were responsible for the observed circumstances. However, if this assertion were valid, it would imply that wind velocities will continue to decrease. The recent surge can be attributed to inherent climate oscillations. The investigation of global wind patterns unveils the significant influence of critical climatic phenomena on regional temperatures, exerting a substantial effect on wind velocities.

The evaluation of wind energy necessitates a comprehensive assessment of the efficacy of wind energy and its energy density per unit area. A wind turbine (WT) operates by harnessing the kinetic energy of wind flow and turning it into mechanical energy, which is further transformed into electrical energy. Relevant characteristics include wind power, rotor blade size, wind speed, and air density. Moreover, the theoretical power associated with air movement can be defined as the kinetic energy flow rate per unit of time a wind turbine exerts. This quantity can be mathematically expressed using the equation (Borhanazad, 2013).

$$P = \frac{1}{2} pAV^3C_p \tag{3}$$

The variable V represents the monthly wind speed in meters per second (m/s), while Cp denotes the coefficient based on the Betz limit. This coefficient has the potential to attain a maximum value of 59% for all types of wind turbines. The swept area, denoted as A, is calculated using the formula πD^2/4, where D represents the rotor diameter in square meters and π is the mathematical constant approximately equal to 3.1416. On the other hand, the corrected monthly air density, represented by ρ and measured in kilograms per cubic meter, can be approximated using the equation referenced as (Nematollahi, 2017).

$$p = \frac{P}{RT} \tag{4}$$

In the given equation, P represents the monthly average air pressure in units of Newton per square meter (N/m^2). TR is the monthly mean air temperature measured in Kelvin (K). Lastly, R signifies the specific gas constant for air, which has a value of 287 Joules per kilogram Kelvin (J/kg K).

According to the equation presented, it can be observed that when the rotor area in square meters remains constant and the air density remains stable at a specific point, the energy contained within the wind is only contingent upon the wind speed. Subsequently, the equation above can be simplified to provide K.V3, where K denotes a constant encompassing the aggregate of fixed rotor blade area, air mass, and turbine efficiency. The amount of wind energy that can be harnessed is directly related to the cube of the wind speed. The significance of this assertion lies in the fact that even a slight alteration in wind velocity yields a substantial variation in the associated power.

The levelized cost of electricity (LCOE) expenses associated with baseload electricity production technologies exhibit variations across different wind technology scenarios. Specifically, the LCOE for onshore wind technology amounts to US$111.64 per megawatt-hour (MWh), whereas offshore wind technology incurs a higher LCOE of US$214.47 per MWh. Despite the general trend of decreasing costs in renewable energy, it is essential to note that the levelized cost of electricity (LCOE) associated with wind power technology remains comparatively higher than that of traditional electricity generation technologies. The primary factor contributing to the rise in the levelized cost of electricity (LCOE) associated with wind generation technology is the substantial fluctuation in capital expenses, representing over 80% of the LCOE for this particular technology. Nevertheless, it is crucial to consider the advantageous aspect of green energy technology, specifically wind energy, as it is not reliant on traditional fuel sources that are susceptible to substantial price changes and associated dangers.

Furthermore, the utilization of green energy technology can lead to a decrease in the emission of carbon dioxide (CO2) and other harmful pollutants. It can also reduce demand for additional generation capacity and lessen the reliance on the grid, hence minimizing associated losses. The cost of producing wind turbines is relatively high, necessitating manufacturers' attention to criteria like size, length, dependability, and efficiency.

Wind energy sources exhibit characteristics of intermittency and unpredictability. To provide a continuous supply of electricity, it is necessary to integrate wind power systems with other energy sources, such as the public electricity grid. Furthermore, proficient short- and long-term forecasting and meticulous planning and analysis will contribute to attaining an optimal solution.

Manufacturers should prioritize criteria such as size, length, dependability, and efficiency. There is a need to address multiple manufacturing difficulties with wind turbine components, as they are exposed to significant forces and experience substantial joint stresses and failures. For instance, blades, towers, and enclosures must possess the capability to endure extreme temperatures, precipitation, and ice formation while also being adaptable to fluctuating wind velocities. Incorporating a high strength-to-weight ratio in the production of knives necessitates the imperative exploration of novel materials. The industry is facing a more complex issue in terms of cost and risk management of transmission technology due to the growing distance between offshore wind farms and destinations. Thorough preparation and meticulous investigation are essential for attaining optimal solutions. Turbines have the potential to generate noise as a result of the movement of their blades.

Nevertheless, this issue is effectively addressed or significantly mitigated through technological advancements and the appropriate choice of wind power facilities. One additional risk issue associated with wind turbines is lightning strikes, which can damage the turbines' electronic components. Nevertheless, advancements in technology and enhanced regular maintenance can substantially mitigate this issue.

Hydroelectric power accounts for around 20% of the global electricity generation. Based on the findings presented in the report published by the International Hydropower Association, the worldwide installed hydropower capacity had a growth of 1.9 percent, resulting in a total of 1,360 gigawatts (GW) by the year 2021. In 2004, China, Canada, Brazil, the United States, and Russia emerged as the leading global contributors to hydropower production. The Three Gorges, situated on the Yangtze River in China, is the largest hydroelectric power facility globally, boasting an impressive installed power capacity of 22,500 MW. In 2020, hydropower emerged as the most cost-effective renewable energy source for electricity generation. The global levelized cost of electricity (LCOE) for hydropower has exhibited consistent stability, approximately at US$0.04 per kilowatt-hour, for the last ten years (Kirmani, 2021). Once the dam and associated infrastructure are constructed, water flow can be harnessed as a renewable energy resource.

Hydroelectric power plants have the advantage of facile water flow and electricity output adjustment, ensuring uninterrupted energy production. The power mentioned above plants are characterized by their lack of environmental pollution and absence of CO2 gas emissions. The sole ecological impact of these large-scale power plants is limited to the construction phase. The constructed dams possess an extensive lifespan and can provide electricity for several decades (Richter, 2021). The safety of these facilities is highly commendable since they demonstrate consistent and stable energy production without any notable fluctuations. The reservoirs created by hydroelectric dams offer a range of water-based recreational activities, including fishing, water sports, boating, and waterfowl hunting. Within the geographical area in which they are situated, the power plants possess the capacity to function as a consistent and reliable water source for irrigation in agricultural endeavors.

Despite the high initial costs associated with constructing power plants, they are characterized by relatively low maintenance expenses. The construction of a hydroelectric power station situated along a transboundary river can exert influence over the downstream country through which the river flows, potentially leading to inter-country disputes. The occurrence of flooding poses a significant risk, necessitating the relocation of residents to designated safe zones. The extent of flooded agricultural areas and habitats is considerable, and this circumstance contributes to the degradation of the natural environment. The interplay between physical and biological elements within a wild river system has the potential to perturb the ecological equilibrium of aquatic life.

The estimated potential for global geothermal power generation ranges from 70 to 80 gigawatts (GW). According to the latest available data in 2019, the global capacity for geothermal electricity is at 15.4 gigawatts (GW). The United States accounts for 23.9 percent of this total, equivalent to 3.68 GW (Anand, 2012). The current utilization of global electricity-producing capacity stands at a mere 15%. According to the publicly available geological knowledge and technology disclosed by the Geothermal Energy Association (GEA), the use of worldwide geothermal energy potential has only reached 6.9 percent thus far. The Intergovernmental Panel on Climate Change (IPCC) has revealed that the estimated range for geothermal power potential is between 35 gigawatts (GW) and two terawatts (TW) (Kabeyi, 2019). 2020 had a modest growth of approximately 2% in geothermal power generation, accompanied by a capacity expansion of about 200 MW. This growth rate falls below the average growth observed during the preceding five-year period. The implementation of this technology is projected to result in a production increase of roughly 13% (equivalent to approximately 3.6 GW per year) within the time frame of 2021 to 2030, in alignment with the Net Zero Emissions Scenario targeted for 2050. To enhance the efficiency of geothermal energy production, it is imperative to mitigate expenses and minimize pre-development uncertainties.

Several countries, namely El Salvador, Kenya, the Philippines, Iceland, New Zealand, and Costa Rica, generate power from geothermal sources, accounting for more than 15 percent of their total energy production. Indonesia possesses the highest potential for geothermal energy resources globally, estimated at 29,000 MW. The installed capacity of Indonesia amounts to 1,800 MW.

Geothermal power plants offer a consistent, environmentally friendly, secure, and sustainable energy source while minimizing land requirements. Due to its regional nature, this energy source provides economic benefits to the countries involved and generates employment opportunities for the local population. The water requirements for geothermal power plants are minimal, with an average usage of approximately 20 liters per megawatt-hour (MWh). The rate of air-cooling in power plants is measured at 1,370 liters per MWh hour. According to a study conducted by reference MicCalmont, (2017), it has been observed that the water consumption of an oil plant is around 15% lower than that of a nuclear plant.

The study also found that the water consumption of an oil plant is approximately 25% lower. Generally, a geothermal power plant requires an area of 404 square meters for every gigawatt-hour (GWh) of electricity generated. In contrast, a coal-fired power plant necessitates an area of 3,632 square meters per GWh, while a wind farm requires 1,335 square meters per GWh. The energy derived from the geothermal reservoir exhibits continuous availability throughout the year, unaffected by temporal variations such as changing seasons, natural calamities, or political disturbances.

The potential gases that may be emitted vary depending on the reservoir type, including carbon dioxide, sulfur dioxide, nitrogen oxides, hydrogen sulfide, and particulate matter. A coal-fired power station exhibits higher kilogram emissions per MWh than a geothermal unit. The concentrations of carbon dioxide (CO2) are 40 parts per million (ppm) and 994 ppm, while sulfur dioxide (SO2) concentrations are 4.71 ppm and 0.16 ppm. Nitrogen oxide (NOx) concentrations are 1.95 ppm and 0 ppm, but hydrogen sulfide (H2S) concentrations are 0 ppm and 0.08 ppm. Lastly, particulate matter concentrations are 1.01 ppm and 0 ppm. In geothermal power facilities, hydrogen sulfide undergoes regular processing and conversion into elemental sulfur. On the contrary, it is worth noting that oil-fired power plants generate an estimated 814 kilograms of hydrogen sulfide (H2S) per megawatt-hour (MWh). In contrast, natural gas-fired power plants produce approximately 550 kilograms of H2S.

2017, the International Energy Agency (IEA) recognized bioenergy as the preeminent renewable energy source. The International Energy Agency (IEA) has argued that the present rate of bioenergy delivery needs to catch up to the levels necessary for future low-carbon scenarios. Consequently, there is an urgent need for the swift and increased implementation of bioenergy technologies. According to the International Energy Agency's Net Zero scenario, it is projected that the proportion of modern bioenergy in the overall energy supply will rise from 6.6% in 2020 to 13.1% in 2030 and further to 18.7% in 2050 (Whitaker, 2018). According to IRENA's analysis in 2014, a modest input may provide around six exajoules (EJ) of energy from conventional bioenergy by the year 2030. According to a scholarly source, the utilization of contemporary bioenergy is distributed as follows: 86% is allocated for heating purposes, 9% for transportation, and the remaining 5% is dedicated to electricity generation (Alsharif, 2018). The majority of global bioenergy is derived from forest resources. Power plants that utilize biomass as a fuel source can generate consistent and reliable power production, in contrast to the intermittent power generation observed in solar or wind farms.

Bioenergy plays a crucial role in mitigating soil and water pollution and minimizing reliance on waste disposal in landfills and agricultural settings throughout the nation. The process of biogas production does not require the presence of oxygen, hence minimizing the consumption of additional fuel resources. Biogas technology in rural areas presents employment prospects and offers a cost-effective solution. Using landfill gas as a means of energy generation mitigates the impact of the greenhouse effect. The extensive utilization of biogas can be attributed to its significant importance. The recycling of biodegradable garbage encompasses a wide range of forms and employs straightforward methods and technologies. The installation of biogas systems on a small scale is characterized by its ease and minimal investment requirements. Biogas facilities and the utilization of animal waste products provide daily advantages to several agricultural establishments. The cow can generate a sufficient quantity of waste products within a single day, which can be utilized to sustain the illumination of a light bulb throughout the day.

Reducing soil and water contamination leads to a decrease in the utilization of waste items in landfills, landfills, and farms nationwide. Despite the initiation of refining operations, bioenergy still retains some contaminants. Methane exhibits a propensity for explosive reactions upon encountering oxygen, rendering it susceptible to flammability. According to the cited source, it is said that current biopower plants show suboptimal efficiency levels and require the implementation of novel technologies to enhance the cost-effectiveness of this particular energy approach (Bagher, 2015).

Green hydrogen is expected to play a prominent role in the inevitable energy transition that nations worldwide are compelled to undertake in order to attain carbon neutrality and effectively address the challenges posed by climate change. In order to achieve this objective, it is crucial to address emissions in sectors that currently pose challenges for electrification. Green hydrogen emerges as a promising solution in this regard, owing to its numerous inherent advantages.

Despite the numerous advantages associated with green hydrogen, its integration into our current energy portfolio remains limited due to a range of obstacles that necessitate concerted efforts from research institutions, government entities, and private investors in order to be overcome. The production cost of green hydrogen is higher compared to that of gray hydrogen. Nevertheless, the decline in the cost of renewable energies has created a favorable prospect for their expanding competitiveness. The cost of solar electricity has decreased by a factor of 10 over the past decade, while wind energy costs less than half of what it did previously. This significant reduction in prices has created the potential for a corresponding decrease in the price of electricity required for the electrolysis process. The successful execution of this endeavor necessitates substantial financial commitments. According to reports, a substantial sum of USD 300 billion is projected to be required on a global scale in the next years, mostly for the purposes of infrastructure development and research endeavors. However, as indicated in a report published by BloombergNEF (BNEF), the implementation of adequate laws to facilitate its advancement might potentially result in a surge in demand for green hydrogen, reaching a staggering 700 million tonnes by the year 2050. Consequently, investing in the development of green hydrogen presents not only a financial cost but also a substantial economic prospect.

5. CONCLUSION AND FUTURE WORK

Integrating renewable energy sources into power system grids remains a persistent challenge for architects and operators. The RES system and green hydrogen production are considered an advanced option for power generation due to its status as a clean energy source. Recently, there has been a notable surge in Renewable Energy Systems (RES) installation owing to a heightened recognition of its advantageous attributes. Numerous obstacles must be addressed to achieve the commercialization and general adoption of the subject matter. The difficulties at hand are interconnected, encompassing the system's design and sizing, the selection of the most suitable and efficient model considering both technical and financial considerations for a renewable energy sources (RES) and green hyrogen smart grid, and the determination of the equilibrium electricity price for the integration of RES inside the smart grid system. Previous research suggests there are limitations to achieving a balanced power price for integrating photovoltaic (PV) systems in a smart grid. These limitations arise from considering the feasibility of deploying renewable energy source (RES) smart grid systems. Therefore, developing a RES smart grid system model that incorporates both technological and financial considerations is imperative. The following information is beneficial to evaluate the equilibrium power price associated with integrating renewable energy sources (RES) inside a smart grid framework.

REFERENCES

Al-Tameemi, M. A., & Chukin, V. V. (2016). Global water cycle and solar activity variations. *Journal of Atmospheric and Solar-Terrestrial Physics*, *142*, 55–59. doi:10.1016/j.jastp.2016.02.023

Alhamid, M. I., Daud, Y., Surachman, A., Sugiyono, A., Aditya, H. B., & Mahlia, T. M. I. (2016). Potential of geothermal energy for electricity generation in Indonesia: A review. *Renewable & Sustainable Energy Reviews*, *53*, 733–740. doi:10.1016/j.rser.2015.09.032

Alsaleh, M., & Abdul-Rahim, A. S. (2021). Do global competitiveness factors effects the industry sustainability practices? Evidence from European hydropower industry. *Journal of Cleaner Production*, *310*, 127492. doi:10.1016/j.jclepro.2021.127492

Alsaleh, M., & Abdul-Rahim, A. S. (2021). The nexus between worldwide governance indicators and hydropower sustainable growth in EU 28 region. *International Journal of Environmental Research*, *15*(6), 1001–1015. doi:10.1007/s41742-021-00366-6

Alsaleh, M., & Abdul-Rahim, A. S. (2022). Moving toward sustainable environment: The effects of hydropower industry on water quality in EU economies. *Energy & Environment*, *33*(7), 1304–1325. doi:10.1177/0958305X211039452

Alsharif, M. H., Kim, J., & Kim, J. H. (2018). Opportunities and challenges of solar and wind energy in South Korea: A review. *Sustainability (Basel), 10*(6), 1822. doi:10.3390/su10061822

Anand, S., Fernandes, B. G., & Guerrero, J. (2012). Distributed control to ensure proportional load sharing and improve voltage regulation in low-voltage DC microgrids. *IEEE Transactions on Power Electronics, 28*(4), 1900–1913. doi:10.1109/TPEL.2012.2215055

Aravani, V. P., Sun, H., Yang, Z., Liu, G., Wang, W., Anagnostopoulos, G., Syriopoulos, G., Charisiou, N. D., Goula, M. A., Kornaros, M., & Papadakis, V. G. (2022). Agricultural and livestock sector's residues in Greece & China: Comparative qualitative and quantitative characterization for assessing their potential for biogas production. *Renewable & Sustainable Energy Reviews, 154*, 111821. doi:10.1016/j.rser.2021.111821

Bagher, A. M., Fatemeh, G., Saman, M., & Leili, M. M. (2015). Advantages and disadvantages of biogas energy. *Bulletin of Advanced Scientific Research, 1*(5), 132–135.

Berga, L. (2008). Solving dam engineering challenges. *HydroWorld. com.*

Berga, L. (2016). The role of hydropower in climate change mitigation and adaptation: A review. *Engineering (Beijing), 2*(3), 313–318. doi:10.1016/J.ENG.2016.03.004

Bergmann, A., Colombo, S., & Hanley, N. (2008). Rural versus urban preferences for renewable energy developments. *Ecological Economics, 65*(3), 616–625. doi:10.1016/j.ecolecon.2007.08.011

Bilgen, S., Kaygusuz, K., & Sari, A. (2004). Renewable energy for a clean and sustainable future. *Energy Sources. Part A, Recovery, Utilization, and Environmental Effects, 26*(12), 1119–1129. https://www.irena.org/

Chia, W. Y., Chew, K. W., Le, C. F., Lam, S. S., Chee, C. S. C., Ooi, M. S. L., & Show, P. L. (2020). Sustainable utilization of biowaste compost for renewable energy and soil amendments. *Environmental Pollution, 267*, 115662. doi:10.1016/j.envpol.2020.115662 PMID:33254731

De Clercq, D., Wen, Z., Gottfried, O., Schmidt, F., & Fei, F. (2017). A review of global strategies promoting the conversion of food waste to bioenergy via anaerobic digestion. *Renewable & Sustainable Energy Reviews, 79*, 204–221. doi:10.1016/j.rser.2017.05.047

Demirbas, A. J. E. E. S. T. (2000). Recent advances in biomass conversion technologies. *Energy Education Science and Technology, 6*, 19–41.

Dinçer, İ. (2001). Environmental Issues: II-Potential Solutions. *Energy Sources, 23*(1), 83–92. doi:10.1080/00908310151092218

Duca, D., Maceratesi, V., Fabrizi, S., & Toscano, G. (2022). Valorising agricultural residues through pelletisation. *Processes (Basel, Switzerland), 10*(2), 232. doi:10.3390/pr10020232

Duca, D., & Toscano, G. (2022). Biomass energy resources: Feedstock quality and bioenergy sustainability. *Resources, 11*(6), 57. doi:10.3390/resources11060057

Farhad, S., Saffar-Avval, M., & Younessi-Sinaki, M. (2008). Efficient design of feedwater heaters network in steam power plants using pinch technology and exergy analysis. *International Journal of Energy Research*, *32*(1), 1–11. doi:10.1002/er.1319

Feng, G., Xia, G., Chen, B., Xiao, Y., & Zhou, R. (2019). A method for rockburst prediction in the deep tunnels of hydropower stations based on the monitored microseismicity and an optimized probabilistic neural network model. *Sustainability (Basel)*, *11*(11), 3212. https://www.irena.org/hydropower. doi:10.3390/su11113212

Fraas, L. M. (2014). *Low-cost solar electric power*. Springer. doi:10.1007/978-3-319-07530-3

Hall, D. O., Mynick, H. E., & Williams, R. H. (1991). Cooling the greenhouse with bioenergy. *Nature*, *353*(6339), 11–12. doi:10.1038/353011a0

Hammar, L., Ehnberg, J., Mavume, A., Cuamba, B. C., & Molander, S. (2012). Renewable ocean energy in the Western Indian Ocean. *Renewable & Sustainable Energy Reviews*, *16*(7), 4938–4950. doi:10.1016/j.rser.2012.04.026

Hart, M. (2015). *Hubris: The troubling science, economics, and politics of climate change*. Lulu.com.

Holm-Nielsen, J. B., Al Seadi, T., & Oleskowicz-Popiel, P. (2009). The future of anaerobic digestion and biogas utilization. *Bioresource Technology*, *100*(22), 5478–5484. doi:10.1016/j.biortech.2008.12.046 PMID:19217772

Hoogwijk, M., & Graus, W. (2008). *Global potential of renewable energy sources: a literature assessment*. Background report prepared by order of REN21. Ecofys, PECSNL072975.

Horst, G. H., & Hovorka, A. J. (2009). Fuelwood: The "other" renewable energy source for Africa? *Biomass and Bioenergy*, *33*(11), 1605–1616. doi:10.1016/j.biombioe.2009.08.007

Ilari, A., Toscano, G., Foppa Pedretti, E., Fabrizi, S., & Duca, D. (2020). Environmental sustainability of heating systems based on pellets produced in mobile and stationary plants from vineyard pruning residues. *Resources*, *9*(8), 94. doi:10.3390/resources9080094

Ishaq, H., Dincer, I., & Crawford, C. (2022). A review on hydrogen production and utilization: Challenges and opportunities. *International Journal of Hydrogen Energy*, *47*(62), 26238–26264. doi:10.1016/j.ijhydene.2021.11.149

Jiang, T., Shen, Z., Liu, Y., & Hou, Y. (2018). Carbon footprint assessment of four normal size hydropower stations in China. *Sustainability (Basel)*, *10*(6), 2018. doi:10.3390/su10062018

Khatib, T., Mohamed, A., Sopian, K., & Mahmoud, M. (2012). A new approach for optimal sizing of standalone photovoltaic systems. *International Journal of Photoenergy*, *2012*, 2012. doi:10.1155/2012/391213

Kirmani, F., Pal, A., Mudgal, A., Shrestha, A., & Siddiqui, A. (2021). Advantages and disadvantages of hydroelectric power plant. *International Journal of Innovative Science and Research Technology*.

Ko, S., Lautala, P., & Handler, R. M. (2018). Securing the feedstock procurement for bioenergy products: A literature review on the biomass transportation and logistics. *Journal of Cleaner Production*, *200*, 205–218. doi:10.1016/j.jclepro.2018.07.241

Kuriqi, A., Pinheiro, A. N., Sordo-Ward, A., Bejarano, M. D., & Garrote, L. (2021). Ecological impacts of run-of-river hydropower plants—Current status and future prospects on the brink of energy transition. *Renewable & Sustainable Energy Reviews*, *142*, 110833. doi:10.1016/j.rser.2021.110833

Kuriqi, A., Pinheiro, A. N., Sordo-Ward, A., & Garrote, L. (2019). Influence of hydrologically based environmental flow methods on flow alteration and energy production in a run-of-river hydropower plant. *Journal of Cleaner Production*, *232*, 1028–1042. doi:10.1016/j.jclepro.2019.05.358

Kusakana, K., & Vermaak, H. J. (2013). Hybrid renewable power systems for mobile telephony base stations in developing countries. *Renewable Energy*, *51*, 419–425. doi:10.1016/j.renene.2012.09.045

Lee, K. C. (2001). Classification of geothermal resources by exergy. *Geothermics*, *30*(4), 431–442. doi:10.1016/S0375-6505(00)00056-0

Lehner, B., Czisch, G., & Vassolo, S. (2005). The impact of global change on the hydropower potential of Europe: A model-based analysis. *Energy Policy*, *33*(7), 839–855. doi:10.1016/j.enpol.2003.10.018

Li, Z., Du, H., Xiao, Y., & Guo, J. (2017). Carbon footprints of two large hydro-projects in China: Life-cycle assessment according to ISO/TS 14067. *Renewable Energy*, *114*, 534–546. doi:10.1016/j.renene.2017.07.073

Lubosny, Z., & Bialek, J. W. (2007). Supervisory control of a wind farm. *IEEE Transactions on Power Systems*, *22*(3), 985–994. doi:10.1109/TPWRS.2007.901101

Luqman, M., Ahmad, S. R., Khan, S., Ahmad, U., Raza, A., & Akmal, F. (2015). Estimation of solar energy potential from rooftop of Punjab government servants cooperative housing society Lahore using GIS. *Smart Grid and Renewable Energy*, *6*(05), 128–139. doi:10.4236/sgre.2015.65012

Mai-Moulin, T., Hoefnagels, R., Grundmann, P., & Junginger, M. (2021). Effective sustainability criteria for bioenergy: Towards the implementation of the european renewable directive II. *Renewable & Sustainable Energy Reviews*, *138*, 110645. doi:10.1016/j.rser.2020.110645

Malafeh, S., & Sharp, B. (2015). Role of royalties in sustainable geothermal energy development. *Energy Policy*, *85*, 235–242. doi:10.1016/j.enpol.2015.06.023

Manwell, J. F., McGowan, J. G., & Rogers, A. L. (2010). *Wind energy explained: theory, design and application*. John Wiley & Sons.

McCalmont, J. P., Hastings, A., McNamara, N. P., Richter, G. M., Robson, P., Donnison, I. S., & Clifton-Brown, J. (2017). Environmental costs and benefits of growing Miscanthus for bioenergy in the UK. *Global Change Biology. Bioenergy*, *9*(3), 489–507. doi:10.1111/gcbb.12294 PMID:28331551

Meisen, P., & Loiseau, A. (2009). *Ocean energy technologies for renewable energy generation*. Global Energy Network Institute.

Melikoglu, M. (2017). Geothermal energy in Turkey and around the World: A review of the literature and an analysis based on Turkey's Vision 2023 energy targets. *Renewable & Sustainable Energy Reviews*, *76*, 485–492. doi:10.1016/j.rser.2017.03.082

Mohanty, P., Muneer, T., Gago, E. J., & Kotak, Y. (2015). Solar radiation fundamentals and PV system components. In *Solar Photovoltaic System Applications: A Guidebook for Off-Grid Electrification* (pp. 7–47). Springer International Publishing.

Nakomcic-Smaragdakis, B., Stajic, T., Cepic, Z., & Djuric, S. (2012). Geothermal energy potentials in the province of Vojvodina from the aspect of the direct energy utilization. *Renewable & Sustainable Energy Reviews*, *16*(8), 5696–5706. doi:10.1016/j.rser.2012.05.038

Nematollahi, O., & Kim, K. C. (2017). A feasibility study of solar energy in South Korea. *Renewable & Sustainable Energy Reviews*, *77*, 566–579. doi:10.1016/j.rser.2017.03.132

Oh, Y. K., Hwang, K. R., Kim, C., Kim, J. R., & Lee, J. S. (2018). Recent developments and key barriers to advanced biofuels: A short review. *Bioresource Technology*, *257*, 320–333. doi:10.1016/j.biortech.2018.02.089 PMID:29523378

Ohunakin, O. S., Adaramola, M. S., Oyewola, O. M., & Fagbenle, R. O. (2014). Solar energy applications and development in Nigeria: Drivers and barriers. *Renewable & Sustainable Energy Reviews*, *32*, 294–301. doi:10.1016/j.rser.2014.01.014

Panwar, N. L., Kaushik, S. C., & Kothari, S. (2011). Role of renewable energy sources in environmental protection: A review. *Renewable & Sustainable Energy Reviews*, *15*(3), 1513–1524. doi:10.1016/j.rser.2010.11.037

Prakasam, C., & Saravanan, R. (2020). Environmental flow—a mitigation measure for impact of hydropower projects. In *Sustainable Civil Engineering Practices: Select Proceedings of ICSCEP 2019* (pp. 207-213). Springer Singapore.

Prakash, R., & Bhat, I. K. (2012). Life cycle greenhouse gas emissions estimation for small hydropower schemes in India. *Energy*, *44*(1), 498–508. doi:10.1016/j.energy.2012.05.052

Prasad, S., Singh, A., Korres, N. E., Rathore, D., Sevda, S., & Pant, D. (2020). Sustainable utilization of crop residues for energy generation: A life cycle assessment (LCA) perspective. *Bioresource Technology*, *303*, 122964. doi:10.1016/j.biortech.2020.122964 PMID:32061494

Rathore, N. S., & Panwar, N. L. (2007). *Renewable energy sources for sustainable development.* New India Publishing. doi:10.59317/9789390083862

Ravindranath, N. H., & Hall, D. O. (1995). *Biomass, energy and environment: a developing country perspective from India.* Oxford University Press. doi:10.1093/oso/9780198564362.001.0001

Richter, A. (2019). *The Top 10 Geothermal Countries 2018.* Academic Press.

Schiel, K., Baume, O., Caruso, G., & Leopold, U. (2016). GIS-based modelling of shallow geothermal energy potential for CO2 emission mitigation in urban areas. *Renewable Energy*, *86*, 1023–1036. doi:10.1016/j.renene.2015.09.017

Sims, R. E. H. (2003). Bioenergy to mitigate for climate change and meet the needs of society, the economy and the environment. *Mitigation and Adaptation Strategies for Global Change*, *8*(4), 349–370. doi:10.1023/B:MITI.0000005614.51405.ce

Toscano, G., Alfano, V., Scarfone, A., & Pari, L. (2018). Pelleting vineyard pruning at low cost with a mobile technology. *Energies*, *11*(9), 2477. doi:10.3390/en11092477

Toscano, G., Leoni, E., Gasperini, T., & Picchi, G. (2022). Performance of a portable NIR spectrometer for the determination of moisture content of industrial wood chips fuel. *Fuel*, *320*, 123948. doi:10.1016/j.fuel.2022.123948

Toscano, G., Pizzi, A., Pedretti, E. F., Rossini, G., Ciceri, G., Martignon, G., & Duca, D. (2015). Torrefaction of tomato industry residues. *Fuel*, *143*, 89–97. doi:10.1016/j.fuel.2014.11.039

Toscano, G., Riva, G., Pedretti, E. F., & Duca, D. (2013). Effect of the carbon oxidation state of biomass compounds on the relationship between GCV and carbon content. *Biomass and Bioenergy*, *48*, 231–238. doi:10.1016/j.biombioe.2012.11.002

Wani, F., Dong, J., & Polinder, H. (2020). Tidal turbine generators. In *Advances in Modelling and Control of Wind and Hydrogenerators*. IntechOpen. doi:10.5772/intechopen.90433

Whitaker, J., Field, J. L., Bernacchi, C. J., Cerri, C. E., Ceulemans, R., Davies, C. A., DeLucia, E. H., Donnison, I. S., McCalmont, J. P., Paustian, K., Rowe, R. L., Smith, P., Thornley, P., & McNamara, N. P. (2018). Consensus, uncertainties and challenges for perennial bioenergy crops and land use. *Global Change Biology. Bioenergy*, *10*(3), 150–164. doi:10.1111/gcbb.12488 PMID:29497458

Yingcheng, X., & Nengling, T. (2012). System frequency regulation in doubly fed induction generators. *International Journal of Electrical Power & Energy Systems*, *43*(1), 977–983. doi:10.1016/j.ijepes.2012.05.039

Youm, I., Sarr, J., Sall, M., & Kane, M. M. (2000). Renewable energy activities in Senegal: A review. *Renewable & Sustainable Energy Reviews*, *4*(1), 75–89. doi:10.1016/S1364-0321(99)00009-X

Zakhidov, R. A. (2008). Central Asian countries energy system and role of renewable energy sources. *Applied Solar Energy*, *44*(3), 218–223. doi:10.3103/S0003701X08030201

Chapter 2
Performance Metrics Analysis Tool for Solar Energy Plants

Unomasa Kasenina
Namibia University of Science and Technology, Namibia

K. S. Sastry Musti
https://orcid.org/0000-0003-4384-7933
Namibia University of Science and Technology, Nambia

ABSTRACT

With the rise of grid connected independent power producers (IPP), performance analysis of energy generation plants becomes an essential task. This chapter presents design and development of a performance analysis tool using MS Excel for typical IPP with a grid connected solar PV plant. The tool utilizes historical data to determine various energy-metrics, econometrics, and financial metrics. Additionally, the tool can determine various key parameters such as the energy sales to the utility, energy that cannot be sold, and the load that cannot be served. The tool will have the capability of capturing data, doing calculations, and visualizing the data and output from calculations through plots and graphs.

INTRODUCTION

Due to the decreasing levelized costs of solar and wind energies, several renewable energy-based power plants are being added to the energy markets. Various stakeholders in the energy supply industry (ESI) want to know the overall performance of the energy production facilities for various purposes. Even the regulator expects the independent power producers (IPPs) to report on their overall performance with details. Thus, the performance analysis of a solar photovoltaic (PV) plant is crucial to ensuring its efficient operation and maximum energy production. The determination of overall performance and various related parameters can be complex as a lot of data is required for the analysis and several parameters need to be computed. Performance analysis requires a lot of data as input, and that includes historical energy production details, weather data (irradiance, temperature, wind speed), equipment specifications, and historical maintenance data. Key parameters in the performance metrics include performance ratio (PR),

DOI: 10.4018/979-8-3693-1014-4.ch002

capacity factor (CF), energy yield, energy losses, etc. Then financial analysis includes the determination of Return on investment (ROI) and payback period based on actual energy production and system costs (Musti & Kapali, 2021). Due to natural environmental operating conditions, solar energy is indeed intermittent, and thus it is required to determine the weather impact on the plant. For this, data for parameters such as irradiance, temperature, wind speed, etc. is to be collected from the specific geolocation, and then their impact on the performance of the plant needs to be determined (Musti, 2020a).

The most vital factor that plays a role in solar energy production is the geographic location of the solar plant itself. Grid scale solar power plants use several hundreds of photovoltaic panels that are typically installed on rotatable mounts so that they can be rotated based on the movement of the Sun. Naturally a lot of land space is required and thus such plants cannot be located in urban areas (Sastry & Sahadeo, 2015). The plants are connected to the grid through what is called a Point of Common Coupling (PCC). Lengths of the two distances, one from the plant to the PCC and then the other from PCC to the grid; play a significant role in losses incurred in the lines, the amount of energy transferred from the plant to the grid and thus overall revenues to the IPP (Hauwanga & Musti, 2022). Then, the energy payback time of solar energy is usually long, thus solar energy IPPs would need to set up the plants in regions or areas that would maximise the power output of the PV modules. This is because the amount of sunlight received can vary significantly based on the location; and this in turn affects the efficiency of the solar PV modules and overall energy output of the plant. The amount of sunlight received has a considerable effect on solar irradiance, temperature, and daylight hours. These are key parameters that contribute to the plant performance. Other factors that can influence the geolocation of the IPP are the climate, weather, and shading. Thus, location of the plant, location of IPP and the shading etc., are important parameters.

Solar energy or electricity generated using solar power plants is often termed as "clean energy". Although this happens to be so, the extraction, production, transportation, and installation of solar power plants make use of fossil fuels during most of these processes (Sharma & Tiwari, 2013). Hence, solar power plants cannot be termed entirely clean, as the processes leading to the product, in this case, the solar power plants, contribute to greenhouse emissions. However, solar power plants mitigate further depletion of fossil fuels and produce cleaner energy that will in turn mitigate greenhouse gas emissions. The flow of energy through the processes of extraction, production, transportation, and installation of solar power plants is studied through energy metrics (Roy et al., 2022). The energy that is consumed during the manufacturing process is termed "embodied energy" (Rachoutis, 2016; Sharma & Tiwari, 2013). The embodied energy is the overall energy required for the making of the various components in a typical solar power plant. This is determined from the energy used for extraction of various components in their natural form, processing and production of the various parts that make up the solar power plant. The major parts include - solar PV panels, charge controller(s), battery bank(s), and racking and other support structures. Then energy required for transporting and the energy used during setting up the plant also need to be considered (Sharma & Tiwari, 2013). Three major studies are undertaken to evaluate solar power plants – 1) Energy Metrics Analysis; 2) Performance Analysis and 3) Cost Benefit Analysis.

The energy metrics study typically involves analysing a number of important parameters. They include - energy payback time (EPBT), energy production factor (EPF), and life cycle conversion efficiency (LCCE) (Li et. al., 2018; Roy et al., 2022; Fthenakis & Raugei, 2017). Energy metrics analysis will play a vital role in helping the IPPs determine the economic viability of the solar power plant as well as key production evaluation factors. Even for small sized solar power production sites such as the smart benches, performance is measured and displayed using versatile dashboards (Chakwesha and Musti, 2023). Most smart cities now-a-days offer smart infrastructure such as the solar-smart benches where in such dashboards can be seen. The performance analysis of a solar power plant for IPPs is another crucial aspect. The IPPs can assess how grid-connected solar power plants operate in real time with the aid of performance analysis (Inamdar et al., 2018). According to Rituraj and Kadar (2021), a few parameters for assessing the solar power plants' performance are outlined in the International Standard IEC61724-6724, which was established by the International Electro-Technical Commission (IEC). The parameters considered are energy losses, capacity utilisation factor (CUF), performance ratio (PR), inverter efficiency, module efficiency, and the electricity output of the Solar power plant (Li et. al., 2018). The same performance parameters were also used by Vasudev et al. (2018). The third, but very important study for the IPPs is to do a cost-benefit analysis (CBA). According to Waewsak et al. (2023), a CBA is a method for evaluating a project's benefits and drawbacks, including those pertaining to the installation or alteration of energy projects. One of the important parameters in the evaluation of cost-energy ratios is the levelized cost of energy (LCOE). The LCOE will help the IPPs in determining the selling price of energy so that the plant can achieve the financial parity within a set period of time.

At present, there is no simple process or a tool that be used to undertake these studies for a solar power plant. This is the rationale for this chapter, as it will focus on designing and developing a tool using MS Excel that will analyse the performance and energy metrics of a solar power plant for an IPP. The tool can be further expanded for additional functionalities like energy sales to the utility, energy that cannot be sold, and energy forecasting. Some of the key functional features include - the capability of capturing data, performing calculations, visualising the results with plots and graphs. Thus core objectives of this chapter are outlined below.

Objectives

- To apply the available mathematical formulas for performance and energy metrics analysis of grid-tied solar power plants.
- To design and develop an MS Excel tool to calculate the various performance and energy metrics parameters essential to utility and owner.
- To generate various graphical visuals that will interpret the data captured and results obtained.

LITERATURE REVIEW

Energy metrics in solar photovoltaic (PV) plants have been under study for the past few years, and they persist up to date. This is because the world is moving towards green or renewable forms of energy to produce electricity to limit the rising challenges that are caused by the growing emissions of greenhouse gases into the atmosphere (Musti, 2023). This move has resulted in use of other alternatives, such as renewable sources such as solar, wind, geothermal, and hydropower, to produce electricity (Njong et. al., 2022). Specifically, overall costs of various components used in solar energy production have reduced drastically over the time, due to technological advances. At the macro level, lending policies have also changed. Green financing for large electricity production plants is increasingly becoming popular. Liberalization and power sector reforms has attracted investments from several IPPs who established their own plants. In the process, several plants have been established in the world. These developments have led to the study of the performance and energy metric analysis as well as the life cycles costs solar PV plants. As of today, several different algorithms and computational tools have been reported to study the performance and costs. Some of them include DigiSilent, SimPow, PVSyst and BluSol etc. This chapter puts the key elements in analysing performance and costs over a given period of time. MS Excel is chosen as the development platform due to its popularity and simplicity. Besides the study of the solar power plants, overall system analysis (load flow studies, stability analysis and energy usage studies etc) is also important in improving the various operations involved. MS Excel platform provides ample avenues for further development and modifications. For instance, performance studies of solar power plants may eventually require further, additional functionalities like energy sales to the utility, energy that cannot be sold, and energy forecasting (Sastry & Sahadeo, 2015). It is well-known that solar energy can be only produced during day, with an average production times of 5 to 6 hours per day. And solar energy peaks during mid-day. However, in the recent times, increased quantities of solar energy production is resulting in what is called duck-curve phenomenon and this is causing a situation wherein the solar energy could not be utilized (Pitra and Musti, 2021). Naturally, IPPs will be subjected to losses. Thankfully, green Hydrogen is becoming popular and IPPs have ample opportunities to expand their business strategies by setting up onsite hydrogen production units. All these engineering tasks are quite complex and require a number of studies which in turn require several tools.

Mathematical models and computational processes of performance, expansion and cost studies are wide ranging. IPPs need to need to study the plant performance on a continuous basis. In that process, several commercial software products have been developed. However, several challenges do exist in using such readymade products due to a number of reasons. Procurement of such products, training the users, product customization for specific requirements, maintenance over a longtime etc., are proved to be quite expensive and yet, the product is not owned by the organization. In most cases, users/ IPPs have to be dependent on the product vendors for customization when further studies are required. For example, several IPPs are now moving towards production of green hydrogen and naturally such move requires simulation of a series different case scenarios to determine the possible outcomes. It is desirable that IPPs/analysts have tools that are simple and easier to customize. Hence, design and development of a performance analysis tool using MS-Excel is well justified.

On the user front, expectations include the capabilities to capturing data, perform calculations using standard computational processes, and visualising the data with plots and graphs. Currently, there is no simple tool for performance analysis of solar power plants and thus there is a need to develop one using the MS Excel. This tool will serve as an alternative to the licenced products that are commonly used for solar PV plant analysis, like PVsyst, BlueSol and System Advisor Model (SAM). Keeping in view of the above aspects, this chapter makes an effort to present design and development of a customizable, MS-Excel tool from the cconceptualization to the deployment. In doing so, the chapter first illustrates the required technical and mathematical concepts.

Mathematical Models of Various Metrics Analysis

According to Roy et al. (2022), the energy metrics are used to study the energy consumption of a solar PV plant. There are three key parameters that were considered in the analysis of energy metrics, which are the energy payback time (EPBT), energy production factor (EPF), and life cycle conversion efficiency (LCCE). The first step when analysing energy metrics is to calculate the embodied energy of the entire solar power plant (Mann et. al., 2014). This is the energy that is required during the making of each component that makes up the solar PV plant. This accounts for all the energy consumed from the point of extraction, processing, manufacture, transportation, and installation. Additional data such as the generation data of the system is also required. This is essential in determining the total energy output of the system, which will be further used in the mathematical expressions to calculate the EPBT (Mann et. al., 2014). After analysing the embodied energy of the solar power plant, the next step is to calculate the three key parameters that are used to calculate energy metrics using the following mathematical equations (Rachoutis, 2016):

Energy payback time (EPBT):

$$EPBT = \frac{E_{in}}{E_{aout}} \tag{1}$$

Where:

E_{in}= total embodied energy
E_{out}= annual energy output of the solar PV plant

Energy production factor (EPF):

$$EPF = \frac{E_{out}}{E_{in}} \tag{2}$$

Where: E_{out}= total energy output over span life of the solar power plant in KWh

$$E_{out} = E_{aout} \times \eta s_{ys} \tag{3}$$

$\eta s_{ys=}$ lifespan of the system in years

Thus, equation number (2) is then rewritten by substituting equation (3) into (2):

$$EPF = \frac{E_{aout} \times \eta_{sys}}{E_{in}} \tag{4}$$

Equation (4) is further rewritten by substituting equation (1) into (4):

$$EPF = \frac{\eta_{sys}}{EPBT} \tag{5}$$

Life cycle conversion efficiency (LCCE):

$$LCCE = \frac{E_{out} \times E_{in}}{\eta_{sys} \times E_{sol}} \tag{6}$$

Where: E_{sol} is the solar energy input on the photovoltaic array
Equation (6) can be rewritten as:

$$LCCE = \frac{E_{out}}{E_{sol}}\left(1 - \frac{E_{in}}{E_{aout} \times \eta_{sys}}\right) \tag{7}$$

Equation (7) then further simplifies to:

$$LCCE = \eta\left(1 - \frac{EPBT}{\eta_{sys}}\right) \tag{8}$$

Where:

$$\eta = \frac{E_{out}}{E_{sol}} \tag{9}$$

η= solar conversion efficiency

The parameter 'energy payback time' relates the overall energy production and the building time. It is the total time required for the solar power plant to regenerate the same amount of energy that was used to build and develop the plant over its life span (Huang and Yu, 2017; Fthenakis & Raugei, 2017). There are a few additional parameters that impact the energy payback time of a Solar power plant. They include the technologies used during production, during the installation of the system, rate of solar radiation, angle of the system, the efficiency of various components, etc. The following mathematical expression was used to account for the energy payback time:

$$EPBT = \frac{E_{input} + E_{BOS,E}}{E_{output}} \tag{10}$$

Where:

E_{input}= is the main energy consumed by the PV modules in the life span.
$E_{BOS,E}$= is the energy consumption for stability of system mechanisms.
E_{output}= is the average yearly output of the Solar power plant as the main energy.

In the same disposition as Huang and Yu (2017) a study by Fthenakis and Raugei (2017) on environmental life-cycle assessment of photovoltaic systems, discussed the energy payback time (EPBT) of photovoltaic (PV) systems. EBPT was defined as the time it would take renewable energy to produce the equivalent amount of energy it took to manufacture the system as well as manage its end of life. To analyse the EPBT of a renewable energy production plant, the following mathematical expression was used:

$$EPBT\left(years\right)=\frac{E_{mat}+E_{manuf}+E_{trans}+E_{inst}+E_{EOL}}{\frac{E_{agen}}{\eta_G}-E_{O\&M}} \tag{11}$$

Where:

E_{mat} is the primary energy required to produce the materials required for various parts of the Solar power plant.
E_{manuf} is the energy required to manufacture the Solar power plant.
E_{trans} is the energy required to transport the materials used life cycle.
E_{trust} is the energy required to install the Solar power plant
E_{EOL} is the energy required for end-of-life management
E_{agen} is the annual electricity generation
ηG is the life cycle conversion efficiency
$E_{O\&M}$ is the annual energy required for operation and management of the Solar power plant

To evaluate the life cycle conversion efficiency of a Solar power plant, the annual generation data needs to be acquired and then account for the embodied energy of the PV plant (Ahirwar, 2021). Next step is to account for the embodied energy of the PV plant. Various processes and constituents exist for accounting for the embodied energy of the solar PV plant Ahirwar (2021), the embodied energy is divided into materials production energy, installation energy, energy used for maintenance and energy used for administration. Below are the expressions that were considered for the embodied energy of the Solar power plant.

$$Embodied\ energy(E_{in}) = E_{mfg} + E_{trust} + E_{admin} \tag{12}$$

Where: E_{mfg} is the total manufacturing energy

$$E_{mfg} = E_{mpe} + E_{main} \tag{13}$$

The above computation is then followed by the LCCE analysis of the solar power plant. LCCE is the net productivity of a solar power plant over the course of its lifetime relative to solar energy (radiation). For the calculation of the LCCE of a solar power plant, overall solar input energy to the solar power plant needs to be determined using the following formula:

$$E_{sol}\left(kWh\right)=\frac{E_{aout}}{Module\ efficiency} \tag{14}$$

Where: E_{aout} is the annual output energy of the solar power plant

The LCCE is then calculated using the following mathematical expression.

$$LLCE=\frac{E_{aout}\times\eta-E_{in}}{E_{sol}\times\eta} \tag{15}$$

Where: η is the lifetime of the system in years.

In 2021, Rituraj and Kadar presented the performance analysis of a 50 kWp rooftop grid-connected Solar power plant. Their study not only elucidates a comprehensive methodological approach for assessing the performance of grid-connected solar power plant but also offers a succinct and lucid framework for conducting such analyses. This was accomplished by identifying the location where the grid-connected system is installed and obtaining the meteorological data for the plant. The technical specification datasheet for PV modules was then provided. The integration of the PV module into the grid was then stated. A mathematical model was then provided based on the IEC 61724published by the Electro-Technical Commission (IEC). The parameters that were considered for the performance analysis of the grid-tied Solar power plant are energy loss, capacity utilisation factor (CUF), performance ratio (PR), inverter efficiency, module efficiency, and the electricity output of the Solar power plant. Below are the parameters and their mathematical expressions that were considered for the performance analysis of the grid tied Solar power plant:

Energy loss

$$System\ loss(L_S)=Y_A-Y_F \tag{16}$$

$$Total\ losses(L_T)=L_C+L_S \tag{17}$$

Capacity utilisation factor (CUF)

$$CUF=\frac{Annual\ AC\ energy\ output\times 100}{the\ rathed\ power\times 24\times 365} \tag{18}$$

Performance ratio (PR)

$$PR=\frac{Final\ yield}{Reference\ yield} \tag{19}$$

Inverter efficiency

$$\eta_{inv} = \frac{P_{AC}}{P_{DC}} \tag{20}$$

Module efficiency

$$\eta_{PV} = \frac{P_{DC}}{G_t \times A_m} \times 100 \tag{21}$$

The electricity output of the Solar power plant

$$E = A \times \eta P_V \times \mathrm{R} \times PR \tag{22}$$

Where: E = Energy (kWh), A = Total solar panel area (m^2), R = Average annual solar irradiation

In alignment with the findings presented by Rituraj and Kadar (2021), a study conducted by Omkar et al. (2015) pertaining to the performance evaluation of a 50 kWp rooftop solar power plant yielded similar results. The congruence in results underscores the consistency of observations across distinct investigations in this domain. Omkar et al. (2015), however, considered a few different parameters in the mathematical model that they presented. The performance evaluation was subdivided into two categories: performance analysis and system analysis. The performance analysis parameters that were considered are the final Yield (Y_F), reference yield (Y_F), performance ratio (PR), and capacity factor (CF). The system analysis parameters that were considered are array yield (Y_A), capture loss (L_C), and system loss (L_S). Below are the parameters and their mathematical expressions that were considered:

Performance parameters:

Final Yield (Y_F)

$$Y_F = \frac{System\ Energy\ Output}{System\ Rated\ Power\,(DC)} \tag{23}$$

Reference yield (Y_R)

$$Y_R = \frac{Total\ In-Plane\ Solar\ Insulation}{Reference\ Irradiance} \tag{24}$$

Performance ratio (PR)

$$PR = \frac{Final\ yield\,(Y_F)}{Reference\ yield\,(Y_R)} \tag{25}$$

Capacity factor (CF)

$$CF = \frac{Actual\ energy\ output\ of\ the\ PV\ system}{8760 \times P_r} \tag{26}$$

Where P_r is the rated power of the system.

With the above equations, overall system analysis can be taken up in terms of Array yield, capture losses and finally the system losses.

Array yield (Y_A)

$$Y_A = \frac{E_A\left(kWhAC\right)}{P_r\left(K\omega DC\right)} \tag{27}$$

Capture loss (L_C)

$$L_C = Y_R - Y_A \tag{28}$$

System loss (L_S)

$$L_S = Y_A - Y_F \tag{29}$$

In 2023, Khamharnphol et al. presented the techno-economic evaluation of a 100 kWp solar rooftop Solar power plant designed for deployment across five hospitals situated in the Central Southern region of Thailand. Part of their study examines the cost-benefit analysis (CBA) associated with the solar power plant. According to Khamharnphol et al. (2023), CBA entails a systematic methodology aimed at evaluating the merits and drawbacks of a specific project, particularly in the perspective of energy projects. This analytical approach aids in making well-informed and data-driven decisions by carefully evaluating the potential benefits and costs associated with the project under consideration. CBA can help IPPs determine the financial viability of their project and help them make informed decisions about the current state and future state of their project. CBA was divided into four subsection which are benefit cost ratio (BCR), net present value (NPV), internal rate of return (IRR) and payback period (PBP). BCR represents the relationship between the overall project benefits and the total project costs across a defined time frame. It serves as a valuable indicator of the projected rate of return for investors involved in the project and may also offer insights into potential risks. The NPV of a project represents the disparity between the present value of all cash inflows and outflows over a specified time frame. Solar energy production projects are deemed economically feasible when the NPV is greater than zero and vice versa. The IRR is synonymous with the discount rate "*r*" and represents the rate at which the NPV of a project becomes zero. In other words, the IRR is the interest rate at which the cash inflows and outflows of the project balance out. A higher IRR indicates that the project yields greater net cash inflows for the company. The PBP represents the duration necessary to ascertain the project's profitability and recoup the initial investment. Typically measured in years, a shorter payback period from project commencement signifies the viability and profitability of the venture. This metric aids in assessing the project's financial attractiveness and potential for returns on investment. For each of the subsections mathematical models were provided for each of them. Below are the mathematical models that were provided:

$$BCR = \frac{\sum_{t=0}^{T} \frac{B_t}{(1+r)^t}}{\sum_{t=0}^{T} \frac{C_t}{(1+r)^t}} \quad (30$$

$$NPV = \sum_{t=0}^{T} \frac{B_t}{(1+r)^t} - \sum_{t=0}^{T} \frac{C_t}{(1+r)^t} \quad (31)$$

$$0 = \sum_{t=0}^{T} \frac{C_t}{(1+r)^t} \quad (32)$$

where C_t cash flow in period t and r = internal rate of return.

$$PBP = \frac{Investment}{Net\ annual\ cashflow} \quad (33)$$

Another vital aspect presented by Khamharnphol et al. (2023) is the economic analysis on levelized cost of energy or electricity (LCOE). LCOE represents a metric used to evaluate the cumulative expenses associated with electric power generation over a defined period. In the context of solar PV projects, this metric encompasses the total costs of the Solar power plant, including installation expenses, ongoing maintenance costs, and component replacement expenditures that arise throughout the system's lifespan (typically ~25 years). LCOE is an important indicator for IPPs as it enables comprehensive financial analysis, facilitates informed decision-making, and ensures the success of their power generation projects in a competitive and dynamic energy market. The mathematical model that was used to account for the LCOE of a solar power plant is:

$$LCOE = \frac{\sum_{t=1}^{n} \frac{I_t + M_t + F_t}{(1+r)^t}}{\sum_{t=1}^{n} \frac{E_t}{(1+r)^t}} \quad (34)$$

where I_t, M_t, F_t, and E_t are the investment costs in year t, the maintenance costs in year t, the fuel expenditures in year t and the electric power production in year t, respectively. Here, r denotes the discount rate and n shows the lifespan, in years, of the Solar power plant. Investigations into the levelized cost of electricity for organic photovoltaics have yielded noteworthy findings. Notably, a study conducted by Mulligan et al. (2015) introduced a mathematical model that bears striking resemblance to the one put forth by Khamharnphol et al. (2023). However, the methodological approach was different and there were a few changes in the breakdown of the mathematical model. The LCOE equation that was considered by Mulligan et al. (2015) is as follows:

$$LCOE = \frac{\sum_{t=0}^{n} \frac{C_t}{(1+r)^t}}{\sum_{t=0}^{n} \frac{E_t}{(1+r)^t}} = \frac{C_{NPV}}{E_{NPV}} \tag{35}$$

The mathematical model approach used by Mulligan et al. (2015) breaks down the LCOE into two important variables, C_{NPV} and E_{NPV}. Steps were then outlined to obtain the two variables. First step is to carry out the energy calculation, this is given by the mathematical expression:

$$E_t = I_{max} \times \frac{\eta}{100} \times 0.365h \tag{36}$$

After obtaining E_t the next step is to find E_{NPV}.

$$E_{NPV} = \sum_{t=0}^{n} \frac{E_t}{(1+r)^t} \tag{37}$$

$$C_{NPV} = \sum_{t=0}^{n} \frac{C_t}{(1+r)^t} \tag{38}$$

In 2021, Abdelhady presented the performance and cost evaluation of a solar dish power plant, particular emphasis was placed on conducting a sensitivity analysis of key financial metrics, including the levelized cost of electricity (LCOE) and net present value (NPV). The works of Abdelhady (2021) provided a similar mathematical model as that provided by Mulligan et al. (2015), however there are certain changes that can be seen in the expression. Below is the mathematical expression that was used to evaluate the LCOE:

$$LCOE = \frac{\sum_{t=1}^{n} \frac{I_t + M_t}{(1+r)^t}}{\sum_{t=1}^{n} \frac{E_t}{(1+r)^t}} \tag{39}$$

$$E_t = E_0\left(1 - \frac{DR}{100}\right)^t \tag{40}$$

Studies on Existing Tools

The domain of solar energy analysis has witnessed an influx of studies comparing various tools used for designing and optimizing photovoltaic systems. Babul et. al (2021) conducted energy and economic studies on grid-tied systems. On the other hand, designing and optimization of the solar power plants include extensive simulation studies. Such studies normally consider key parameters such as the Maximum Power Point Tracking (MPPT) and DC/AC ratio. The research revealed intriguing insights into the sensitivity of these tools to temperature changes, the impact of data entry methods, and the role of Maximum Power Point Tracking (MPPT) algorithms. PVsyst is favoured by a good number of authors due to its ease of design and extensive database, while acknowledging MATLAB's strengths in customizable inverter and panel designs, along with its diverse MPPT algorithms (Bhandari et.al, 2015; Matheus & Musti, 2023) .

In a contrasting exploration, Khan et. al., (2023) undertook a comparative analysis of the tools RET Screen and SAM. Their focus extended to a feasibility study for a 600 kW Solar power plant Installation in Riyadh, Saudi Arabia. The study emphasized the efficiency of monocrystalline silicon modules in the Gulf Cooperation Council (GCC) region. The results showcased minimal deviation between the two software tools (< 1.2 % deviation), reinforcing their reliability in conducting feasibility studies for solar power plants. Furthermore, the capacity factor of the optimized Solar power plant was increased to 17.6%.

The third study takes a broader approach, evaluating 14 popular photovoltaic design and optimization tools. The investigation, conducted at the Faculty of Sciences and Mathematics in Niš, compared simulated results with experimental data. SolarPro's simulation results exhibited the smallest deviation from the experimental average daily solar irradiation per annum. Concerning the performance ratio, PVGIS 5 demonstrated superior performance, achieving the lowest simulation deviation results at 17.9%, in stark contrast to HOMER Grid, which recorded a higher deviation of 52.2%. Analysing the capacity factor showed that the highest annual deviation of the simulation results compared to the experimental results was obtained by HOMER Grid (63.96%), while the lowest deviation was obtained by PVGIS 5 (0.2%). Furthermore, PVGIS displayed the most accurate simulation results in terms of annual electricity production when compared to experimental measurements, solidifying its position as the premier PV analysis tool among the 14 evaluated. The study also highlighted the influence of solar databases on result accuracy (Milosavljević, Kevkić, & Jovanović, 2022).

Information from the performance analysis can be utilized in various planning and operational tasks in the ESI by various stakeholders (Li, 2018). For instance, newer plants can be planned in terms of their size, nature of the fuel and the geolocation (Li, 2018; Musti Merwe, 2022; Musti, 2023b). Load estimation and Demand respond are the essential aspects that can influence the capacity addition (Chhetri et. al., 2014; Musti, 2023a). Though levelized energy costs vary widely, renewable energy sources are the clear winners due to their lower costs and environment friendly nature. Solar PV plants have been actively built all over the world, even several countries have built floating type solar plants in water bodies (Matheus & Musti, 2023). Thus, a lot of cheaper and greener capacity is being added to the grids. However, it is interesting to note that the duck curve phenomenon can interfere with the solar energy production levels, especially in terms of overall utilization (Pitra and Musti, 2021). It is quite possible that excess levels of solar energy production can lead to certain operating conditions where in the solar PV plants may have to curtail their production (Musti, 2023a; Musti 2023b). In such a case, the IPP is subjected to financial loss. Though a few research articles point to the use of energy storage technologies, such a move incurs additional investment.

From the above, it is clear that there is a significant interest in capacity addition in the renewable segment and in particular the solar energy. Planning is heavily dependent on the overall load and the demand response.

METHODOLOGY

Microsoft Excel is used to design and develop a simple MS-Excel tool for incorporating the data, mathematical models, and visualisation of performance analysis. Evaluation of energy metrics is divided into three main categories: energy payback time, energy production factor, and life cycle conversion efficiency. Then each category is considered, and various parameters will be determined. The overall performance analysis is carried out, and the results will be presented using tables and graphs. Figure 1 shown above shows the work breakdown structure of the MS Excel-based tool that is to be developed. The MS Excel-based tool, as seen above, is going to be divided into four main categories: energy metrics analysis, performance analysis, econometric analysis, and additional features. Each of the main branches will be further divided into subdivisions according to their respective branches. The MS Excel-based tool will have a homepage that comes up when the user opens the tool and the same is shown in figure 2. The homepage will consist of a start button that will be required to be pressed by the user in order to proceed with the tool. The tool will not proceed or start unless the start button is pressed. Once the start button is pressed, the user will be brought to the main page.

Under the main page, the user will have three analysis options that they can choose from, namely energy metrics analysis, performance analysis, and econometric analysis. Upon clicking on the button that directs to the analysis of any of the names above, the user will then be required to enter data that is required to carry out calculations or generate visuals. Once the required data is entered, the tool will then carry out calculations and present the user with results, graphical interpretations, or plots where required. Details regarding the incorporation of above discussed mathematical models into the spreadsheet are avoided due to the scope of this chapter.

To demonstrate the effectiveness, the tool is now applied to a specific location for which the values of all the parameters are readily available. The following sections provide the case study, results and discussions.

CASE STUDY AND RESULTS

The geographical site of the system under study is located in Gobabis, Namibia. The site is situated at a latitude of -22.45° S, longitude of 18.97° E and is at an altitude of 1440m. Meteonorm 8.1 is used to obtain the meteorological data of the solar power plant under study. Gobabis has a yearly horizontal global average of 6.42 kWh/m^2/day, horizontal diffuse average of 1.56 kWh/m^2/day, extraterrestrial average of 9.31 kWh/m^2/day, clearness index (ratio) average of 0.689, ambient temperature average of

Figure 1. Breakdown structure part of the MS Excel-based tool considered for this study

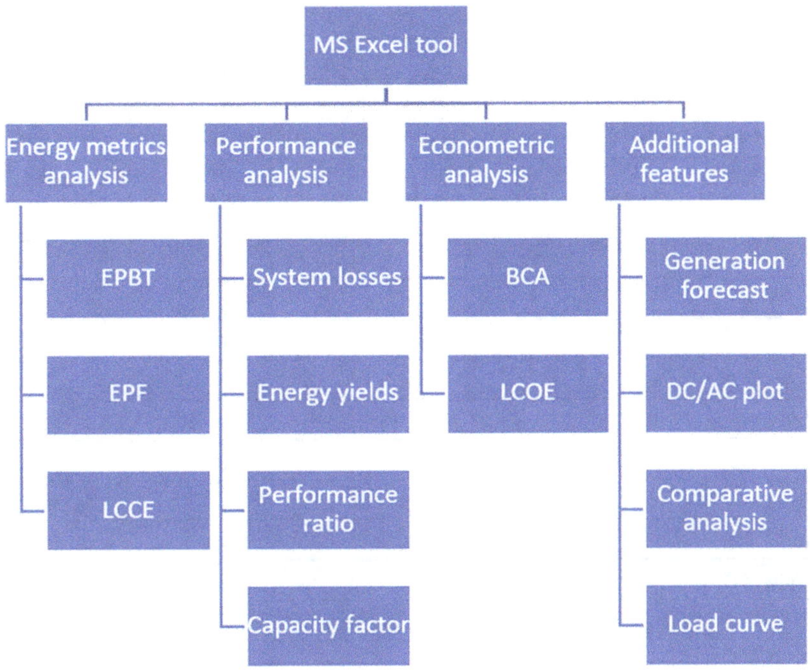

20.4°, and wind velocity average of 2.6 m/s. The plant is grid tied and hence it feeds its energy that it generates to the national grid. The technical specifications of the solar PV plant are shown in table 1.

There will be several components in the solar PV production system. Each component consumes some portion of energy for its own operation. Thus, energy lost is considered as loss. Table 2 provides a listing of various losses along with the percentages. These values were more or less standard in the sense, as they were used by many earlier authors and thus appear as default in the developed spreadsheet as well.

The meteorological data of the solar PV plant was recorded into the tool, using the meteorological data entered into the tool. The tool then visualises the meteorological data captured as shown in figure

Figure 2. Home page of the tool

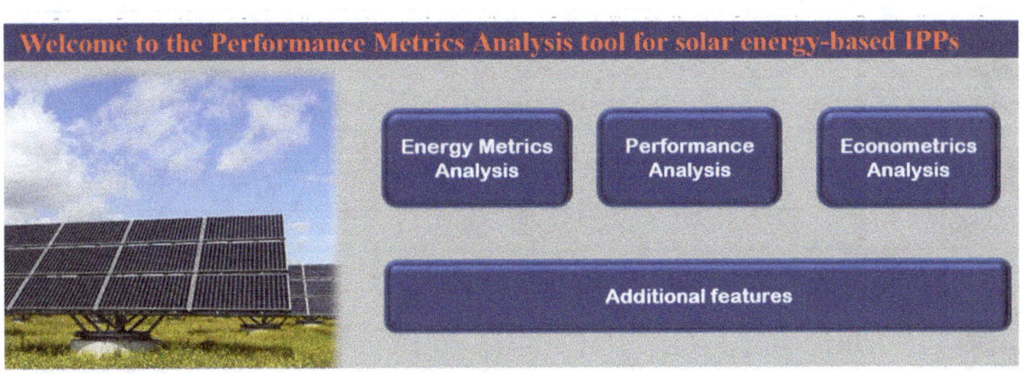

Table 1. Plant details specifications of Komuhuka solar PV plant

Plant Details	
Name of PV plant	Komuhuka
Location	Gobabis, Namibia
Capacity of solar PV plant (kWp)	1000
Module Make	Canadian Solar
Type of Module	Poly Crystalline
Module Efficiency (%)	15%
Single Module Wattage (Wp)	320
No. of Modules	5360
Area of single module (m^2)	1.3
Total solar panel area (m^2)	6968
Inverter Make	ABB
Inverter type	Central Inverter
Inverter capacity (kW)	1000
Grid Inter-connection Voltage (kV)	11
Tilt Angle (β)	18
Azimuth (φ)	180
Installation type	Ground Mount
Inter-Row Pitch (m)	6.5

3. There are a few more visuals to display the meteorological data, that will be produced by the tool, however they are not shown here to avoid multiple images in the captured data.

The next part is assessing the performance of the solar PV plant. This process requires entering the daily plant data by entering the plants daily data captured of the GII, the DC energy from the solar arrays and the energy fed to the grid. For Komuhuka solar PV plant the energy fed to the grid for each month

Table 2. Losses details of Komuhuka solar PV plant

Losses Details	
Inverter losses	7%
Temperature losses	8%
DC Cable losses	2%
AC cable losses	2%
Shadings	3%
Losses due to weak irradiation	3%
Losses due to dust	2%
Soiling losses	2%
Manufacturing tolerances	5%
Other losses	0%

Figure 3. GHI vs. ambient temp. graph of Komuhuka solar PV plant

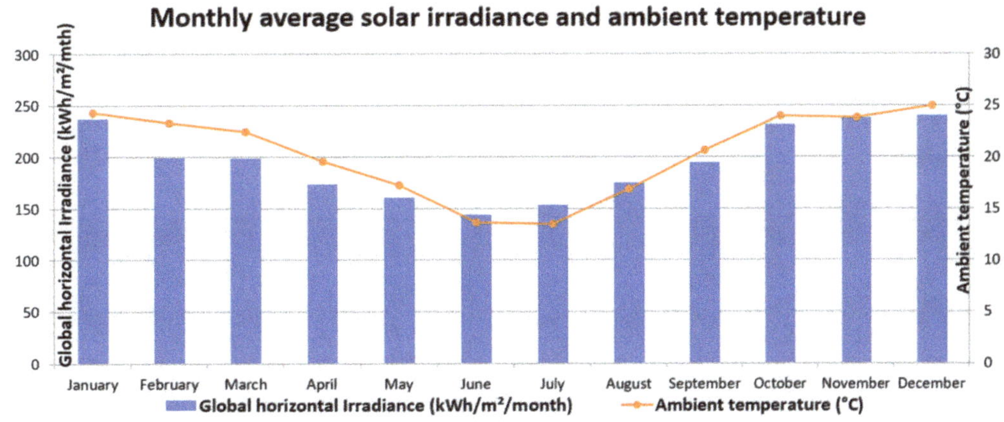

was recorded into the MS Excel tool. The MS Excel tool then uses this data and the technical datasheet to evaluate the performance metrics of the solar PV plant. Based on the captured data and the results produced by the tool, figures 4 and 5 shows the monthly parameters related to energy fed into the grid and the yields.

Figures 6 shows the much required monthly performance ratios and the figure 7 shows the capacity utilization factors. Another important point that needs to be noted, the seasons in Namibia are different from other parts of the world. Therefore, energy yields and even the loads can vary as per the user needs which are mostly influenced by local climatic conditions.

Figure 8 shows the monthly wise capture losses and system losses. And figure 9 shows the most important aspect – actual energy produced versus energy forecasted.

Figure 4. Monthly energy generated by the solar PV plant

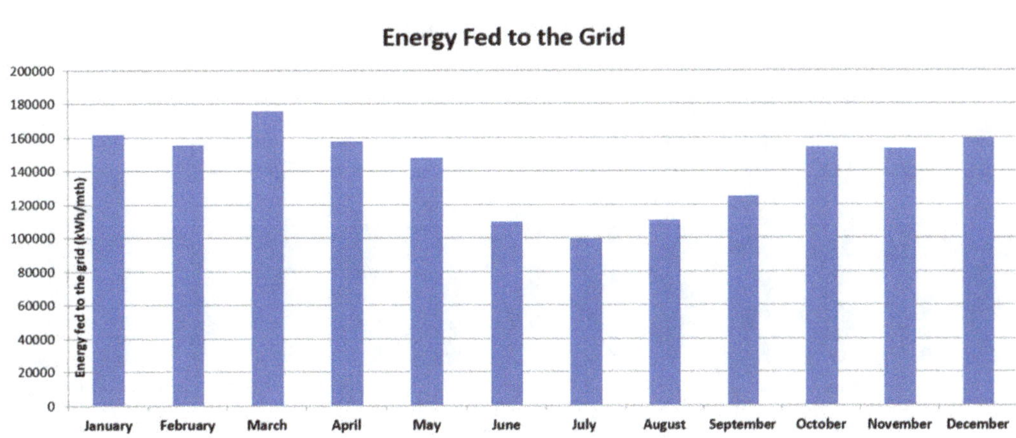

Figure 5. Monthly energy yields of the solar PV plant

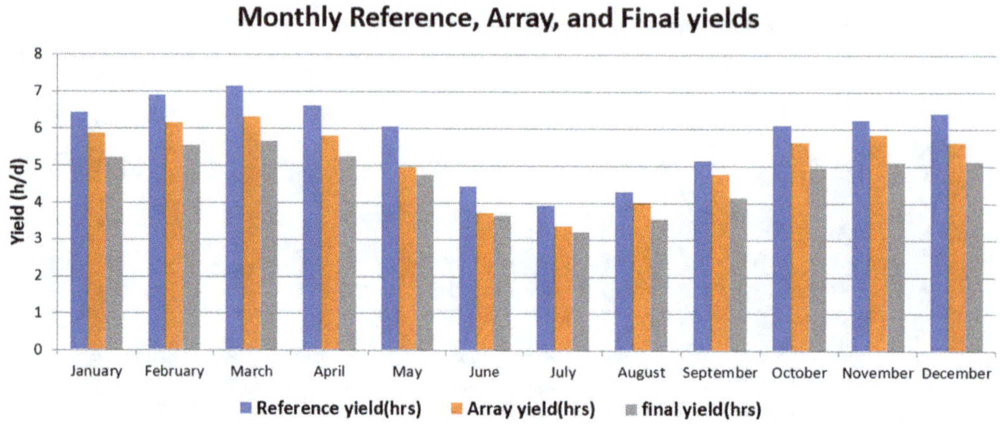

Figure 6. Monthly energy performance ratio of the solar PV plant

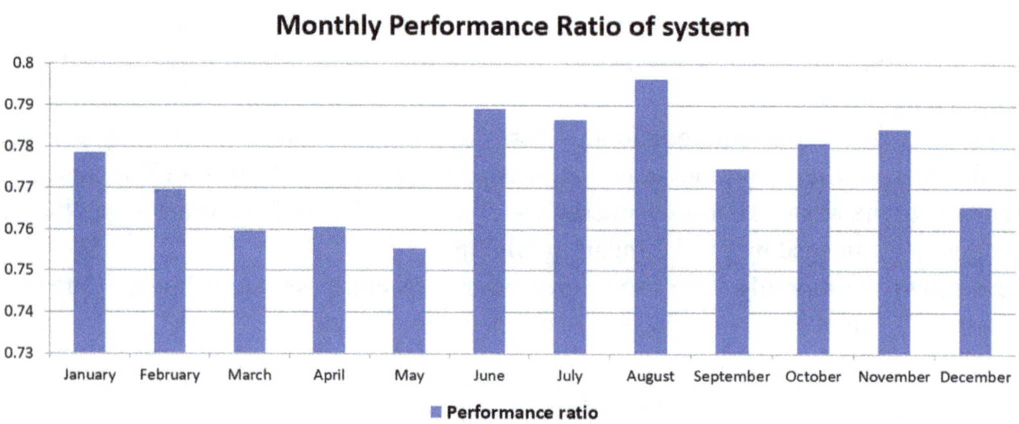

Figure 7. Monthly CUF of solar PV plant

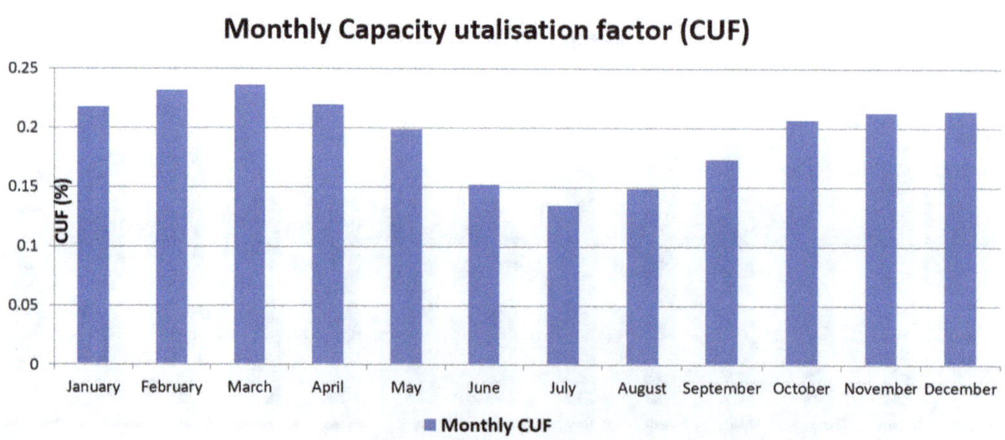

Figure 8. Monthly losses of the solar PV plant

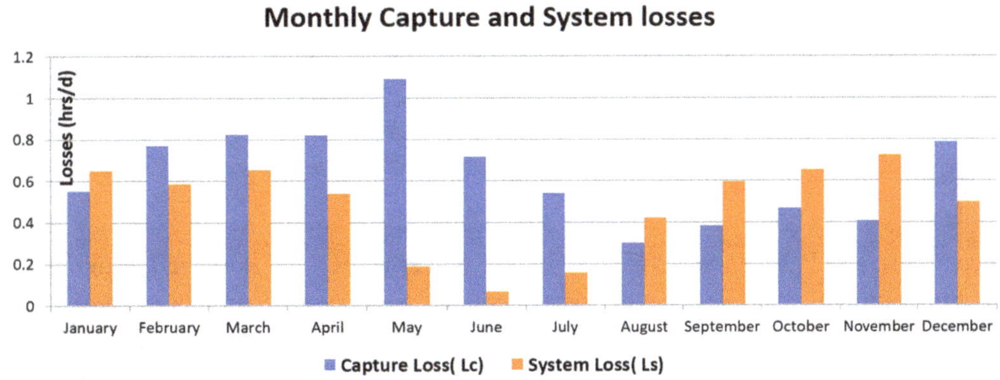

DISCUSSION

From the case study and the results, it can be seen that, the tool is effective in capturing the data of different parameters and perform the required computations. One aspect needs to be noted, the design and development as presented in this chapter have the aim of providing a simple interface for the users, which can be customized as and when required. Importantly, the above chapters bring the idea of developing a tool for assisting the IPPs in evaluating the performance of their plants. As a first attempt, one specific location in Namibia is considered to apply the tool to the context. Comparing the results obtained from this study to that of the work of Roy et al. (2022) on the performance analysis of solar power plant, it can be seen that the location receives relatively higher quantities of sunshine throughout the year and a greater GHI is recorded that resulted in higher energy output.

As explained earlier, solar power based IPPs are now moving towards adding onsite hydrogen production plants and for that, a lot of additional features need to be added. Using this tool, it is possible to apply the data of different years to obtain both the output and visuals for different time durations as

Figure 9. Monthly generated energy vs forecasted energy

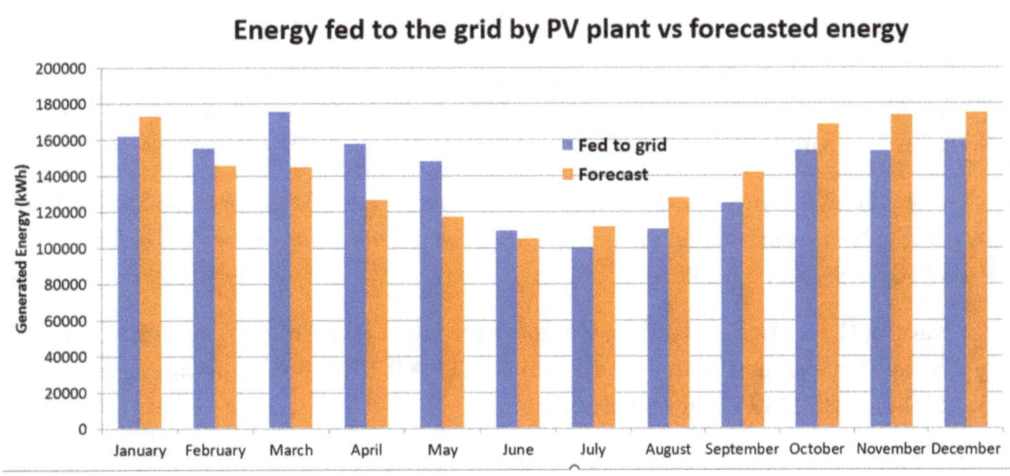

well, with some diligence and customization. While the MS-Excel as a platform offers simple interface, there are a few disadvantages do exist. The data cannot be shared seamlessly due to its inherent nature being a desktop platform. However, future works can consider moving the concept onto web-enabled platforms where data and visuals can be shared seamlessly.

CONCLUSION

Performance, operational and energy metrics of solar power plants are illustrated in detail. Design and development of a simple, yet effective MS-Excel tool and its successful application etc., are presented. A case study was carried out on a 1000 kWp grid-tied solar power plant that is assumed to be on the outskirts of Gobabis, a town located to the east of Namibia for testing the tool. The results obtained from the tool indicated the overall performance ratio of the system to be 77.49%, the utilisation capacity factor to be 19.51%, and the DC/AC ratio to be 1.101. The annual reference yield, array yield, and final yield of the system are 5.79 hrs/day, 5.16 hrs/day, and 4.68 hrs/day, respectively. The annual average capture and system losses are 0.64 hrs/day and 0.47 hrs/day, respectively. A relationship was established between the ambient temperature, global horizontal irradiance (GHI), and losses in the solar power plant. It was noticed that the higher the temperature and GHI, the greater the losses experienced in the system.

REFERENCES

Ahirwar, H. (2021). Life Cycle Conversion Efficiency Analysis of Roof-Top Solar Photovoltaic System. *International Journal of Innovative Research in Electrical, Electronics. Instrumentation and Control Engineering*, *9*(5), 2321–5526. doi:10.17148/IJIREEICE.2021.9558

Babul, A. K., Rajput, S. K., Singh, H., & Yadaw, R. C. (2021). Energy and Economic Analysis of Grid-Type Roof-Top Photovoltaic (GRPV) System. *Lecture Notes in Electrical Engineering*, *710*, 399–408. doi:10.1007/978-981-15-8815-0_34

Bhandari, K. P., Collier, J. M., Ellingson, R. J., & Apul, D. S. (2015). Energy payback time and energy return on energy invested of solar photovoltaic systems: A systematic review and meta-analysis. *Renewable & Sustainable Energy Reviews*, *47*, 133–141. doi:10.1016/j.rser.2015.02.057

Chakwesha, R. T., & Musti, K. S. (2023). Designing a Futuristic Solar Smart Bench for Smart Cities: Towards Developing the Futuristic Urban Infrastructure. In Intersecting Health, Livability, and Human Behavior in Urban Environments (pp. 42-58). IGI Global. doi:10.4018/978-1-6684-6924-8.ch003

Chhetri, R., Wangchuk, K., & Sastry, M. K. S. (2014). Home Energy Audit: A Case Study of Phuentsholing, Bhutan. *IEEE International Conf. on Communication Systems and Network Technologies (CSNT)*, 1022-1025. 10.1109/CSNT.2014.208

Çinici, O. K., Karaca, B., & Acir, A. (2023). Comparison Study in Terms of The Results of Photovoltaic Solar Energy Systems Designed with PVsyst and MATLAB Simulink Software. *Politeknik Dergisi*, *26*(4), 1635–1649. doi:10.2339/politeknik.1266464

Fthenakis, V., & Raugei, M. (2017). Environmental life-cycle assessment of photovoltaic systems. In *The Performance of PV Systems* (pp. 209–232). Modelling, Measurement and Assessment. doi:10.1016/B978-1-78242-336-2.00007-0

Hauwanga, A. S., & Musti, K. S. S. (2022). Evaluation of Energy Losses and Costs Based on the Location of PCC in Smart Grids. *IEEE 11th Int. Conf. on Communication Systems and Network Technologies (CSNT),* 76-82. 10.1109/CSNT54456.2022.9787642

Huang, D., & Yu, T. (2017). Study on Energy Payback Time of Building Integrated Photovoltaic System. *Procedia Engineering*, *205*, 1087–1092. doi:10.1016/j.proeng.2017.10.174

Inamdar, S., & Singh, S. (2018). Performance Evaluation of 5kWp grid connected polycrystalline Si-based PV Plant. *National Power Systems Conf,* 1-6. 10.1109/NPSC.2018.8771846

Kamanja, L., Komarala, V., Dutta, V., Waita, S., & Wachira, K. (2022). *Techno-Economic Analysis of a Rooftop Grid-connected PV Solar System: A case study of JK University of Agriculture and Technology (SAJOREC).* IEEE PES/IAS PowerAfrica. doi:10.1109/PowerAfrica53997.2022.9905282

Khamharnphol, R., Kamdar, I., Waewsak, J., Chiwamongkhonkarn, S., Khunpetch, S., Kongruang, C., & Gagnon, Y. (2023). Techno-Economic Assessment of a 100 kWp Solar Rooftop PV System for Five Hospitals in Central Southern Thailand. *International Journal of Renewable Energy Development*, *12*(1), 77–86. doi:10.14710/ijred.2023.46864

Khan, S. U.-D., Wazeer, I., & Almutairi, Z. (2023). Comparative Analysis of SAM and RETScreen Tools for the Case Study of 600 kW Solar PV System Installation in Riyadh, Saudi Arabia. *Sustainability (Basel)*, *15*(6), 5381. Advance online publication. doi:10.3390/su15065381

Li, C. (2018). Comparative performance analysis of grid-connected PV power systems with different PV technologies in the hot summer and cold winter zone. *International Journal of Photoenergy*, *2018*, 1–9. Advance online publication. doi:10.1155/2018/8307563

Li, T., Roskilly, A. P., & Wang, Y. (2018). Life cycle sustainability assessment of grid-connected photovoltaic power generation: A case study of Northeast England. *Applied Energy*, *227*, 465–479. doi:10.1016/j.apenergy.2017.07.021

Mann, S. A., de Wild-Scholten, M. J., Fthenakis, V. M., van Sark, W. G. J. H. M., & Sinke, W. C. (2014). The energy payback time of advanced crystalline silicon PV modules in 2020: A prospective study. *Progress in Photovoltaics: Research and Applications*, *22*(11), 1180–1194. doi:10.1002/pip.2363

Matheus, M. N., & Musti, K. S. (2023). Design and Simulation of a Floating Solar Power Plant for Goreagab Dam, Namibia. In Human Agro-Energy Optimization for Business and Industry (pp. 1-26). IGI Global. doi:10.4018/978-1-6684-4118-3.ch001

Milosavljević, D., Kevkić, T., & Jovanović, S. (2022). Review and validation of photovoltaic solar simulation tools/software based on case study. *Open Physics*, *20*(1), 431–451. doi:10.1515/phys-2022-0042

Musti, K. S., & Kapali, D. (2021). Digital Transformation of SMEs in the Energy Sector to Survive in a Post-COVID-19 Era. In *Handbook of Research on Strategies and Interventions to Mitigate COVID-19 Impact on SMEs* (pp. 186–207). IGI Global. doi:10.4018/978-1-7998-7436-2.ch009

Musti, K. S. S. (2020). Quantification of Demand Response in Smart Grids. *IEEE Int. Conf. INDISCON*, 278-282. 10.1109/INDISCON50162.2020.00063

Njong, V. A., Musti, K. S., & van der Walt, J. P. (2022), Modeling and Simulation of Solar District Heating Systems. *IEEE 11th Int. Conf. on Communication Systems and Network Technologies (CSNT)*, 125-131. 10.1109/CSNT54456.2022.9787573

Omkar, K., Srikanth, M. v., Swaroop, K. P., & Rama Rao, P. V. V. (2015). Performance evaluation of 50KWp rooftop solar PV plant. *2015 International Conference on Industrial Instrumentation and Control, ICIC 2015*, 761–765. 10.1109/IIC.2015.7150844

Pitra, G. M., & Musti, K. S. S. (2021). Duck Curve with Renewable Energies and Storage Technologies. *IEEE 13th Int. Conf. (CICN)*, 66-71. 10.1109/CICN51697.2021.9574671

Rachoutis, E., & Koubogiannis, D. (2016). Energy Payback Time of a Rooftop Photovoltaic System in Greece. *IOP Conference Series. Materials Science and Engineering*, *161*(1), 012092. Advance online publication. doi:10.1088/1757-899X/161/1/012092

Rajput, P., Singh, Y. K., Tiwari, G. N., Sastry, O. S., Dubey, S., & Pandey, K. (2018). Life cycle assessment of the 3.2 kW cadmium telluride (CdTe) photovoltaic system in composite climate of India. *Solar Energy*, *159*, 415–422. doi:10.1016/j.solener.2017.10.087

Rajput, S. K., & Dheer, D. K. (2022). Integration of 100-kWp PV with low-voltage distribution power system in composite climate: Performance and energy metrics analysis. *International Journal of Ambient Energy*, *43*(1), 8176–8192. doi:10.1080/01430750.2022.2092775

Rituraj, R., & Kadar, P. (2021). Performance Analysis of 50 KWp Rooftop Grid-Connected PV System. *SISY 2021 - IEEE 19th International Symposium on Intelligent Systems and Informatics, Proceedings*, 177–182, 10.1109/SISY52375.2021.9582513

Roy, S. K., Rajput, S. K., Prakash, R., & Dheer, D. K. (2022). Performance and Energy Metrics Analysis of Stand Alone Rooftop PV plant. *IEEE International Conf. on Emerging Frontiers in Electrical & Electronic Tech, ICEFEET.* 10.1109/ICEFEET51821.2022.9848352

Sastry, M. K. S., & Sahadeo, A. (2015). Distributed Cloud Computing Based GIS Solution for Electrical Power Utility Asset Management. *IEEE International Conf. on Computational Intelligence and Communication Networks (CICN)*, 832-838. 10.1109/CICN.2015.169

Sastry Musti, K. S. (2023a). Energy Sector Growth and Sustainability: Role of Government. In Leadership and Governance for Sustainability (pp. 186-203). IGI Global. doi:10.4018/978-1-6684-9711-1.ch010

Sastry Musti, K. S. (2023b). Multicriteria Decision Analysis for Sustainable Green Financing in Energy Sector. In *Green Finance Instruments, FinTech, and Investment Strategies. Sustainable Finance*. Springer. doi:10.1007/978-3-031-29031-2_1

Sastry Musti, K. S., & Van der Merwe, M. (2022). A Novel MS Excel Tool for Multi-Criteria Decision Analysis in Energy Systems. In *Optimal Planning of Smart Grid With Renewable Energy Resources* (pp. 83–109). IGI Global. doi:10.4018/978-1-6684-4012-4.ch003

Sharma, R., & Tiwari, G. N. (2013). Life cycle assessment of stand-alone photovoltaic (SAPV) system under on-field conditions of New Delhi, India. *Energy Policy*, *63*, 272–282. doi:10.1016/j.enpol.2013.08.081

Tripathy, M., Joshi, H., & Panda, S. K. (2017). Energy payback time and life-cycle cost analysis of building integrated photovoltaic thermal system influenced by adverse effect of shadow. *Applied Energy*, *208*, 376–389. doi:10.1016/j.apenergy.2017.10.025

Vasudev, K. P., Mathew, M., Anand, A., & Hossain, J. (2018). Performance Analysis of a 48 kWp Grid connected Rooftop Photovoltaic System. *2018 4th International Conference for Convergence in Technology (I2CT)*. 10.1109/I2CT42659.2018.9058318

Waewsak, J., Khamharnphol, R., Khunpetch, S., Kamdar, I., Chiwamongkhonkarn, S., Kongruang, C., & Gagnon, Y. (2023). Techno-Economic Assessment of a 1 MW Solar PV Rooftop System at Thaksin University (Phatthalung Campus), Thailand. ASEAN Journal of Scientific and Technological Reports, 26(1), 59–75. doi:10.55164/ajstr.v26i1.248159

Wang, R., Lam, C. M., Hsu, S. C., & Chen, J. H. (2019). Life cycle assessment and energy payback time of a standalone hybrid renewable energy commercial microgrid: A case study of Town Island in Hong Kong. *Applied Energy*, *250*, 760–775. doi:10.1016/j.apenergy.2019.04.183

Yadaw, R. (2020). Energy and Economic Analysis of Grid-Type Roof-Top Photovoltaic (GRPV). *System*. Advance online publication. doi:10.1007/978-981-15-8815-0_34

Zhang, Y., Zhu, N., Zhao, X., Luo, Z., Hu, P., & Lei, F. (2023). Energy performance and enviroeconomic analysis of a novel PV-MCHP-TEG system. *Energy*, *274*, 127342. Advance online publication. doi:10.1016/j.energy.2023.127342

KEY TERMS AND DEFINITIONS

Econometrics: Is the analysis and measurement of relationships between financial parameters using statistical techniques, mathematics, and theoretical frameworks.

Energy Forecasting: Is the process of estimating future energy output, demand, or consumption using data from the past, trends, and numerous contributing factors.

Energy Supply Industry (ESI): Is a complicated industry made up of various segments where several organisations and businesses can be found and operated. The generating, transmission, and distribution systems are among the key segments.

Independent Power Producer (IPP): A private entity that generates electricity typically from renewable sources such as solar, wind, hydro, or biomass and sells the electricity either directly to consumers or to utilities.

Levelized Cost of Electricity (LCOE): Is an economic measure which indicates the expense of producing electricity from a specific energy source or technology over the lifespan of a power plant.

Power Purchase Agreement (PPA): Is an agreement between two parties defining the terms and conditions for the sale and purchase of electricity over a certain period of time. The parties are often a power producer (such as a developer of renewable energy) and a power purchaser (such as a utility company or an enterprise entity).

Return on Investment (ROI): Is a financial metric that is used to determine if a project is profitable or not, by measuring the returns generated from the project over the project's cost.

Visualisation: Is an illustration or graphical depiction of data, information, or concepts. It involves presenting abstract or difficult-to-understand concepts in a more engaging and comprehensible way by using charts, graphs, diagrams, pictures, and other visual elements.

Chapter 3
Long-Term Analysis of the Usability and Performance of Photovoltaic Systems

Assia Koussa
https://orcid.org/0000-0003-2953-9037
Université des sciences et technologies Houari Boumediene, Algeria

Sabrina Boudra
Université des sciences et technologies Houari Boumediene, Algeria

Djohra Saheb
https://orcid.org/0000-0003-2953-9037
Centre de Développement des Energies Renouvelables, Algiers, Algeria

Mustapha Koussa
Centre de Développement des Energies Renouvelables, Algiers, Algeria

Abdelhak Razagui
Centre de Développement des Energies Renouvelables, Algiers, Algeria

Houdaifa Ounis
Centre de Développement des Energies Renouvelables, Algiers, Algeria

ABSTRACT

The PVF-Chart method consists of a combination of correlations and fundamental expressions for the hourly calculations of solar radiation at a given location. It uses long-term monthly average solar radiation and ambient temperature to predict the annual performance of a photovoltaic array. The chapter aims to analyse the long-term performances of a PV system used to cover an individual household situated in different Algerian climatic zones. The obtained results are compared to those presented in the literature and from which it can be concluded that the developed software program permits to calculate the PV system performances with a good exactitude which confirm its uses as a tool to analyse the long-term performance of PV system at any climate where only the monthly mean daily weather data is available. From these later, the hourly, the daily, and the annual performance of a photovoltaic array are predicted. For the preliminary assessment of the energy-producing of the studied PV project, the LCOE was calculated for three interest rates and two lifetimes.

DOI: 10.4018/979-8-3693-1014-4.ch003

INTRODUCTION

The imperative to increase renewable resources, seen as the primary alternative to fossil fuels, has become a crucial strategy in addressing electricity shortages following the depletion of traditional energy sources and mitigating environmental issues. Renewable energy encompasses natural power sources capable of self-replenishment, including solar, wind, geothermal, tidal or hydroelectric, and biomass. According to the International Energy Agency, renewable electricity is poised to contribute to nearly 95% of the increase in global power capacity through 2026, with solar (photovoltaaic) PV alone accounting for more than half (IEA, May 2022).

To enhance the efficiency and reliability of energy derived from photovoltaic systems, evaluating the long-term performances related to solar energy conversion systems is essential. A significant challenge in this evaluation lies in the unavailability of hourly measured data, often limited to monthly mean daily weather data. To address this issue, the utilizability and the PV F-Chart method emerge as the most suitable solutions.

The PV F-Chart method estimates the long-term average performance of various photovoltaic systems, including utility feedback systems, battery storage systems, and systems without feedback or storage. Developed by Klein and Beckman (Klein S. A., 1981) & (Klein S. A., 1985) based on earlier work (Siegel, 1981), (Evans D. L., 1978.), (Evans D. L., 1980), extended by (Clark D. R., 1984) & (Clark D. R., 1983), (Oh S., 2013.), (Assad, 2021), (Xiao Z. e., 2023) and this method incorporates a computationally simplified algorithm for evaluating the utilizability function to predict long-term performance. The utilizability function traces back to works such as those of (Klein S. A., 1978), (Liu, 1963), (Whillier A., 1953a), and (Whillier, 1953b), (Evans D. L., 1982), (Haberl, 2004), (Oh S. a., 2016), (Ray, 2020), and (Mazzeo, 2023) which have developed a modified and somewhat simplified general method for calculating monthly average daily utilizability.

In this context, a MATLAB program was developed based on these two methods to calculate the long-term performance of a photovoltaic system. The program's parameters were compared with those presented in (Duffie, 1991) and (Koussa, 2014). Furthermore, the program predicts the hourly, daily, and annual performance of a photovoltaic array.

The developed program was utilized to evaluate the long-term performances of a PV system covering an individual household in different sites in Algeria. To highlight the effects of climatic parameters and site latitude on the system's long-term performances, sites belonging to different Algerian climatic zones (Hot-summer Mediterranean climate (Csa), Hot and dry arid climate (Gwh), Tropical and Subtropical Desert Climate (Bwh), and Cold semi-arid climates (Bsk)) were considered. For the preliminary assessment of energy production in the considred PV project, an economic study was conducted, calculating the Levelized Cost of Energy (LCOE) for three interest rates and two lifetimes.

THE METHODOLOGY DESCRIPTION

The Monthly Hourly Array Electricity Output

The monthly hourly array electricity output Ei for the hour h can be determined by using the following equation:

$$\overline{E_i} = S_c \overline{I_i} \overline{\eta_i} \tag{1}$$

Where

S_c: photovoltaic cell array area (m^2) ;

$\overline{I_i}$: Monthly average hourly solar radiation on tilted surface (MJ/m^2);

$\overline{\eta_i}$: Monthly average array efficiency for the hour h.

The Monthly Average Hourly Solar Radiation on Tilted Surface

The monthly – average hourly solar radiation on tilted surface $\overline{I_i}$ can be evaluted as follow:

$$\overline{I_i} = \overline{K_T}\,\overline{H_0}\left[\left(r_i - \frac{\overline{H_d}}{\overline{H}} r_d\right) R_b + \frac{\overline{H_d}}{\overline{H}} r_d \left(\frac{1+\cos\beta}{2}\right) + \rho_g r_i \left(\frac{1-\cos\beta}{2}\right)\right] \tag{2}$$

Where

$\overline{K_T}$ is the monthly – average daily clearness index

$\overline{H_0}$ is the monthly – average daily extraterrestrial solar radiation (Mj/m^2)

r_i is the ratio of total radiation in an hour to total in a day (%)

$\overline{H_d}$ is the monthly – average daily diffuse solar radiation (Mj/m^2)

$\overline{H}$ is the monthly – average daily solar radiation (Mj/m^2)

r_d is the ratio of diffuse radiation in an hour to diffuse in a day (%)

R_b is the ratio of beam radiation on titled surface to that on a horizontal surface (%)

β is tilt angle of the array from horizontal (°), and

ρg is the ground reflectance (%).

The Monthly Average Array Efficiency $\overline{\eta_i}$

The monthly average array efficiency ($\overline{\eta_i}$) for the hour h is found by the following equation:

$$\overline{\eta_i} = \eta_{mp,ref}\eta_e\left[1 + \frac{\mu_{mp}}{\eta_{mp,ref}}\left(\overline{T}_{a,i} - T_{ref}\right) + \frac{\mu_{mp}\overline{I_T}}{\eta_{mp,ref}}\frac{\tau\alpha}{U_L}\left(1-\eta_{mp,ref}\right)Z_i\right] \tag{3}$$

Where

$\eta m_{p,ref}$ is the maximum power point efficiency from measurement at reference condition,

ηe is the efficiency of any power conditioning equipment,

μm_p is the temperature coefficient at the maximum power point efficiency measurement,

$\overline{T}_{a,i}$ is the average ambient temperature for the hour i (° F),

T_{ref} is the temperature from measurement at reference condition (° F),

τ is the transmittance

α is the absorptance,
U_L is collector overall heat loss coefficient, and
Z_i Can be expressed empirically by the following equation:

$$Z_i = \left(\frac{\overline{I_0}}{\overline{I_T}}\right)(a_1 b_1 + a_2 b_2 + a_3 b_3) \quad (4)$$

Where

$$a_1 = R_b^2 + \rho(1-\cos\beta)R_b + \rho^2\frac{(1-\cos\beta)^2}{4}$$

$$a_2 = R_b(1+\cos\beta - 2R_b) + \frac{\rho(1+\cos\beta-2R_b)(1-\cos\beta)}{2}$$

$$a_3 = \left[\frac{(1-\cos\beta)}{2} - R_b\right]^2$$

$$b_1 = -0.1551 + 0.922\overline{k_T}$$

$$b_2 = 0.1456 + 0.0544\ln\overline{k_T}$$

$$b_3 = k_T\left(0.2769 - 0.3184\overline{k_T}\right)$$

The utilizability function Ø, is defined using the following expression:

$$\varnothing = \begin{cases} 0 & if\ X_c \geq X_m \\ (1 - X_c/X_m)^2 & if\ X_m = 2 \\ \left| |g| - \left[g^2 + (1+2g)\left(1 - X_c/X_m\right)^2\right]^{1/2} \right| & otherwise \end{cases} \quad (5)$$

Where

$$X_m = 1.85 + \frac{0.169\overline{R_h}}{\overline{K_T^2}} - \frac{0.0696\cos\beta}{K_T^2} - \frac{0.981\overline{K_T}}{\cos\delta^2}$$

R_h is the ratio of monthly-average hourly radiation on the tilted surface to that on a horizontal surface ($\frac{\overline{I_T}}{\overline{I}}$, where $\overline{I}$ is the monthly – average hourly radiation)

$\overline{K_T}$ is the monthly –average hourly clearness index ($\overline{I}/\overline{I_0}$),

δ is the declination

$g = (X_m - 1) / (2 - X_m)$

X_c is the monthly-average critical radiation ratio, (L_i/E_i where L_i is the load for the hour i and E_i is the electrical energy output of the array)

expressed by the following relation:

$$X_c = \frac{\overline{L_i}}{\overline{\overline{E_i}}} = \frac{\overline{L_i}}{S_c \eta_i G(\beta, \gamma)_i}$$

The Excess Electrical Energy

The hourly excess electrical energy produced by the retained PV system can be estimated in term of the utilizability by the following equation:

$$E_{exc}^i = E_i \varnothing \tag{6}$$

So, the hourly energy used to cover the electrical load, is as follows:

$$E_{load}^i = E_i (1 - \varnothing) \tag{7}$$

While the monthly- average daily excess electrical energy produced by the considred PV system is given by:

$$G_{exc}^d = \sum_{i=1}^{24} E_{exc}^i \tag{8}$$

and the monthly- average daily energy used to cover the electrical load, is expressed by:

$$G_{load}^d = \sum_{i=1}^{24} E_{load}^i \tag{9}$$

Therefore, the monthly – average fraction of the load carried by the PV system without batteries is given by the following equation:

$$F_0 = \frac{G_{load}^d}{\sum_{i=1}^{24} L_i} \tag{10}$$

The monthly – average fraction of the load carried by the PV system with batteries is given by the following equation:

$$\Delta F_s = 1 - F_o \tag{11}$$

To calculate ΔF_s, the term So is defined which is a dimensionless parameter and represents the fraction of electrical energy produced by the photovoltaic system, not consumed and which is stored in the batteries. It is quantified by the following relationship:

$$S_o = \frac{\eta_B G_{ex}^d}{\sum_{i=1}^{24} Li} \tag{12}$$

Where η_B is the efficiency of the storage batteries used.

So is a parameter characterizing the storage battery's state of charge.

So, if $S_o << \frac{Q_b}{\sum_{i=1}^{24} L_i}$, this indicates that the state of charge of the storage system is at its low level and at this charge limit, So and ΔFs approach zero value.

However if, $S_o >> \frac{Q_b}{\sum_{i=1}^{24} L_i}$, the excess energy G_{exc}^d available is greater than the average load per hour and can be stored while controlling the parameter ΔFs. Thus, as the loading of the accumulators F0 increases and approaches its limit value, it should be maintained below the unit value.

On clear days, an optimaly sized photovoltaic system produces a sufficient amount of electrical energy. This will cover all the electrical load during the day period and the storage batteries will only be used during the night period. This leads to adding another parameter imposing the upper charge limit of the storage system and which is defined as the ratio of the available capacity Qb and the maximum possible charge that can be stored in a battery. This is expressed by the following relationship:

$$\Delta F \max = \min \left[\Delta Fs, \frac{Qb}{\sum_{i=1}^{24} L_i} \right] \tag{13}$$

To simplify the use of the equation (13), Clark et al by (Clark D. R., 1984) developed an empirical relationship to calculate the additional coverage provided by the storage system and which allows the two limits required by their state of charge to be respected. This relationship is given by the following equation:

$$\Delta Fs = \frac{So + \Delta F\,\max - \left[\left(So + \Delta F\,\max \right)^2 - 4\ P\ So\ \Delta \mathrm{Fmax} \right]^{0.5}}{2\ \mathrm{P}} \tag{14}$$

With

$$P = 1.315 - 0.1059 \frac{F_o \sum_{i=1}^{24} L_i}{Q_b} - \frac{0.1847}{\overline{K_T}} \tag{15}$$

METHODOLOGY OF THE CALCULATION

To calculate the monthly electrical output and solar fraction for a PV Power System Array, the following equations are utilized. To evaluate the developed software (C), an example presented in (Duffie, 1991) (DB) is used for comparison, as well as the one provided by (Koussa, 2014) (KM). In this particular example, a PV power system array with four modules is considered, characterized by a Nominal Operating Cell Temperature (NOCT) of 46°C, a reference temperature of 25°C, a maximum power efficiency of 10.4%, a single module area of 0.427m^2, a temperature coefficient of the maximum power point efficiency of -0.0005, an efficiency of maximum power point electronics of 0.9, an array tilt angle of 35°C, an array azimuth of 0°C, a usable battery capacity of 1200 Wh, a battery efficiency of 0.8, a system location at a latitude of 40°C, and the month is March. The monthly-average daily and hourly horizontal radiation, diffuse radiation, and global solar radiation incident on the PV solar collector are computed by the developed software. The ground reflectance is set at 0.2, and the load remains constant and equal to 100 W between 9 AM and 3 PM and zero at other times. The monthly-average hourly temperatures, starting at 6:00 a.m., are 4, 5, 5, 6, 8, 10, 11, 12, 12, 11, 10, and 9°C. Additionally, Kt is 0.448, δ is -2.42 degrees, ωs is 87.97 degrees, and β is 40 degrees. Table 1 presents the hourly results obtained by the developed software, (Duffie, 1991), and (Koussa, 2014). In this table, Gi, Zi, Ef, E. PV, Xc, phi, Eexe, El, and Us respectively represent the global solar radiation on the tilted surface, the array efficiency, the PV electrical output, the critical radiation ratio, the utilisability function, the excess energy, the load, and the energy consumed by the load.

The presented results are relative to a standalone system and collector type is not a tracking one. The selection is done in the developed software as is shown in figures 1 and 2.

Utilizing the aforementioned inputs, the developed PV F-Chart calculates the electricity produced by the system. The hourly values predicted for March performance, including total solar radiation, monthly average array efficiency, monthly average excess energy, monthly average load, monthly average PV electrical output, and the monthly-average fraction of the load carried by the PV system without batteries, are as follows:

- Monthly average solar radiation (MWh): 15.5935
- Monthly average array efficiency (%): 0.0973

Table 1. Results comparison between the developed software (Duffie, 1991; Koussa, 2014) (C): Results obtained by the simulation using the developed MATLAB program

Time	Gi Mj/m²			Zi			E. PV (Wh)			Xc			phi			Eexc (Wh)			El (Wh)			Us (Wh)		
	C	KM	DB	C	KM	DB	C	KM	DB	C	KM	DB	C	KM	DB	C	KM	DB	C	KM	DB	C	KM	DB
4-5	0	0	0	0.000	0	0	0	0	0	0.000	0.000	0	0.000	0.000	0	1.000	1.000	1	0	0	0	0	0	0
5-6	0	0	0	0	0	0	0	0	0	0	0	0	0	0	0	1	1	1	0	0	0	0	0	0
6-7	0.151	0.149	0.15	2.134	2.256	2.228	0.101	0.103	0.102	7.286	7.32	7	0	0	0	1	1	1	7.286	7.32	7	0	0	0
7-8	0.6 03	0.596	0.60	1.822	1.903	1.906	0.099	0.100	0.099	28.23	28.21	28	0	0	0	1	1	1	28.22	28.21	28	0	0	0
8-9	1.124	1.110	1.11	1.651	1.731	1.728	0.095	0.097	0.096	50.88	51.06	50	0.00	0	0	1	1	1	50.87	51.06	50	0	0	0
9-10	1.631	1.610	1.61	1.533	1.603	1.604	0.092	0.094	0.096	71.27	71.87	71	1.403	1.391	1.412	0087	0.089	0.087	6.190	6.42	6	100	100	100
10-11	2.032	2.006	2.00	1.457	1.525	1.525	0.090	0.091	0.090	86.35	87.09	86	1.158	1.168	1.166	0.176	0.162	0.159	15.17	14.15	14	100	100	100
11-12	2.255	2.226	2.22	1.420	1.484	1.486	0.088	0.090	0.089	94.37	94.72	93	1.060	1.066	1.082	0.220	0.197	0.193	20.76	18.7	18	100	100	100
12-13	2.255	2.226	2.22	1.420	1.484	1.486	0.088	0.089	0.088	93.89	94.24	93	1.065	1.066	1.077	0.218	0.195	0.181	20.42	18.36	18	100	100	100
13-14	2.032	2.006	2.00	1.457	1.524	1.525	0.089	0.090	0.088	85.480	85.38	84	1.17	1.178	1.189	0.171	0.153	0.149	14.60	13.04	13	100	100	100
14-15	1.631	1.610	1.61	1.533	1.603	1.604	0.096	0.091	0.090	70.229	69.83	69	1.424	1.442	1.454	0.081	0.077	0.075	5.771	5.39	5	0	0	0
15-16	1.124	1.101	1.11	1.651	1.731	1.728	0.094	0.094	0.093	49.916	49.64	49	0	0	0	1	1	1	49.91	49.64	49	0	0	0
16-17	0.603	0.596	0.60	1.822	1.903	1.906	0.097	0.098	0.096	27.713	27.57	27	0	0	0	1	1	1	27.71	27.88	27	0	0	0
17-18	0.151	0.149	0.15	2.134	2.256	2.228	0.104	0.101	0.100	7.447	7.16	7	0	0	0	1	1	1	7.447	7.16	7	0	0	0
18-19	0.00	0	0	0.00	0	0	0.08	0	0	0.00	0	0	0	0	0	1	1	1	0	0	0	0	0	0

Figure 1. Selection of system type in the PVF-chart program

```
****************System Type********************
*            1. utility feedback system          *
*            2. Battery Storage System           *
*            3. stand alone system               *
*            4. Get user file                    *
*            5. Quit                             *
*************************************************
```

Figure 2. Selection of collector type in the PVF-chart program

```
***************Collector Type*******************
*            1. No tracking                      *
*            2. Horizontal axis                  *
*            3. Vertical axis                    *
*            4. Two axis                         *
*            5. Quit                             *
*************************************************
```

- Monthly average excess energy (Wh): 254.290 (C), 247 (Koussa, 2014), and 242 (Duffie, 1991).
- Monthly average to the load (Wh): 418.770 (C), 427.04 (Koussa, 2014), and 422 (Duffie, 1991).
- Monthly average load (Wh): 600.000 (C), (Koussa, 2014) & (Duffie, 1991).
- Monthly average PV electrical output (Wh): 673.060 (C), 646 (Koussa, 2014), & 664 (Duffie, 1991).

Therefore, the fraction of the load carried by the PV system without batteries, F0 in %, is 69.80 (C), 71.17 (Koussa, 2014) & 70.3 (Duffie, 1991).

The fraction of the energy stored in the battery (%) is 30.20 (C), 34.9 (Koussa, 2014), & 32.3 (Duffie, 1991).

The fraction of the load covered by the batteries (%) is 24.02 (C), 23.69 (Koussa, 2014), & 23.2 (Duffie, 1991).

The fraction of the load covered by the overall PV system (%) is 93.82 (C), 94.86 (Koussa, 2014), & 93.5 (Duffie, 1991).

From the results obtained for the 12 months of the year, the developed program allows the calculation of the yearly average values of each of the considered parameters. Interestingly, the results presented in Table 1 and above show good agreement between all three methods with very small relative differences, validating the results obtained employing the developed PV F-Chart software.

CASE STUDY: DESCRIPTION OF THE CONSIDERED SYSTEM

To evaluate and compare the performance of the considered system, a photovoltaic installation with an electrical load consisting of several pieces of equipment was selected. The synoptic diagram of the proposed installation is given in Figure 3.

As noted above, this method requires to know the following information:

Figure 3. Configuration of the proposed solar system

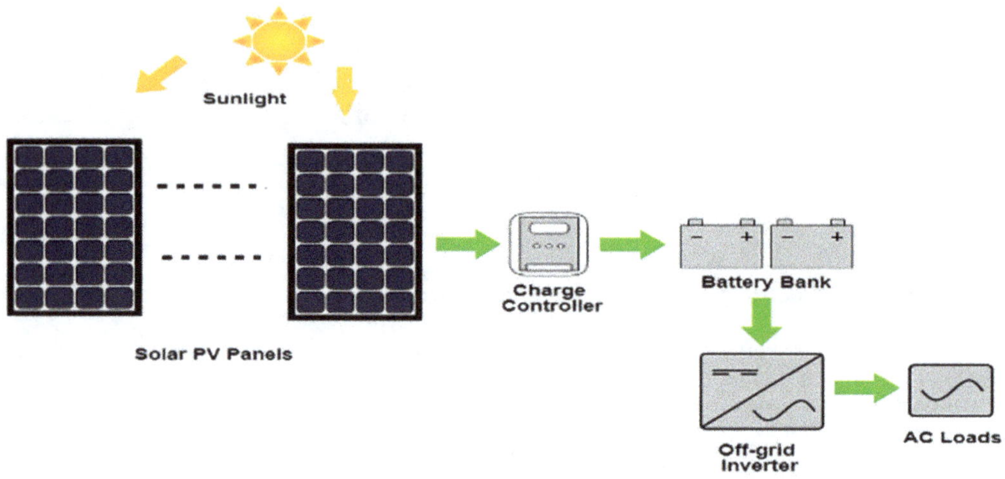

- The geometric and optoelectronic characteristics of the photovoltaic module;
- The Electrical load;
- The Monthly averages per hour of the overall incident irradiation on the surface of the photovoltaic collector;
- The monthly average per hour of the air temperature.

Characteristics of the Used Photovoltaic Module

In this study, the SEP 330 Sunceco PV module was selected as generator, whose specifications are given in Table 2.

Estimation of the Monthly Average Daily of the Global and Diffuse Solar Radiation on Inclined Surfaces

The reconstitution of the monthly averages per day and per hour of the global and diffuse irradiations on a horizontal surface is ensured by the methodology presented in Section 2. The obtained results are depicted in Figures 4 and 5 .

On the other hand, employing the method outlined in Section 2.b, the program enables the calculation of the global irradiation on the designated surfaces, as illustrated in Figure 6.

The Considered Sites

To underscore the influence of climatic parameters and site latitude on the long-term performance of the selected system, multiple sites from various climatic zones in Algeria are taken into account. The selection of these sites is grounded in the availability of meteorological data. The respective geographical coordinates and classifications of these sites are detailed in Table 3 below:

Table 2. Characteristics of the used photovoltaic module

Model	SOLAR CELLS POLY-CRYSTALLINE 156 × 156 MM 72 PCS. (6×12) – 4 BUS BARS
Rated Power (Wp)	330
Max Power Point Vmpp (V)	37.23
Open Circuit Voltage Voc (V)	44.71
Short circuit current Isc (A)	8.06
Maximum System Voltage (V DC)	1000 V (iec), 600 V (UL)
Cell Efficiency	17.46 %
Module Efficiency	15.46 %
Number of By-pass Diodes	6
Maximum Series Fuse	15 A
Temperature Coefficient of Pmax	- 0.45 % / °C
Temperature Coefficient of Voc	- 0.34 % / °C
Temperature Coefficient of Isc	- 0.05 % / °C
Nominal Operating Cell Temperature	47 ± 2°C

Figure 4. The daily evolution of the global irradiations on a horizontal surface

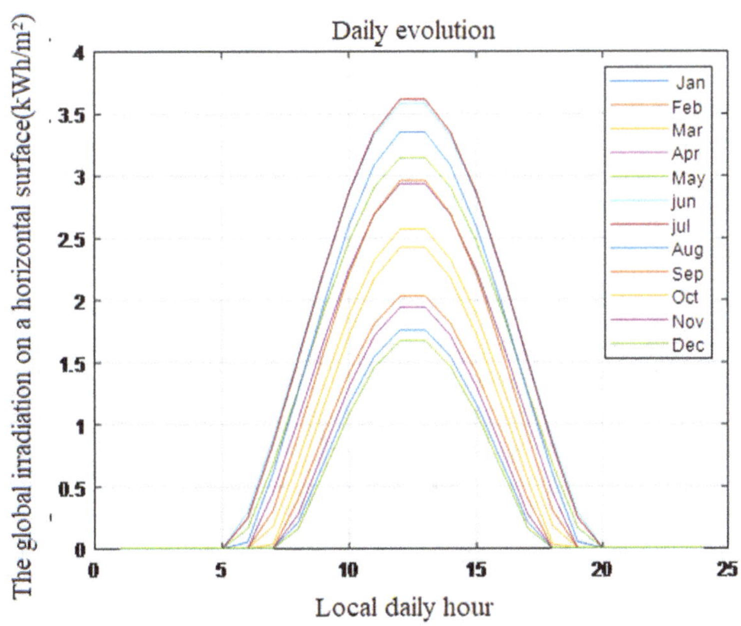

Electrical Load

In the current investigation, the selected representative household is intentionally isolated from the conventional energy distribution network and is fully equipped with all necessary devices to ensure comfort for its occupants. Moreover, the assumption is made that the residence will be consistently oc-

Figure 5. The daily evolution of the diffuse irradiations on a horizontal surface

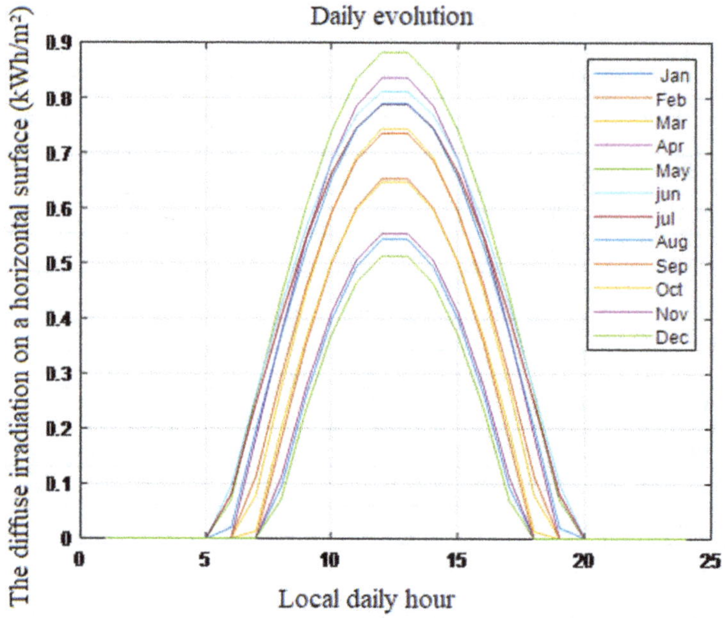

Figure 6. The daily evolution of the global irradiations on an inclined surface

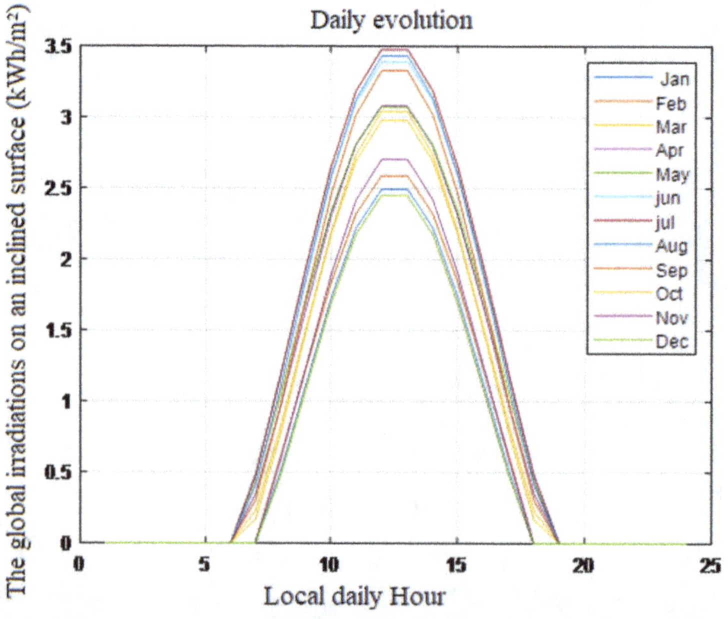

cupied throughout the entire year, and all domestic equipment operates at a standard voltage of 230V and frequency of 50Hz. The specific characteristics of this chosen household are outlined below:

Table 3. Geographical coordinates and climatic characteristics of the considered sites (Chai, 2015)

Site	Latitude in [deg.]	Longitude in [deg.]	Altitude in [m]	Climatic Classification
Ghardaïa	32.4	3.8	468.5	Hot and dry arid climate (Gwh)
Timimoune	29.15	0.28	312	Hot and dry arid climate (Gwh)
Adrar	27.82	0.18	263.9	Hot and dry arid climate (Gwh)
Bechar ville	31.64	2.25	777	Hot and dry arid climate (Gwh)
In Salah	27.12	2.28	268	Hot and dry arid climate (Gwh)
Mascara	35.3	0.18	474	Hot-summer Mediterranean climate (CSA)
Miliana	36.30	2.41	715	Hot-summer Mediterranean climate (Csa)
Annaba	36.83	7.82	3	Hot-summer Mediterranean climate (Csa)
Batna	35.55	6.18	1051.9	climat subtropical humide chaud sans saison sèche (Cfa)
Bejaia Soumam	36.72	5.07	1.8	Hot-summer Mediterranean climate (Csa)
Beni Abbes	30.13	2.17	499.4	Hot-summer Mediterranean climate (Csa)
Beni Saf	35.30	1.38	68	Hot-summer Mediterranean climate (Csa)
BISKRA	34.8	5.73	87.6	Hot and dry arid climate (Gwh)
Constantine	36.28	6.62	694	Hot-summer Mediterranean climate (Csa)
Dar El beidha	36.72	3.25	24.4	Hot-summer Mediterranean climate (Csa)
Djanet	24.55	9.47	1054	Hot and dry arid climate (Gwh)
Djelfa	34.69	3.25	1140	Cold semi-arid climates(BSk)
Chlef	36.22	1.33	103	Hot-summer Mediterranean climate (Csa)
El Beidh	33.67	1	1341	Cold semi-arid climates(BSk)
El Golea	30.57	2.87	397	Hot (subtropical) deserts climate (BWh)
El Kreider	34.15	0.07	1000	Hot (subtropical) deserts climate (BWh)
Essenia Oran	36.63	0.60	90	Hot-summer Mediterranean climate (Csa)
Mostaganem	35.81	0.12	137	Hot-summer Mediterranean climate (Csa)
M'Sila	35.68	4.50	441	Tropical and Subtropical Desert Climate (Bwh)
Saida	34.86	0.16	750	Cold semi-arid climates(BSk)
Setif	36.18	5.25	1033	Hot-summer Mediterranean climate (Csa)
Skikda	36.88	6.92	13	Hot-summer Mediterranean climate (Csa)
Tebessa	35.48	8.13	812.6	Cold semi-arid climates(BSk)
Tamenrasset	22.76	5.52	1377	Hot and dry arid climate (Gwh)
Tindouf	27.67	8.13	403	subtropical desert climate (GWh)
Tlemcen Zenata	35.00	1.45	248.5	Hot-summer Mediterranean climate (Csa)
Tougourth	33.12	6.13	85	Hot and dry arid climate (Gwh)

Type: Individual
Number of Rooms: 4 rooms, kitchen, bathroom, and hallway.

Figure 7 illustrates the daily load profile of the power demand for this household. Notably, there is a peak power demand of 2900 W observed around 12 p.m., contributing to a total daily consumption of 9739 Wh/day.

To address the total electrical load associated with the specified household, a photovoltaic (PV) system has been employed. This system comprises the following components:

- A solar panel array,
- A storage system,
- An inverter.

The chosen PV modules have a capacity of 300 Wc. The storage battery system is characterized by a capacity of 200Ah, a voltage of 12 V, and an efficiency of 80%. The overall system operates at a voltage of 24 V. This configuration is designed to effectively harness solar energy, store it in the battery system, and convert it into usable electrical power for meeting the household's energy requirements.

RESULTS AND DISCUSSION

The results obtained by the developed program are presented in the figures 8 and 9.

In the Figure 8 are represented the collected global solar radiation and Electrical power output values in solar photovoltaic (PV) panels.

On other hand in Figure 9 are presented the stored slectrical energy, used to cover load, part of the electrical load covered only by the panels, the electrical required load and the electrical load covered by the PV system. From the first result presented in Figure 8, it's remarked that the energy efficiency

Figure 7. The daily household load profile

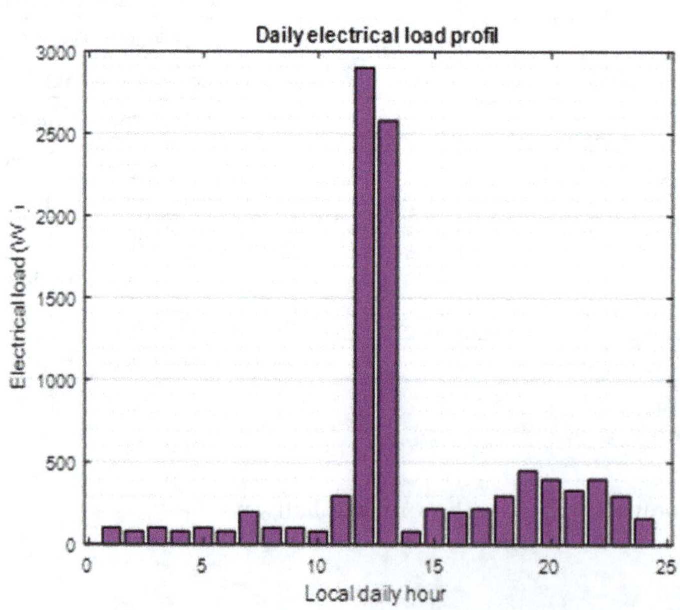

of most present-day photovoltaic cells is only about 9 to 10.3 percent. This represent the cells ability to convert radiative energy to electrical energy which depends on the solar radiation availability and to the seasonal variation.

Figure 8. The collected global solar radiation and electrical power output values in solar photovoltaic (PV) panels

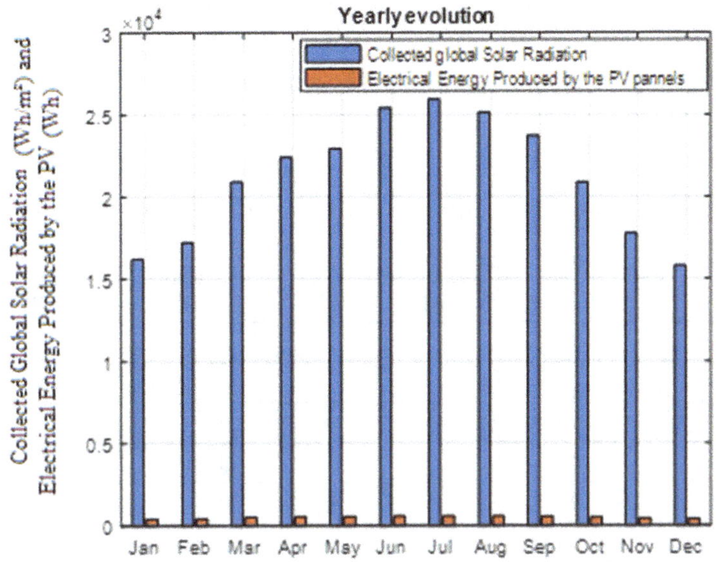

Figure 9. The stored electrical energy and used to cover load, Part of the electrical load covered only by the panels and the Electrical required

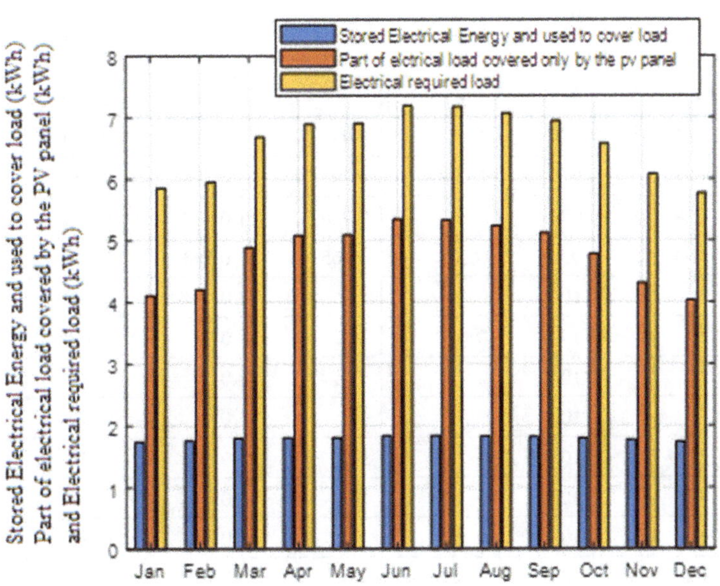

Using the above inputs, PV F-Chart calculates electrical energy produced by the system as shown in Table 4 and Table 5. Table 4 shows the hourly values predicted for December performance, including: the ambient temperature, the total solar radiation incident on the collector, the array efficiency, the array power output, the utilizability, the total excess electrical energy, the energy consumed by the load, the fraction of the load carried by the PV system without batteries. The monthly total values are shown in the bottom of the Table were the solar radiation is 15.8 kWh for the month of December, and the percent of load supplied directly by the array is 40%.

Table 5 presents the results obtained through the calculation of monthly and annual performance for the twelve months. The table includes data on the total solar radiation incident on the collector, array power output, total excess electrical energy, energy consumed by the load, and the fraction of the load covered by the PV system. The results reveal that the annual fraction of the load supplied by the array

Table 4. Calculation results of monthly–average fraction of the load carried by the PV system for December

Time (Hour)	T. °C	Gi Mj	Ef. %	E. PV kWh	Phi	Eexe kWh	El kWh	Us kWh	F0 %
0-1	3.75	0	0.103	0	1	0	0.100	0	0
1 - 2	3.78	0	0.103	0	1	0	0.080	0	0
2-3	3.07	0	0.103	0	1	0	0.100	0	0
3-4	2.3	0	0.104	0	1	0	0.080	0	0
4- 5	1.88	0	0.104	0	1	0	0.100	0	0
5-6	1.99	0	0.104	0	1	0	0.080	0	0
6-7	2.63	0	0.104	0	1	0	0.200	0	0
7-8	3.72	0.484	0.1	0.313	0.723	0.226	0.100	0.868	86.8
8-9	5.10	1.088	0.097	0.680	0.858	0.582	0.100	0.967	96.7
9-10	6.56	1.693	9.4	1.025	0.923	0.945	0.080	0.788	98.5
10-11	7.91	2.182	9.1	1.288	0.777	0.100	0.300	0.287	95.7
11-12	9.00	2.455	9.0	1.427	0	0	2.900	1.427	49.2
12-13	9.69	2.455	9.0	1.423	0	0	2.580	1.422	55.1
13- 14	9.94	2.182	9.0	1.275	0.938	1.196	0.080	0.079	98.9
14 -15	9.72	1.693	9.2	1.008	0.791	0.798	0.220	0.211	95.7
15-16	9.12	1.088	9.5	0.666	0.720	0.480	0.200	0.186	93.1
16-17	8.22	0.484	9.8	0.307	0.465	0.143	0.220	0.164	74.6
17-18	7.17	0	10.2	0	1	0	0.299	0	0
18-19	6.15	0	10.2	0	1	0	0.450	0	0
19-20	5.28	0	10.2	0	1	0	0.400	0	0
20-21	4.68	0	10.3	0	1	0	0.333	0	0
21-22	4.36	0	10.3	0	1	0	0.400	0	0
22-23	4.17	0	10.3	0	1	0	0.300	0	0
23-24	3.82	0	10.3	0	1	0	0.160	0	0
Month12	**5.6**	**15.8**	**10**	**9.410**	**0.62**	**5.371**	**9.862**	**4.039**	**40**

is 49%. Therefore, 51% of the electricity must be sourced from batteries or the utility to meet the total load of the system at the El Beidh site.

The temperature data mapping over a ten-year period is shown in the Figure 10. The red-garnet shaded areas indicate above average temperatures, while the light yellow shaded areas indicate lower than average temperatures those in orange indicate average temperatures. The maximum temperature is observed in the In Salah area with an annual average temperature of 26°C. The minimum average temperature is 14°C observed on the following sites: Sétif, Djelfa Batna and El Beidh which are high plateau sites.

On account of its geographical location, Algeria holds one of the highest solar potentials as shown in the Figure 11. Following this evaluation, it can be concluded that Algeria represents the most important solar potential of the entire Mediterranean basin, with 7896 MJ/year for solar photovoltaic. The distribution of solar potential by climatic region at the level of the Algerian territory is represented in the Figure 11 according to the sunshine received annually: it is noticed that in the coastal zones, the variation of the solar potential is between 215 to 230 MJ, on the highlands, it spans from 235 to 255 MJ and in the Sahara it is from 260 to 280 MJ where the maximum value is obtained on the Adrar site.

The knowledge of the fundamental concepts and particularly the global radiation on the ground and the temperature, enabled the determination of the solar energy produced by a set of photovoltaic modules. So, it's noticed that the maximum, which evaluates to 12.7 kWh, is observed in the south of Algeria while the minimum of 10.2 kWh is accounted for the coastal zones of Algeria.

Also, the same remarks are valid regarding the excess energy produced by the PV generator (Figure 13), such that the maximum is 7.7 kWh in the south and the minimum is 5.9 kWh in the coastal zones of Algeria. This excess energy must be accumulated as rationally as possible in order to be available on days with little sunshine as shown in Figure 13. It can also be injected into the grid or used to produce heat.

Table 5. Calculation results of yearly: Average fraction of the load carried by the PV system for December

Months	Gi Mj	El kWh	F0 %	E.PV kWh	Eexe kWh
Jan	16.21	9.862	41.7	9.669	5.557
Feb	17.24	9.862	42.6	10.158	5.955
Mar	20.91	9.862	49.55	12.121	7.234
Apr	22.42	9.862	51.5	12.840	7.758
May	22.92	9.862	51.65	12.961	7.868
Jun	25.43	9.862	54.2	13.844	8.497
Jul	25.95	9.862	54.09	13.768	8.433
Aug	25.15	9.862	53.13	13.370	8.13
Sep	23.71	9.862	51.94	12.894	7.772
Oct	20.81	9.862	48.5	11.785	7.005
Nov	17.77	9.862	43.7	10.377	6.068
Dec	15.81	9.862	41	9.409	5.371
Year	254.40	118.344	49	143.1960	85.6480

Figure 10. The temperature data mapping in °C

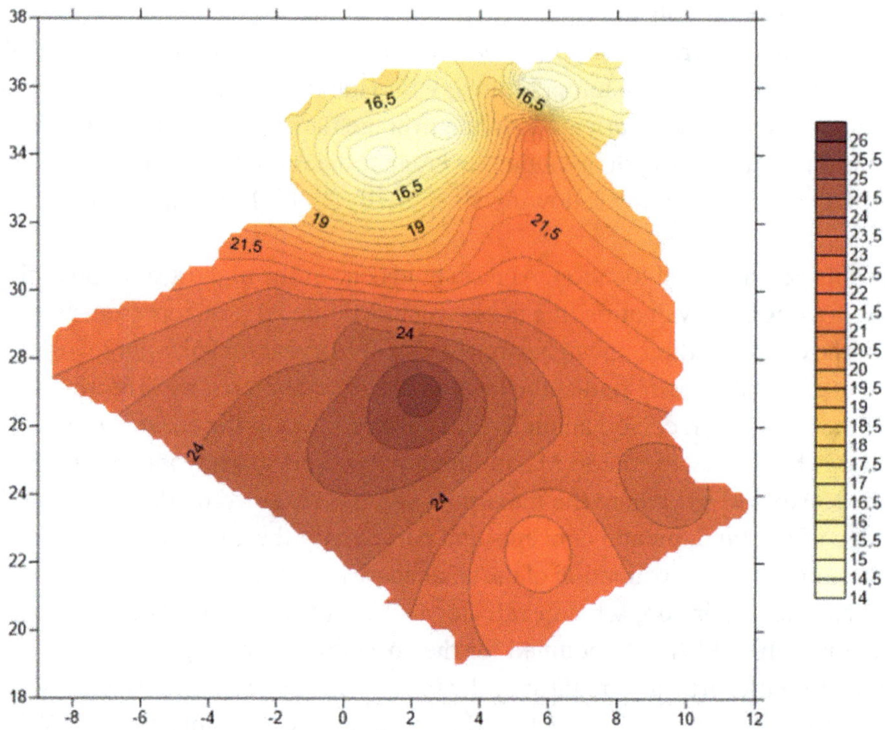

Figure 11. The global radiation data mapping in Mj

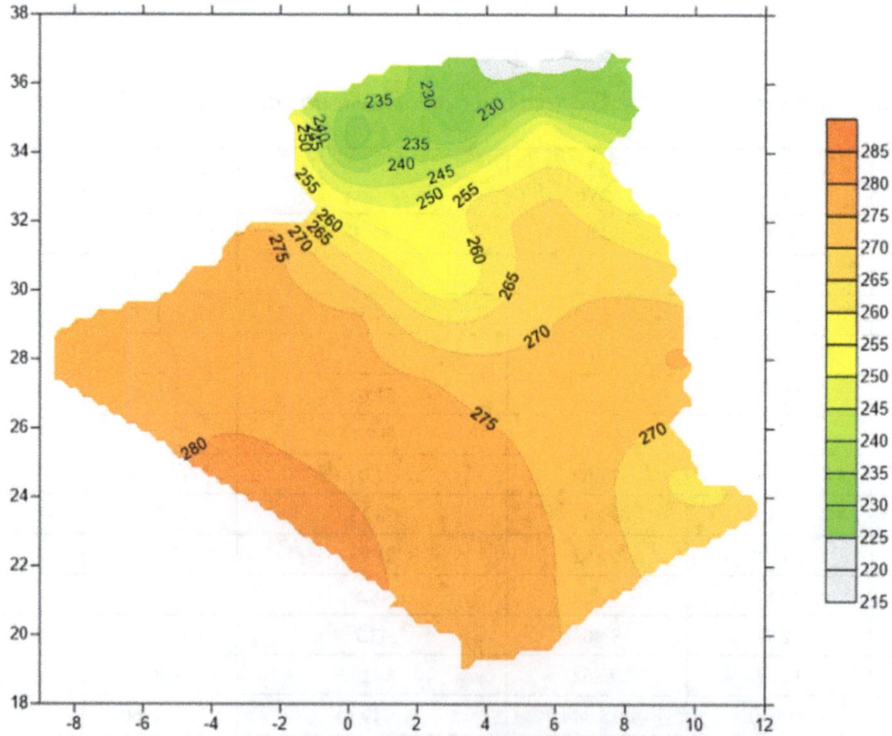

Figure 12. Produced power by the PV system in Wh

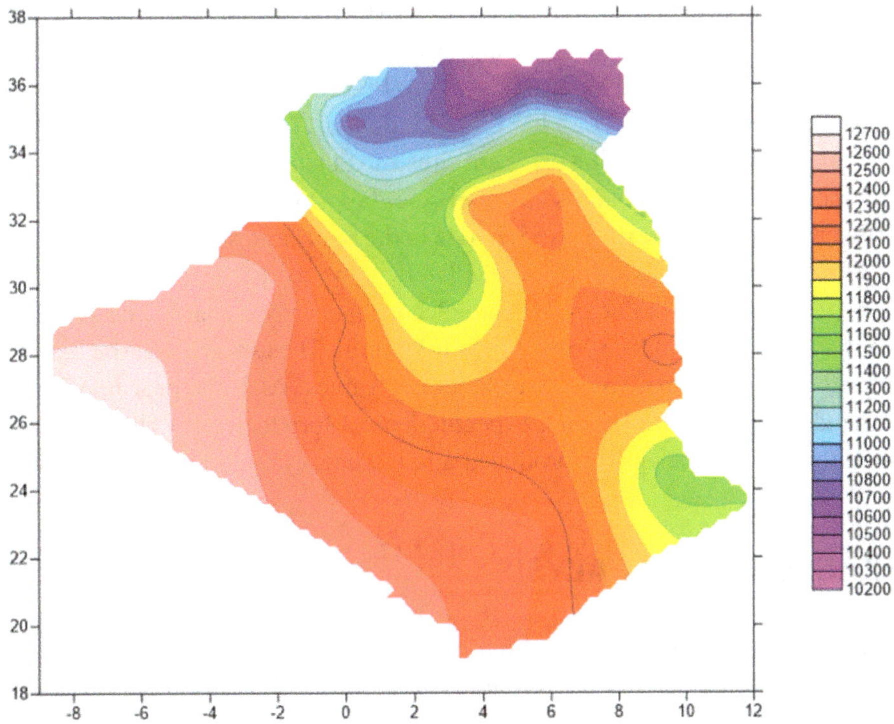

Figure 13. The excess energy data mapping in Wh

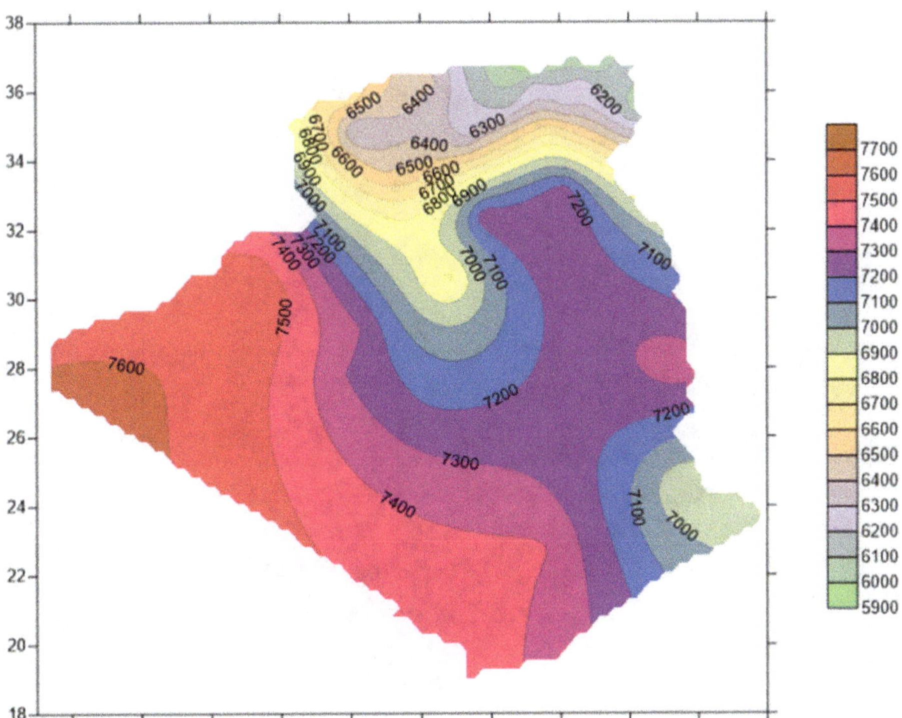

The Figure 14 below represents the electrical load covered by the storage system. It's observed that more the PV production is ensured, the less recourse to the storage system takes place, which is confirmed by comparing the results of Figure 12 with those of Figure14.

Regarding the coverage of the load by the PV system (Figure 15), it's noted that the maximum is 5.05 kWh in the south and the minimum is 4.25 kWh in Skikda and Annaba in the coastal zones of Algeria.

The utilizability variations considering different sites and during various months are analysed in terms of the average temperature, solar radiations and coefficient of global losses. The results presented in the Figure 16 show that the usability amounts to 57.7-61% in the Sahara climatic zones and Pre-Sahara, in the Highlands and in the Coastal climatic zones. These later can be considered as the most favourable zones for the conversion of the available solar energy into electrical energy through the solar panels. The study of the effect on the system performance considering utilizability has shown that the load distribution has a significant impact on the obtained average utilizability values. Also the profile of the electrical load shows a low demand in the high-load hours, expected from 13pm to 14pm. This means that there is an excess of production between the energy produced and the energy that satisfies the demand.

THE TECHNO-ECONOMIC ANALYSIS

The techno-economic analysis examines research, development, and deployment areas with a focus on benefits, costs, risks, timeframes, and uncertainties (Lai, et al., 2019), (Vögele, 2022) and (Ibrahim, 2024). The levelized cost of electricity (LCOE), or levelized cost of energy is calculated as the ratio be-

Figure 14. The load covered by the storage system

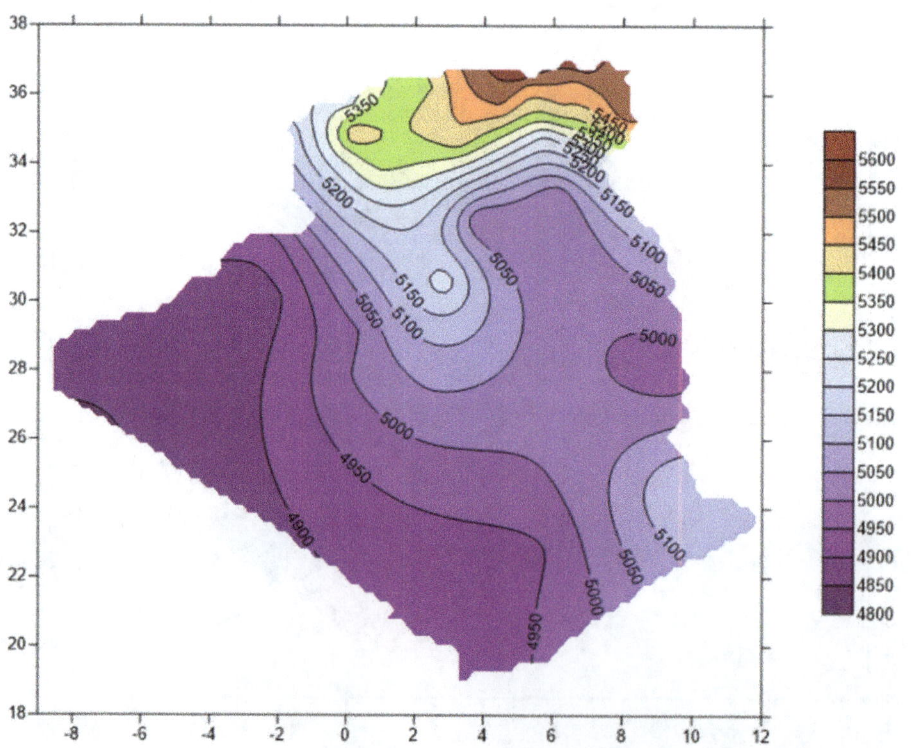

Figure 15. The electrical load covered by the PV system

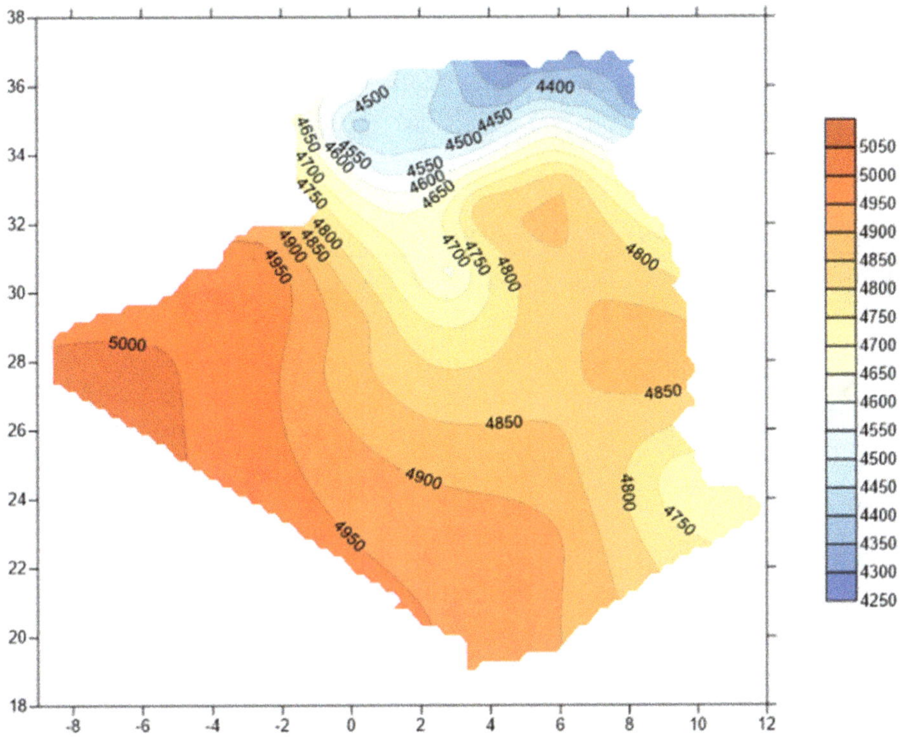

Figure 16. The utilizability variations

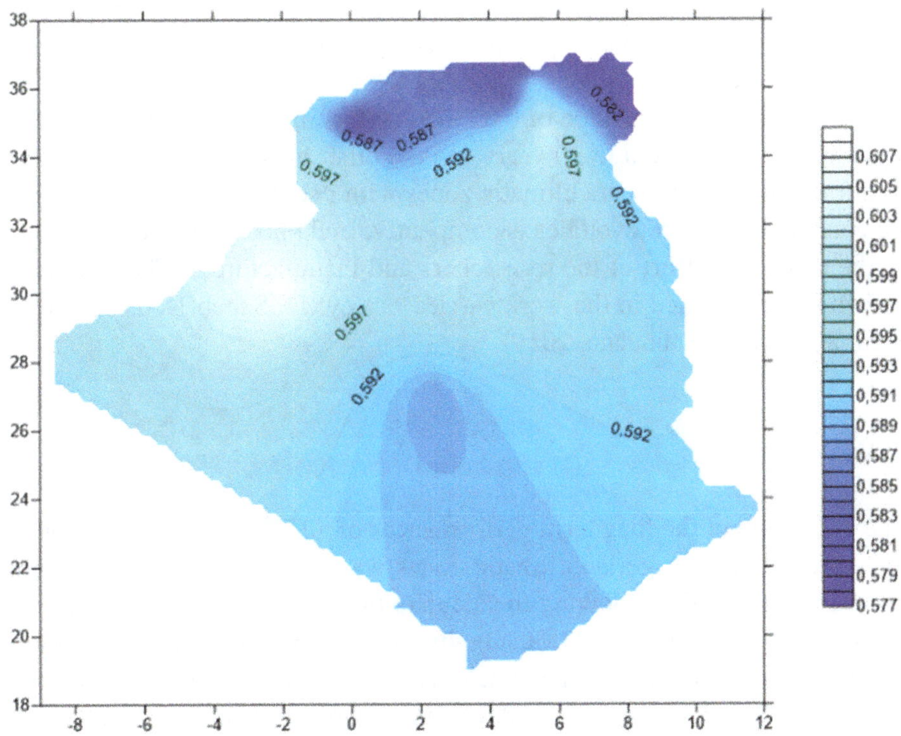

tween all the discounted costs over the lifetime of an electricity generating plant divided by a discounted sum of the actual energy amounts delivered (Lai C. S., 2017), (Abdelhady, 2021)) (Kabeyi, 2023) and (Castro-Santos, 2024) and Inputs to LCOE are chosen by the estimator.

An energy system typically operates over a long lifetime; for a PV system it is around 25 years Thus, LCOE includes a discount rate that converts future cash flows into their present value. The LCOE is calculated as:

$$LCOE = \frac{Sum\ of\ cost\ over\ lifetime}{Sum\ of\ electrical\ energy\ produced\ over\ lifetime}$$

$$LCOE = \frac{I_0 + \sum_1^n \frac{M_t}{(1+r)^t}}{\sum_1^n \frac{E_t}{(1+r)^t}} \tag{16}$$

I0: is the capital cost (€)
M_t: operation and maintenance (O&M) cost (€)
n: system lifetime in years
r: is the discount rate (%)
E_t: is the electrical energy generated (kWh)

The economic study is elaborated for three discount rates and two system lifetimes. The results presented in Figure 17, show that the solar energy price is higher in the Coastal climatic zones than in the Highlands and in the Sahara ones. It is also shown that the solar energy price is inversely proportional to the project lifetime, the maximum value of 0.0336 €/kWh being observed for Annaba in the Coastal climatic zones with r set to 8% and n to 20 years. On the other hand, the minimum value of 0.0248 €/kWh is observed for Adrar in the Sahara climatic zones with r set to 2% and n to 25 years. From these results it can be concluded that photovoltaics is competitive and photovoltaic energy can become more competitive by the conjugate effort of the researchers and manufacturers. These obtained results and conlusions confirm those obtained in the work carried out by (D. Saheb-Koussa, 2009) (Maammeur, 2017), (Chabachi, 2022), and (Sahouane, 2019).

CONCLUSION

In this study, the evaluation of the long term performances of a PV system used to cover an individual household situated in different Algerian climatic zones was considered. To achieve the objective, a MATLAB program was developed allowing to calculate the long-term performance of a photovoltaic system. The different parameters obtained by using this program were compared with those presented in (Duffie, 1991) and (Koussa, 2014), it can be concluded that the obtained results showed a good agreement between all three methods with very small relative differences, which validates the developed PV

Figure 17. The solar energy price

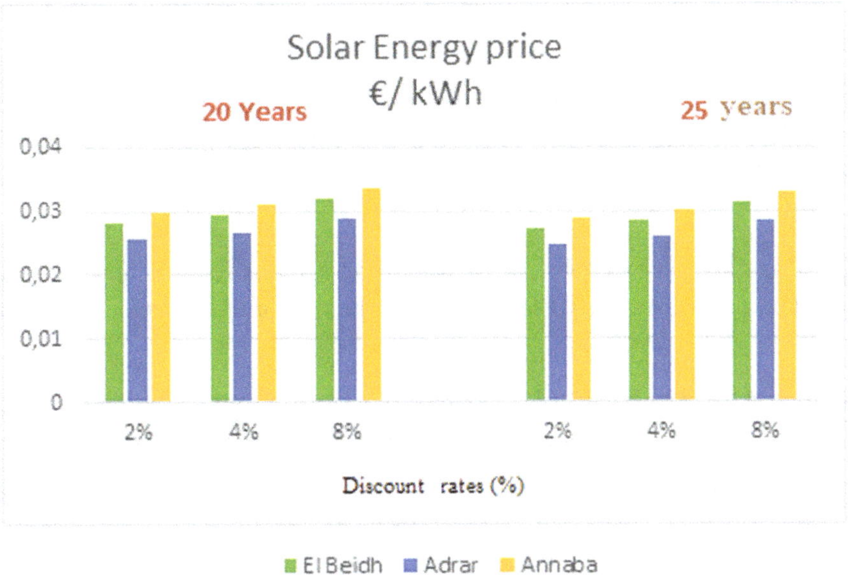

F-Chart software results. Furthermore, the hourly, the daily and the annual performance of a photovoltaic array were predicted

In order to highlight the effect of the climatic parameters as well as that of the site latitude on the long-term performances of the considered system, the majority of the sites belonging to different Algerian climatic zones (Hot-summer Mediterranean climate (Csa), Hot and dry arid climate (Gwh), Tropical and Subtropical Desert Climate (Bwh) and Cold semi-arid climates(BSk) are considered and related data mapping performed. For the preliminary assessment of the energy-producing of the considred PV project, an economic study is carried out in which the LCOE was calculated for three interest rate values (2%, 4% and 8%) and two lifetime ones (20 and 25 yrs) . It is found that the solar energy price is higher in the Coastal climatic zones, the maximum value of 0.0336 €/kWh being observed then follow the Highlands climatic zones with an intermediate price value. And last, the minimum value of 0.0248 €/kWh is observed for Adrar in the Sahara climatic zones. Furtheremore, it is found that the solar energy price is inversely proportional to the project lifetime.

The future study will be the subject of the environmental analysis of a photovoltaic system for each of the sites.

REFERENCES

Abdelhady, S. (2021). Performance and cost evaluation of solar dish power plant: Sensitivity analysis of levelized cost of electricity (LCOE) and net present value (NPV). *Renewable Energy*, *168*, 332–342. doi:10.1016/j.renene.2020.12.074

Assad, M. a. (2021). *Design and performance optimization of renewable energy systems*. Academic Press.

Castro-Santos, L. e., Filgueira-Vizoso, A., Costoya, X., Arguilé-Pérez, B., & Ribeiro, A. S. (2024). Economic viability of floating wave power farms considering the energy generated in the near future. *Renewable Energy*, *222*, 119947. doi:10.1016/j.renene.2024.119947

Chabachi, S. e., Necaibia, A., Abdelkhalek, O., Bouraiou, A., Ziane, A., & Hamouda, M. (2022). Performance analysis of an experimental and simulated grid connected photovoltaic system in southwest Algeria. *International Journal of Energy and Environmental Engineering*, *13*(2), 831–851. doi:10.1007/s40095-022-00474-9

Chai, M. M. (2015). Evaluation of Global Solar Radiation Models for Inclined Surfaces. *Proceedings of the 2015 International Conference on Electrical, Automation and Mechanical Engineering,* (pp. 472-475). Academic Press.

Clark, D. R., Klein, S. A., & Beckman, W. A. (1983). Algorithm for evaluating hourly radiation utilizability function. ASME. *Journal of Solar Energy Engineering*, *105*(3), 281–287. doi:10.1115/1.3266379

Clark, D. R., Klein, S. A., & Beckman, W. A. (1984). A method for estimating the performance of photovoltaic systems. *Solar Energy*, *33*(6), 551–555. doi:10.1016/0038-092X(84)90010-0

Duffie, J. W. (1991). Solar Engineering of Thermal Process. John Wiley & Sons Inc.

Evans, D. L. (1978.). *Combined photovoltaic/ thermal system studies.* Arizona State University – Tempe.

Evans, D. L. (1980). *Simplified design guide for estimating photovoltaic flat array and system performanc.* Sandia: SAND 80-7185.

Evans, D. L., Rule, T. T., & Wood, B. D. (1982). A new look at long term collector performance and utilizability. *Solar Energy*, *28*(1), 13–23. doi:10.1016/0038-092X(82)90219-5

Haberl, J. S. (2004). *Literature Review of Uncertainty of Analysis Methods*. Energy Systems Laboratory.

Ibrahim, N. A.-e., Wan Alwi, S. R., Abd Manan, Z., Mustaffa, A. A., & Kidam, K. (2024). Climate change impact on solar system in Malaysia: Techno-economic analysis. *Renewable & Sustainable Energy Reviews*, *189*, 113901. doi:10.1016/j.rser.2023.113901

IEA. (2022). *Renewable Energy Market Update*. Paris: IEA. https://www.iea.org/reports/renewable-energy-market-update-may-2022

Kabeyi, M. J., & Olanrewaju, O. A. (2023). The levelized cost of energy and modifications for use in electricity generation planning. *Energy Reports*, *9*, 495–534. doi:10.1016/j.egyr.2023.06.036

Klein, S. A. (1978). Calculation of flat-plate collector utilizability. *Solar Energy*, *21*(5), 393–402. doi:10.1016/0038-092X(78)90171-8

Klein, S. A. (1981). *TRNSYS - A transient system simulation program*. Medison: Solar Energy Laboratory.

Klein, S. A. (1985). *PV F-Chart User's Manual: DOS Version*. www.fchart.com

Lai, C. S., Locatelli, G., Pimm, A., Tao, Y., Li, X., & Lai, L. L. (2019). A financial model for lithium-ion storage in a photovoltaic and biogas energy system. *Applied Energy*, *113179*, 113179. Advance online publication. doi:10.1016/j.apenergy.2019.04.175

Lai, C. S., & McCulloch, M. D. (2017). Levelized cost of electricity for solar photovoltaic and electrical energy storage. *Applied Energy*, *190*, 191–203. doi:10.1016/j.apenergy.2016.12.153

Liu, B. Y. (1963). A rational procedure for predicting the long term performance of flat plate solar energy collectors. *Solar Energy*, *7*, 53. doi:10.1016/0038-092X(63)90006-9

Maammeur, H. H., Hamidat, A., Loukarfi, L., Missoum, M., Abdeladim, K., & Nacer, T. (2017). Performance investigation of grid-connected PV systems for family farms: Case study of North-West of Algeria. *Renewable & Sustainable Energy Reviews*, *78*, 1208–1220. doi:10.1016/j.rser.2017.05.004

Oh, S. (2013). *Origins of analysis methods in energy simulation programs used for high performance commercial buildings*. PhD Diss.

Oh, S., & Haberl, J. S. (2016). Origins of analysis methods used to design high-performance commercial buildings: Solar energy analysis. *Science and Technology for the Built Environment*, *22*(1), 87–106. doi:10.1080/23744731.2015.1090277

Ray, B. e., Shah, R., Islam, M. R., & Islam, S. (2020). A new data driven long-term solar yield analysis model of photovoltaic power plants. *IEEE Access : Practical Innovations, Open Solutions*, *8*, 136223–136233. doi:10.1109/ACCESS.2020.3011982

Saheb-Koussa, D., Haddadi, M., & Belhamel, M. (2009). Economic and technical study of a hybrid system (wind–photovoltaic–diesel) for rural electrification in Algeria. *Applied Energy*, *86*(7-8), 1024–1030. doi:10.1016/j.apenergy.2008.10.015

Sahouane, N. e., Dabou, R., Ziane, A., Neçaibia, A., Bouraiou, A., Rouabhia, A., & Mohammed, B. (2019). Energy and economic efficiency performance assessment of a 28 kWp photovoltaic grid-connected system under desertic weather conditions in Algerian Sahara. *Renewable Energy*, *143*, 1318–1330. doi:10.1016/j.renene.2019.05.086

Siegel, M. D., Klein, S. A., & Beckman, W. A. (1981). A simplified method for estimating the monthly-average performance of photovoltaic systems. *Solar Energy*, *26*(5), 413–418. doi:10.1016/0038-092X(81)90220-6

Vögele, S. P.-J., Poganietz, W.-R., Kleinebrahm, M., Weimer-Jehle, W., Bernhard, J., Kuckshinrichs, W., & Weiss, A. (2022). Dissemination of PV-Battery systems in the German residential sector up to 2050: Technological diffusion from multidisciplinary perspectives. *Energy*, *248*, 123477. doi:10.1016/j.energy.2022.123477

Whillier, A. (1953a). The Utilization of Solar Energy in South Africa. *Journal of South Africa Institute of Mechanical Engineers*, *2*(9), 260–266.

Xiao, Z. e., Huang, X., Liu, J., Li, C., & Tai, Y. (2023). A novel method based on time series ensemble model for hourly photovoltaic power prediction. *Energy*, *276*, 127542. doi:10.1016/j.energy.2023.127542

Yusuf, A., & Ballikaya, S. (2023). Performance analysis of concentrated photovoltaic syste using thermoelectric module with phase change material. *Journal of Energy Storage*, *59*, 106544. doi:10.1016/j.est.2022.106544

Chapter 4
Effect of Solar Energy Technology on Green Hydrogen Production

Ilyès Nouicer
Renewable Energy Development Center, Algeria

Bessem Kaghouche
Abdelhafid Boussouf University Centre of Mila, Algeria

Chaouki Ghenai
University of Sharjah, UAE

Meziane Fares
Renewable Energy Development Center, Algeria

Kabouche Nourdine
Renewable Energy Development Center, Algeria

ABSTRACT

Hydrogen can be produced in a clean, emission-free manner using solar technologies. Solar energy technology, specifically photovoltaic energy paired with PEM (proton exchange membrane), is required to complete the electrochemical water splitting process. The actual challenge of integrating renewable energy sources into PEM is developing the best possible set of operating parameters. Hydrogen generation flow in this study depends on a number of variables. Using two distinct solar energy methods with equivalent energy production, the authors discovered that the flow of hydrogen production may be altered. The best results were recorded in Illizi (Afra), where the annual power consumption was measured at 896.65 kWh and the flow rate was calculated to be around 4.59 L/h. The global annual loss under this setup is rather small at 10.6%.

1. INTRODUCTION

According to the global energy outlook by 2050, electricity, hydrogen and syngas could represent 50% of the energy mix (McKinsey 2022). However, the vast majority of the resources used in this energy

DOI: 10.4018/979-8-3693-1014-4.ch004

mix come from fossil fuels, which have negative consequences on the environment and speed up the progression of global warming. According to the International Energy Agency's 2021 assessment, approximately 62% of the globe's electricity is generated through fossil fuel sources (International Energy Agency 2022).

Emissions from the energy sector can be reduced by the widespread use of renewable energy sources like wind and solar power. According to the information found in reference (International Renewable Energy Agency (IRENA) 2022a), their total capacity topped 845 GW and 942 GW in the year 2021.

Based on the levelized cost of electricity (LCOE) solar energy technology (USD$0.040/kWh) is the cheapest renewable energy than onshore wind energy (USD$0.043/kWh) and can replace fossil fuel energy (Temiz and Dincer 2022)(International Renewable Energy Agency (IRENA) 2022b).

However, the unpredictable nature of solar and wind power plants is one of the numerous technological obstacles that make the integration of renewable energy sources into the traditional national power system more difficult. A feasible approach to amplify the proportion of renewable energy involves creating an autonomous system for producing green hydrogen, powered by solar and wind plants. Hydrogen produced in this way has the potential to contribute to the nation's energy security (Renewable Energy Agency 2021) and the reduction of emissions in hard-to-decarbonize industries including steel manufacture, transportation, fertilizer production, chemical production, heating, and so on.

Ultra-pure green hydrogen has been created from water using a PEM electrolyzer in several demonstration projects and research, both industrial and commercial (Norazahar et al. 2023)(Majumdar et al. 2023). Among the advantages of this technology:

1. It works at extremely high current densities of the order of 2 A/cm^2.
2. It produces very pure hydrogen of the order of 99.99%.
3. A very high efficiency of around 80-90% (Wang, Cao, and Jiao 2022).
4. An acceptable lifetime (Weiß et al. 2019)(Möckl et al. 2022).

Because of its layout, PEM is preferable to other electrolysis units because it is easier to implement and generate new catalysts and cathodes materials that can boost efficiency at low cost (Baroutaji et al. 2023). Direct integrated photovoltaic (PV) electrolysis systems have seen a significant uptick in published research that focuses on optimizing system sizes, operational strategies, and the incorporation of auxiliary components such energy storage systems. Moreover, the quest for appropriate materials for enhancing solar hydrogen production efficiency in these systems has emerged as a prominent research avenue, attracting considerable interest. Previous investigations have explored this domain extensively, with expectations of further publications in the future, given the substantial potential this research avenue holds (Kellenberger et al. 2020).

Jingyong Cai et al (Cai et al. 2023) studied the performance of PV-PEMWE and PV/T-PEMWE with the same area and optimized connection mode under various condition. The best and the optimized connection based on efficiency was the indirect connection followed by the optimized connection and the direct connection. In the work of Eveline Kuhnert et al (Kuhnert et al. 2024), under real-life condition with single cell PEMWE, two accelerated stress tests were designed. First, a protective current was used to compensate shutdown periods by a battery assisted system. Second, at night during periods of inactivity where the cell is maintained at open circuit voltage. A PV profile accelerated over time which represents the power imbalances is used to accelerate stress test cell. Based on electrochemical and physical characterization, a lower degradation rate has been remarked with the first solution.

This study focuses on the influence of solar irradiation and ambient temperature on hydrogen production flow using two different solar energy technologies with the same characteristic of energy production.

2.0 METHODS AND MODELISATION

2.1 Models Used for Global Solar Radiation Estimation

The generalized *Liu & Jordan* model (Nouicer et al. 2021)(Liu and Jordan 1960) was employed to calculate the global solar radiation for surfaces both horizontal (RGH) and inclined (RGI). This study aimed to estimate solar irradiation under varying conditions. In Figure 1-A and 1-B, the visual representations offer insights into the global average monthly solar irradiation and average temperature patterns, providing a comprehensive understanding of these factors for the two cities in Algeria as shown in Table 1 (for Constantine and Afra). The analysis bridges the estimation of solar radiation with real-world climatic parameters, enriching our comprehension of solar energy dynamics.

2.2 Modeling of Photovoltaic Modules

This research was conducted to investigate the performance of horizontal and optimally tilted photovoltaic (PV) modules at two distinct locations in Algeria: Constantine (Airport) and Illizi (Afra), as depicted in Table 1 and Figure 2. The study involves modeling the system design for both locations, considering horizontal orientation as well as optimized tilt angles.

The validation methodology for the model was previously introduced in the work by (Nouicer et al. 2019)(Nouicer et al. 2023).

Figure 1. (a) Monthly average solar irradiation (horizontal and inclined) and (b) mean temperature in the two selected cities

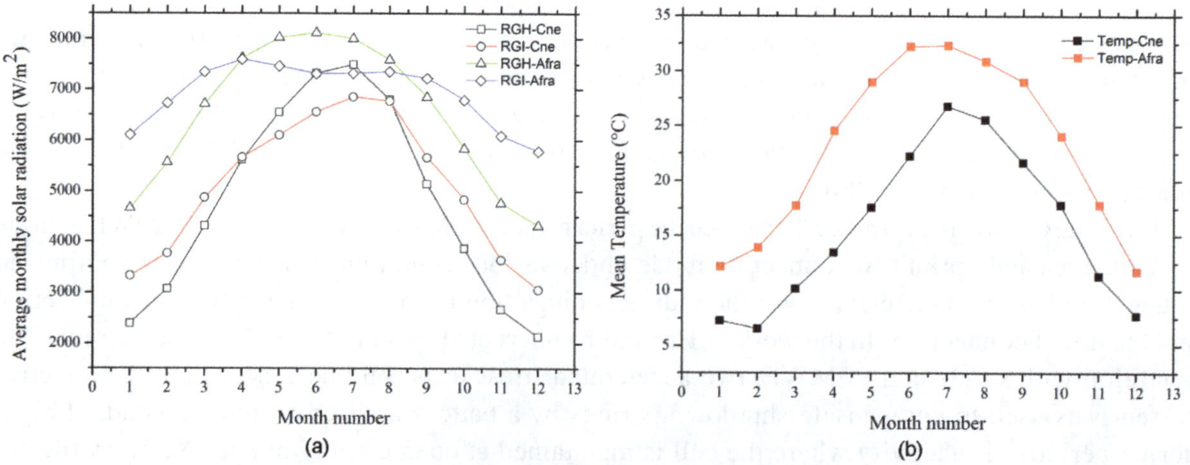

Table 1. Two Algerian sites selected for this investigation

City	Latitude (°)	Longitude (°)	Altitude (m)
Constantine	36,28	6,62	690.40
Illizi (Afra)	25,33	7,80	1240

2.3 Design of Solar PV/PEM Green Hydrogen System

A schematic representation of proposed PV/PEM systems based on proton exchange membrane is given in **Figure 3** includes: two kinds of PV module: the first one is: SUN POWER MAXEON 3 (SPR-MAX3-400-COM (Reil et al. 2013)) and the second one is: Canadian Solar (CS6R-405MS (CanadianSolar n.d.)), MPPT, PEM electrolyzer system, and storage tanks.

3.0 RESULTS AND DISCUSSIONS

3.1 Hydrogen Production Flow Estimation From One Cell of PEM Electrolyzer

By utilizing the photovoltaic module's seasonal variation in power output, the demand for hydrogen production can be met (Nouicer et al. 2023). By leveraging the temporal variations in solar energy output, the photovoltaic system can contribute to the generation of hydrogen as a viable energy storage solution. This process involves capturing electricity generated over the year and utilizing it for electrolysis to produce hydrogen gas. Consequently, this approach offers a sustainable means of addressing energy storage

Figure 2. Algerian map and location of the two selected cities

Figure 3. Schematic diagram of the solar hydrogen production system with two types of PV modules

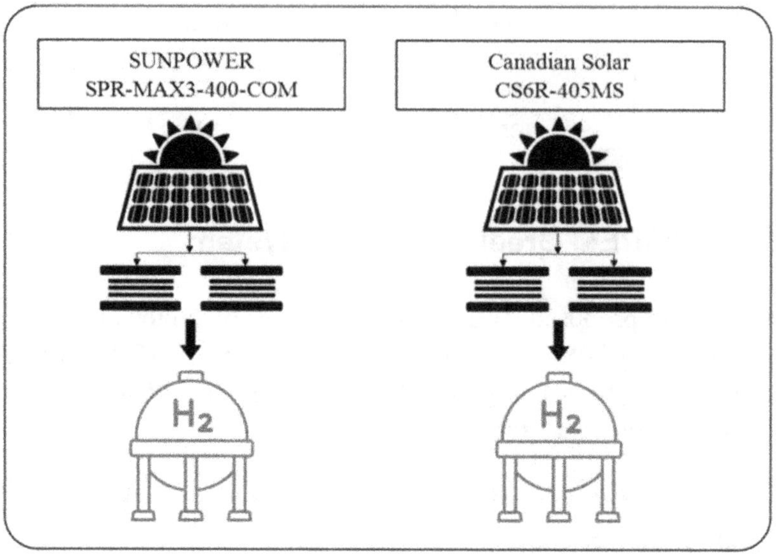

challenges, enhancing the utilization of renewable energy sources, and ensuring a stable energy supply even when solar irradiance is variable. The integration of photovoltaic with hydrogen production capitalizes on renewable energy synergies, contributing to a more resilient and adaptable energy ecosystem.

Hydrogen production rates from two different locations in Algeria using horizontal and optimized photovoltaic (PV) modules are summarized annually in Table 2. Several critical characteristics must be fully understood for precise determination of PV module power output and its subsequent application to a Proton Exchange Membrane (PEM) electrolyzer. These encompass critical factors such as the photo current (I_{Ph}), the diode's reverse saturation current (I_0), the series resistance (R_s), the shunt resistance (R_p), and the ideality factor (a), collectively instrumental in shaping the behavior of the PV system. Additionally, the specific classification of the PV module employed significantly shapes the operational dynamics and overall efficacy of the system. The complex interplay of these several components

Table 2. Various PV/PEM combinations for GRH and GRI at different annual power produced by photovoltaic with two kinds of PV module

City	Type of PV	Angle (°)	Power (kWh)	Hydrogen Flow (L/h)	GAL (%)
Constantine	CS6R400	0	663.02	2,69	12.32
	CS6R400	36	703.10	2,94	11.14
	SPR400	0	367.66	1,53	12.50
	SPR400	36	391.98	1,63	12.34
Afra	CS6R400	0	844.62	4,10	12.01
	CS6R400	25	896.65	4,59	10.60
	SPR400	0	504.58	2,10	11.99
	SPR400	25	537.51	2,24	11.86

highlights the critical need of accurate parameter identification in maximizing hydrogen generation using PV-powered electrolysis. As a result, this work makes a significant impact on the development of renewable energy paradigms.

The best flow in this case study is in Illizi (Afra) with annual power consumption 896.65 kWh and an annual hydrogen volumetric flow rate of about 4.59 L/h with minimum global annual loss (GAL =10.6%). Our findings align with those of commercial electrolyzes and other investigations (Www. fuelcellstore.com n.d.). The energy consumption of commercial electrolysis stacks is between ; 4.4 – 5.0 kWh/Nm^3 (Buttler and Spliethoff 2018) and 5.0-6.5 kWh/Nm^3 (Zhu et al. 2023), but in our case the energy consumption of PEM electrolysis stacks is between: 2.98 – 4.66 kWh/Nm^3.

4. CONCLUSION

The accurate determination of hydrogen production flow hinges upon the knowledge of multiple critical parameters. Variables like ambient temperature and sun radiation have a crucial role in the efficiency with which power from an optimum photovoltaic (PV) module is transferred to the Proton Exchange Membrane (PEM) electrolyzer. Additionally, the operational efficiency of the PEM electrolyzer can be influenced by diverse parameters such as membrane thickness, cathode, and anode pressures, among others.

To establish the reliability of the proposed model, validation was conducted, involving the comparison of the model's predictions with experimental data from a manufacturer's photovoltaic module. This procedure demonstrates the reliability and precision of the resulting model.

Interestingly, the results show that hydrogen generation in the Illizi (Afra) region is significantly more important than in the Constantine site. When comparing two PV module technologies that generate electricity in exactly the same way, this difference becomes readily obvious. The geographical disparity in hydrogen generation underscores the importance of site-specific factors and underlines the broader implications for optimizing renewable energy systems tailored to distinct regions.

REFERENCES

Baroutaji, A., Arjunan, A., Robinson, J., Abdelkareem, M. A., & Olabi, A. G. (2023). Additive Manufacturing for Proton Exchange Membrane (PEM) Hydrogen Technologies: Merits, Challenges, and Prospects. *International Journal of Hydrogen Energy*.

Buttler, A., & Spliethoff, H. (2017, September). Current Status of Water Electrolysis for Energy Storage, Grid Balancing and Sector Coupling via Power-to-Gas and Power-to-Liquids: A Review. *Renewable & Sustainable Energy Reviews*, *82*, 2440–2454. doi:10.1016/j.rser.2017.09.003

Cai, J., Weng, C., Zhang, R., Li, Q., Zhang, T., & Shi, Z. (2023). Comparative Analysis on the Dynamic Operation Performance of Photovoltaic/Thermal Powered Proton Exchange Membrane Water Electrolysis Cogeneration System (PV/T-PEMWE) under Different Connection Modes. *Renewable Energy*, *219*(P2), 119566. doi:10.1016/j.renene.2023.119566

CanadianSolar. (n.d.). *CS6R-405MS Datasheet*. Retrieved September 7, 2023 (https://static.csisolar.com/wp-content/uploads/2020/06/27131418/CS-Datasheet-HiKu6_CS6R-MS_v2.0_EN.pdf)

International Energy Agency. (2022). *International Energy Agency (IEA) World Energy Outlook 2022.* Https://Www.Iea.Org/Reports/World-Energy-Outlook-2022/Executive-Summary

International Renewable Energy Agency (IRENA). (2022a). *Renewables 2022: Global Status Report.* IRENA.

International Renewable Energy Agency (IRENA). (2022b). *World Energy Transitions Outlook 2022.* IRENA.

Kellenberger, A., Duca, D., Vaszilcsin, N., & Craciunescu, C. M. (2020). Electrochemical Evaluation of Niobium Corrosion Resistance in Simulated Anodic PEM Electrolyzer Environment. *International Journal of Electrochemical Science, 15*(11), 10664–10673. doi:10.20964/2020.11.47

Kuhnert, E., Mayer, K., Heidinger, M., Rienessel, C., Hacker, V., & Bodner, M. (2023, September). Impact of Intermittent Operation on Photovoltaic-PEM Electrolyzer Systems: A Degradation Study Based on Accelerated Stress Testing. *International Journal of Hydrogen Energy, 55*, 683–695. doi:10.1016/j.ijhydene.2023.11.249

Liu, B. Y. H., & Jordan, R. C. (1960). The Interrelationship and Characteristic Distribution of Direct, Diffuse and Total Solar Radiation. *Solar Energy, 4*(3), 1–19. doi:10.1016/0038-092X(60)90062-1

Majumdar, A., Haas, M., Elliot, I., & Nazari, S. (2023). Control and Control-Oriented Modeling of PEM Water Electrolyzers: A Review. *International Journal of Hydrogen Energy, 48*(79), 1–21. doi:10.1016/j.ijhydene.2023.04.204

McKinsey. (2022). Global Energy Perspective 2022 McKinsey's Global Energy Perspective Is a Collaboration between Energy Insights and Adjacent Practices. *Executive Summary.*

Möckl, M., Ernst, M. F., Kornherr, M., Allebrod, F., Bernt, M., Byrknes, J., Eickes, C., Gebauer, C., Moskovtseva, A., & Gasteiger, H. A. (2022). Durability Testing of Low-Iridium PEM Water Electrolysis Membrane Electrode Assemblies. *Journal of the Electrochemical Society, 169*(6), 064505. doi:10.1149/1945-7111/ac6d14

Norazahar, N., Khan, F., Rahmani, N., & Ahmad, A. (2023). Degradation Modelling and Reliability Analysis of PEM Electrolyzer. *International Journal of Hydrogen Energy.*

Nouicer, Yaiche, Menia, Meziane, & Kabouche. (2021). *Estimation of Hydrogen Production in Three Cities in the North of Algeria.* Academic Press.

Nouicer, I., Khellaf, A., Menia, S., Yaiche, M. R., Kabouche, N., & Meziane, F. (2019). Solar Hydrogen Production Using Direct Coupling of SO2 Depolarized Electrolyser to a Solar Photovoltaic System. *International Journal of Hydrogen Energy, 44*(39), 22408–22418. doi:10.1016/j.ijhydene.2018.11.106

Nouicer, I., Menia, S., Kabouche, N., Ghenai, C., & Meziane, F. (2023). Production : Case Study in Constantine, Algeria. *2023 14th International Renewable Energy Congress (IREC)*, 1–4.

Reil, Baumann, Althaus, & Gebhard. (2013). SunPower Maxeon 3 Datasheet. *Conference Record of the IEEE Photovoltaic Specialists Conference,* 1537–41.

Renewable Energy Agency. (2021). *International* (Vol. 56). Estadísticas De Energía Renovable.

Temiz, M., & Dincer, I. (2022). Development of Solar and Wind Based Hydrogen Energy Systems for Sustainable Communities. *Energy Conversion and Management*, *269*(June), 116090. doi:10.1016/j.enconman.2022.116090

Wang, T., Cao, X., & Jiao, L. (2022). PEM Water Electrolysis for Hydrogen Production: Fundamentals, Advances, and Prospects. *Carbon Neutrality*, *1*(1), 1–19. doi:10.1007/s43979-022-00022-8

Weiß, A., Siebel, A., Bernt, M., Shen, T. H., Tileli, V., & Gasteiger, H. A. (2019). Impact of Intermittent Operation on Lifetime and Performance of a PEM Water Electrolyzer. *Journal of the Electrochemical Society*, *166*(8), F487–F497. doi:10.1149/2.0421908jes

Zhu, T., Qi, M., Yin, M., Dang, J., Zhang, X., & Zhao, D. (2023). On the Hazards of Proton Exchange Membrane (PEM) Water Electrolysis Systems Using System Theoretic Process Analysis (STPA). *Process Safety and Environmental Protection*, *180*(November), 1118–1131. doi:10.1016/j.psep.2023.10.062

Chapter 5
Green Hydrogen:
A Clean Energy Solution for Electricity and Heat

Ayşenur Öztürk Aydın
https://orcid.org/0000-0001-9421-8091
Atatürk University, Turkey

Gamze Bozkurt
Erzurum Technical University, Turkey

Elif Daş
https://orcid.org/0000-0002-3149-6016
Atatürk University, Turkey

ABSTRACT

Hydrogen (H_2) is an energy carrier that offers both some of the benefits of fossil fuels and is clean with a low carbon footprint. Production of H_2 from the water electrolysis using renewable energy is called green hydrogen and is the preferred pathway for future H_2 production. In order to decarbonize the energy sector, green hydrogen must be adapted to existing systems using different technologies. Fuel cells, gas turbines (GTs), and combined heat and power (CHP) systems are the primary applications for hydrogen in power generation. CHP systems are highly efficient and can be employed in buildings, industrial facilities, and district energy systems to provide electrical and thermal energy. Heat pumps integrated with fuel cells are another example of energy-efficient technology. Mixing hydrogen in a specific volumetric ratio improves the resource utilization of the present natural gas heating systems. This chapter focuses on green hydrogen practices in electrical and thermal energy generation applications.

INTRODUCTION

Governments and the media warn that there will be severe problems with energy supply to citizens in all countries and that the situation is concerning. Energy supply has been a concern due to the widespread

DOI: 10.4018/979-8-3693-1014-4.ch005

usage of hydrocarbon-based fossil fuels, which are depleting. Fossil fuels are nonrenewable resources, and recent natural occurrences around the world have been linked to the detrimental impacts of fossil fuels on the climate. This condition increases psychological stress and lowers people's quality of life. Among the different alternatives proposed by scientists and experts, hydrogen technology is the most widely accepted. For this reason, the terms "hydrogen society" and "hydrogen economy" have recently gained prominence.

Hydrogen societies rely on reducing the use of fossil fuels in favor of emission-free sources and hydrogen-based generation to decarbonize energy systems. The hydrogen economy is described as a new industry that will create new jobs, contribute to reducing energy consumption and emissions, and stimulate economic growth (Dou et al., 2017). The hydrogen economy aims to generate significant amounts of power and hydrogen using primary energy sources such as solar, nuclear, and even fossil fuels (Sherif et al., 2014). Industrial and political milestones have arisen for the hydrogen economy, which are linked and will shift in relative importance over time. The following milestones are being considered: a) hydrogen from coal; b) hydrogen as a transportation fuel; c) the exergetically efficient stationary fuel cell; d) hydrogen and fuel cell-induced industry structure change; e) political framework conditions for investment planning security (Winter, 2005). Many countries are developing hydrogen strategies or roadmaps to encourage their citizens to accept hydrogen technology and ensure the growth of the hydrogen economy.

Hydrogen is an energy carrier with the highest energy content per weight. Hydrogen is a unique energy carrier due to its excellent qualities, which are detailed below:

- It can both produce and convert electricity with high efficiency.
- It is produced using water, which is a plentiful raw material on the planet.
- It is a renewable fuel.
- It can be stored in gaseous form for large-scale storage, in liquid form for air and space transportation, and as metal hydrides for cars and small-scale storage.
- Long-distance transportation can be accomplished by using pipelines or tankers.
- Catalytic combustion, electrochemical conversion, and hydration are all efficient methods for converting it into other types of energy.
- It emits no greenhouse gases (GHG) or other pollutants into the environment during manufacture, storage, transportation, or consumption (Sherif et al., 2014).
- So far, hydrogen has been produced using coal, crude oil, or natural gas, resulting in an unclean production technique with GHG emissions. Today, hydrogen synthesis by electrolysis of water using only renewable electricity is called "green hydrogen" and is encouraged.

Green hydrogen plays an important role in the decarbonization of the energy industry, particularly in electricity and heat generation, due to its inherent sustainability and environmental benefits. Green hydrogen is a versatile energy carrier that may store extra electricity generated by renewable energy sources (RESs) during peak production. This stored energy can be efficiently converted into electricity using fuel cells or turbines, improving grid stability and reliability when RESs are less productive or unavailable due to wind and solar power's intermittent nature. Fuel cells are at the forefront of using green hydrogen to generate electricity by aiding the electrochemical reaction of hydrogen and oxygen, leaving only water vapor as a byproduct. This technology is not only highly efficient, but it also provides a clean and quiet alternative to traditional combustion-based power generation. Aside from fuel cell technology, gas turbines are a feasible alternative for utilizing hydrogen in power generation. This device generates electricity by combusting compressed air and hydrogen at high temperatures. The resulting heated gases expand via a turbine, causing a connecting shaft to rotate, which drives a generator. This mechanism effectively transfers the chemical energy of the fuel into mechanical energy, and eventually into electrical energy. Combined heat and power (CHP) or cogeneration systems generate both electricity and heat from the same fuel source (Hwang et al., 2023). A complete CHP system comprises of a power generation unit (which includes a prime mover and a generator), a heat recovery unit, and a heating system. This technology efficiently catches and uses the heat generated during electricity production for heating. As shown in these cases, incorporating green hydrogen into electricity and heat generation can considerably cut carbon emissions, increase energy sustainability, and create a greener, more resilient power system for the future.

Domestic buildings account for 9.9% of the total 36.1 Gt of CO_2 emissions produced globally in 2022 (Liu et al., 2023). Furthermore, space and water heating account for 78% of the total energy consumed by houses (Longoria et al., 2021). The domestic heat demand varies according to size, building structure, and climatic circumstances (Dodds et al., 2015). The electrification of fossil fuel-based heating systems is essential to meet thermal energy demand while also lowering CO_2 emissions from heating. For stand-alone electricity and heat delivery, as shown in Figure 1, hybrid energy systems are alternative options, which include green hydrogen production and storage methods derived from RESs (Lacko et al., 2014). Although CHP systems are made of fossil fuels, incorporating hydrogen with a high energy density into them would improve energy efficiency. CHP systems can be classified into two types: heat engine-based and electrochemical-based hydrogen CHP systems. In the second category, fuel cell CHP systems achieve a total electrical and thermal energy efficiency of 90%. There are various methods based on RESs for manufacturing hydrogen for use as fuel in CHP systems, including water electrolysis, photocatalytic hydrogen generation, thermochemical cycles, and biomass hydrogen production. Because RESs are intermittent, some of the hydrogen produced at a convenient time is stored to avoid delays in hydrogen production and respond to energy demand, particularly during peak hours (Yu et al., 2023).

Heat pumps are also an effective solution for obtaining heat from power. They are used in low-temperature (75-100 °C) thermal applications for buildings, district heating, and industries because their coefficient of performance (COP) drops as temperature increases due to the moving parts (Pimm et al., 2023). In terms of operating principles, heat pumps are classified into two types: compression heat pumps and absorption heat pumps, which are fueled by electricity and high-grade heat sources, respectively (Chen et al., 2023). In hydrogen-integrated hybrid systems for carbon-free heating, hot water produced from hydrogen combustion or the excess heat emitted in the electrolyzer provides the thermal energy; the current generated by hydrogen-powered fuel cells supplies the electrical energy to heat pumps (Savu et al., 2021). The growing demand for air conditioning, particularly in hot climates due to global warming, has created a desire for alternative cooling systems that incorporate ecologically benign working fluids and efficient energy consumption (Chauhan et al., 2022). As a result, instead of the standard vapor-compression refrigeration cycle (VCRC), thermal-based sorption cooling systems with absorption or adsorption cooling have gained popularity. These systems use solar-based and low-grade waste heat, making them more energy efficient (Hassan et al., 2023). Hydrogen was employed as a combustion fuel in the vapor absorption cycle refrigeration (VACR) system instead of propane in recreational vehicles (Zun et al., 2023).

Using an existing gas network for long-distance, large-volume hydrogen transportation and distribution is less expensive than using trucks or tube trailers and results in less energy loss. The viability of

Figure 1. A hybrid energy system for electricity and heat delivery
Reproduced from Ref. (Boulmrharj et al., 2020) (with different figure caption) from the MDPI according to the open access article distributed under the terms and conditions of the Creative Commons Attribution (CC BY) license (https://creativecommons.org/licenses/by/4.0/)

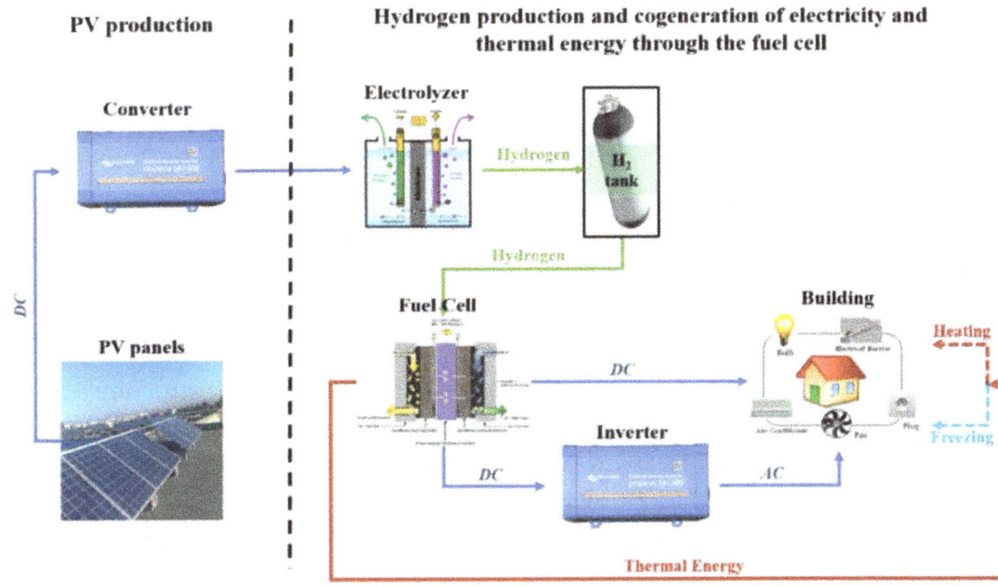

blending hydrogen with natural gas should be established based on the method of production, hydrogen supply and demand, and existing pipeline materials and specifications. Mixing hydrogen with natural gas in appropriate ratios reduces the risk of fire or explosion caused by ignition or leakage. Hythane gas, which is a mixture of 10-30% hydrogen and 90-70% methane, can be made available to customers for domestic use within the framework that assures the safety and integrity of the local gas network (Chae et al., 2022). The practice of hydrogen in industrial processes requiring high heat (iron, steel, cement, glass, aluminum, etc.) is also one of the applications documented in literature.

This book chapter offers hydrogen's multifaceted role in the energy sector, covering its entire life-cycle from production to utilization. With a primary focus on green hydrogen, we explore its potential applications in both electrical and thermal energy generation.

MAIN FOCUS OF THE CHAPTER

What Is Hydrogen?

Hydrogen is present in the universe long before the earth, planets, sun, Milky Way, and chemical elements like oxygen, sodium, iron, and gold. It is also the most plentiful element (in the form of compounds) in the cosmos, serving as the major fuel source for the Sun and other stars. The sun consumes approximately 600 million tons of hydrogen every second (Hordeski, 2008). Hydrogen accounts for 75% of the universe's mass and more than 90% of all atoms. Hydrogen is crucial for stars and planets, making it a critical element for the universe's survival (Zhang et al., 2014; Züttel et al., 2008). It is also the first element in the periodic table.

Hydrogen is a technologically and economically essential element composed of one proton and one electron, with no neutrons. Two hydrogen atoms combine to produce a stable H_2 molecule with a single, strong covalent bond (Eq. 1):

$$2H^{+} + 2e^{-} \rightarrow H_{2(1atm)} \quad (1)$$

The isotope of hydrogen with one neutron is deuterium (D or 2H), and the isotope with two neutrons is tritium (T or 3H). At standard temperature and pressure, hydrogen is a colorless, tasteless, odorless and easily flammable gas. Hydrogen's physical properties, like temperature, pressure, liquid density, and gas density, are different at the triple point, boiling point, and critical point (Fichtner & Idrissova, 2009). In addition to producing H_2, the hydrogen element is present in the structure of many vital compounds, including water, hydrocarbons, proteins, and DNA. Furthermore, all structures, such as cells, viruses, bacteria, and tissues that play a significant role in the universe and for people, contain hydrogen as a basic element (Zhang et al., 2014). Hydrogen, an energy carrier, can be produced from a variety of sources, including hydrocarbons and RESs such as sun, wind, hydro, geothermal, biomass, and nuclear energy. It is also easily produced from synthesized hydrogen carriers such as methanol, ammonia, and synthetic fuels (Veneri, 2011).

Hydrogen Production

Hydrogen is produced globally from several renewable and nonrenewable sources, with varied costs and CO_2 emissions. Figure 2 provides a general summary of hydrogen production methods.

Hydrogen production is classified according to production technology into different colors, such as black, brown, grey, blue, green, pink, yellow, and white (Ajanovic et al., 2022; Panic et al., 2022). It appears that the color scale for hydrogen production will gradually expand. The literature contains different viewpoints on colors, but the issue is complex and perplexing. Perhaps a more appropriate way

Figure 2. Hydrogen production methods

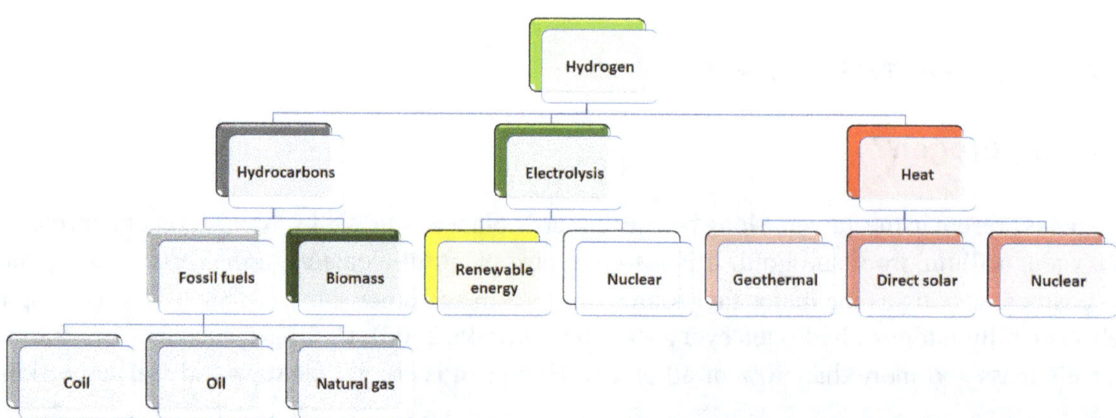

would be to categorize hydrogen production into two types: technologies that emit GHG emissions and those that do not. If we were to explain the production technologies one by one based on colors, as is now offered in literature:

- **Brown/Black H_2 Production:** Brown and black hydrogen colors represent hydrogen produced from lignite (brown) and bituminous coal (black). This is a popular hydrogen production technology since coal is the fossil energy source with the most reserves worldwide and is relatively inexpensive. China produces large amounts of hydrogen by gasifying coal due to high natural gas prices and large coal reserves (Arcos & Santos, 2023). However, mining and gasification of coal to produce hydrogen has numerous severe environmental consequences, including land degradation, soil erosion, dust, effects on biodiversity, waste dumps, and abandoned mines (Leon, 2008).
- **Grey H_2 Production:** Grey hydrogen is produced from natural gas through steam reforming, partial oxidation, or autothermal reformation. Because these processes create hydrogen with a mixture of carbon monoxide (CO), CO_2, and unconverted methane (CH_4), additional processing is required to purify the hydrogen. The steam-methane reforming reaction given in the equations (Eq. 2, Eq. 3, Eq. 4) below is a frequently used method that produces hydrogen and synthesis gas.

$$CH_4 + H_2O \rightarrow CO + 3H_2 \tag{2}$$

$$CO + H_2O \rightarrow CO_2 + H_2 \text{ (water-gas shift reaction)} \tag{3}$$

$$CH_4 + 2H_2O \rightarrow CO_2 + 4H_2 \text{ (overall reaction)} \tag{4}$$

The partial oxidation reaction of natural gas (Eq. 5) performed in high-pressure reactors includes steps such as removing the large amount of heat produced, converting CO to H_2, eliminating CO_2, and purifying hydrogen.

$$CH_4 + \frac{1}{2}O_2 \leftrightarrow CO + H_2 \tag{5}$$

Autothermal reforming is a mixture of steam reforming and partial oxidation that produces synthesis gas consisting of hydrogen, CO, and CO_2 (Hočevar & Summers, 2008).

- **Pink H_2 Production:** Pink hydrogen production methods vary depending on whether nuclear energy is employed as a thermal or electrical energy source. These may include thermochemical water splitting and low- and high-temperature water electrolysis. Many countries are taking significant moves toward nuclear energy. This will also affect hydrogen production. However, existing world uranium resources are sufficient for about 90 years. Limited Uranium-235 supply and technical constraints hinder large-scale nuclear energy expansion, potentially impeding long-term nuclear-based hydrogen production (Panic et al., 2022).

- **Yellow H_2 Production:** There is some confusion in the literature between pink hydrogen and yellow hydrogen. However, hydrogen production from electrolysis with mains electricity is commonly known as yellow production (Ajanovic et al., 2022).
- **White H_2 Production:** Like the production of yellow hydrogen, the synthesis of white hydrogen is a relatively new process in the literature. That is why "white hydrogen" is a confusing term. It is naturally occurring, stored in natural reserves, or used as geological hydrogen (Panic et al., 2022). Some sources recognize white hydrogen as a byproduct of chemical processes (Incer-Valverde et al., 2023).
- **Blue H_2 Production:** The blue hydrogen method refers to the process of producing hydrogen from fossil fuels (steam reforming) by capturing the CO_2 emitted. Because it is not possible to capture all of the CO_2 in this production method, the emissions are low (Nami et al., 2022). Furthermore, capturing and storing CO_2 incurs additional expenditures, and the efficiency of collection systems varies greatly. Although the long-term advantages of blue hydrogen are unknown, it is considered a hydrogen production method that could enable a global transition to green hydrogen production (Panic et al., 2022).
- **Green H_2 Production:** Green hydrogen is the synthesis of hydrogen using renewable electricity from water electrolysis powered by RESs. Green hydrogen that contains no carbon, is rapidly commercializing in many countries; however, it is still prohibitively expensive compared to fossil-based hydrogen production, so it cannot yet become widespread in daily life (Yu et al., 2021). Green hydrogen is expected to be used in three pillars of the industry: transportation, heating, and energy. The first example of green hydrogen applications in industry is its use in chemical synthesis to meet the world's hydrogen demand, which is responded to by gray hydrogen produced by reforming fossil fuels. The second application is as a fuel in fuel cell vehicles, as well as in natural gas pipelines for heating purposes. The third long-term application example is its use as a fuel in aviation applications and in the acquisition of high-quality heat for industries (Oliveira et al., 2021).

Although green hydrogen production is costly, it is recommended as a way to solve climate-related challenges. Solar and wind energy are RESs that are ideal for producing hydrogen through water electrolysis. The sole disadvantage of these energy sources is that they are dependent on weather conditions. Electrolysis of water is an electrochemical process, and its basic reaction is as shown in Eq. 6 (Kumar & Lim, 2022):

$$H_2O + Electricity_{237.2\text{kJ/mol}} + Heat_{48.6\text{kJ/mol}} \rightarrow H_2 + \frac{1}{2}O_2 \quad (6)$$

Electrolysers

Electrolysers are used to perform the electrolysis process. Hydrogen is not only produced from electricity, but it is also a source of electricity. Technically, the efficiency of converting electricity into hydrogen is lower (Karayel et al., 2023; Karayel et al., 2022). Electrolysers that produce hydrogen by electrolysis of water are classified according to their electrolytes and operating conditions. Nowadays, hydrogen is produced using different electrolysis methods, including low-temperature technologies like alkaline electrolysis cells (AECs), proton exchange membrane electrolysis cells (PEMECs), and high-temperature solid oxide electrolysis cells (SOECs). These technologies have various advantages and disadvantages. Low-temperature electrolysers are more advanced than high-temperature electrolysers, while high-temperature technologies are more efficient. Another classifications are: a) alkaline water electrolysis; b) anion exchange membrane (AEM) water electrolysis; c) PEM water electrolysis; and d) solid oxide water electrolysis (Kumar & Lim, 2022). There have been studies in the literature using AEC, PEMEC, and SOEC electrolysers to calculate various hydrogen production costs.

Converting Green Hydrogen Into Useful Energy

Electrical Energy

As a fuel, hydrogen can be used in a variety of ways to generate energy in an efficient and environmentally friendly manner. One of the most notable approaches is to use hydrogen fuel cells, which electrochemically convert hydrogen and oxygen into electricity while emitting only water as a byproduct (Daş et al., 2020). This clean and sustainable technique has the potential to provide a consistent source of electrical energy for both industrial and residential uses, meeting the growing need for power while lowering GHG emissions and fostering a more sustainable energy future.

Fuel cells work on a similar concept to reverse electrolysis, converting fuel and an oxidizing agent (usually hydrogen and oxygen) into electricity, heat, and water. In a fuel cell, hydrogen gas is fed into the anode, and oxygen or air is fed into the cathode. The catalyst at the anode splits hydrogen molecules into protons and electrons in a process known as oxidation. Protons pass through an electrolyte membrane to reach the cathode, while electrons are forced to travel through an external circuit, resulting in an electrical current. At the cathode, oxygen reacts with the protons and electrons to form water molecules. This controlled electrochemical reaction efficiently generates electrical energy while emitting only water vapor as a byproduct (Corbo et al., 2011). There are several types of fuel cells, including proton exchange membrane or polymer electrolyte membrane (PEM) fuel cells, alkaline fuel cells (AFCs), phosphoric acid fuel cells (PAFCs), direct methanol fuel cells (DMFCs), solid oxide fuel cells (SOFCs), and molten carbonate fuel cells (MCFCs), each with specific applications and advantages, ranging from portable power sources to large-scale stationary power generation systems (Lucia, 2014). Table 1 lists the types of fuel cells along with their main characteristics (Corbo et al., 2011; EG&G Technical Services, 2004; Głuchy et al., 2014). They are generally classified according to the electrolyte. The cell's operating temperature ranges from 40 to 1000 °C based on the electrolyte's ion conduction mechanism. Fuel cells are separated into two temperature ranges: low-temperature (up to 200 °C) and high-temperature (600-1000 °C). The medium-temperature range remains idle due to the lack of a viable electrolyte material.

PEMFCs and DMFCs are the most common types of low-temperature fuel cells, with applications ranging from portable electronics to transportation. In particular, PEMFCs are widely used in vehicles and small-scale power generation due to their quick startup times and high power density (Sasaki et al., 2016). On the other hand, DMFCs, while suitable for portable electronics, face challenges due to the toxicity of methanol fuel and efficiency issues, necessitating further research and development. In contrast, high-temperature fuel cells can achieve higher electrical efficiencies than low-temperature counterparts and use a broader range of fuel sources, including hydrocarbons. They are less susceptible to fuel contaminants and can be used in CHP applications, which improves total efficiency (Pasupathi et al., 2016). Furthermore, the higher operating temperatures enable faster electrochemical reactions,

Table 1. Types of fuel cells and their key characteristics

Fuel cell	Electrolyte	Charge Carrier	Catalyst	Fuel	Oxidant	Operation Temperature (°C)	Efficiency %
PEMFC	Polymeric membrane	H^+	Noble metal	H_2	O_2, air	40-90	Up to 60
AFC	Potassium hydroxide	OH^-	Noble/non-noble	H_2	O_2, air	60-220	Up to 70
PAFC	Phosphoric acid	H^+	Noble metal	H_2	O_2, air	160-220	Up to 50
DMFC	Polymeric membrane	H^+	Noble metal	CH_3OH	O_2, air	90-130	Up to 40
SOFC	Perovskites	O^{2-}	Non-noble	H_2, Light HC*	O_2, air	600-1000	Up to 65
MCFC	Li and K carbonates	CO_3^{2-}	Non-noble	H_2, Light HC*	CO_2, O_2, air	600-700	Up to 60

* Light hydrocarbons, predominantly methane or substances resulting from methane reforming reactions
(Corbo et al., 2011; EG&G Technical Services, 2004; Głuchy et al., 2014)

leading to increased power generating capabilities. Despite problems like material durability and thermal management, high-temperature fuel cells offer substantial benefits for efficient and environmentally friendly electricity production in large-scale stationary plants.

Today, the main focus of research and development (R&D) is primarily on PEMFCs for small and medium-sized applications, including DMFCs, while SOFCs are the primary workspace for larger applications. In terms of commercialization, PEMFCs have advanced significantly in recent years and may now be regarded a well-established and mature technology. In contrast, SOFCs are still in the process of gaining broader commercial acceptance. While SOFC technology has shown significant potential, particularly for larger-scale applications and stationary power generation, it is still in early stages of commercialization. Material durability, high operating temperatures, and cost challenges must be addressed to improve competitiveness and open the path for further commercialization. Nevertheless, ongoing research and investment propel SOFC technology, positioning it as a strong contender for future clean and efficient energy production.

In general, fuel cell systems are expensive, and this is an issue that research groups around the world must address. In March 2017, the Hydrogen and Fuel Cell Supergen Hub of the United Kingdom published a white paper that methodically analyzed the current state and future possibilities of hydrogen and fuel cells within future energy systems (Ekins & Shah, 2017). According to the paper, the cost of hydrogen, particularly the cost of fuel cell technology, has decreased dramatically over the last decade. This means that hydrogen and fuel cell technologies are likely to play an important role in one or more industries in the near future. In this perspective, especially considering recent improvements, the transportation sector has been identified as the most prospective market for hydrogen and fuel cell technology. Several recent studies concerning fuel cell electric vehicles (FCEVs) have suggested that as production volumes increase, the total cost of ownership can become competitive with alternative options. The United States has gained the lead in the market of FCEVs, selling 2,750 FCEVs by 2017, more than Japan and Europe combined. The position of China regarding fuel cells may also affect other countries. Fuel Cell Vehicle Technology Roadmap (2016) of China had set a modest target of 5000 FCEVs for 2020; however, this target will reach a scale of millions by 2030 (Staffell et al., 2019).

CHP systems are an alternative to fuel cells for electricity generation since, they can work with an engine or a fuel cell and give high overall efficiency while simultaneously producing electricity and heat (Ekins & Shah, 2017). CHP is regarded as the largest and most established market for fuel cells. PEM,

SOFC, MCFC, and PAFC are the most common fuel cell technologies for CHP systems. The diversity of these systems provides numerous benefits across a wide range of application areas. Firstly, it offers for greater flexibility in meeting specific energy requirements. PEMFC systems, for example, are well-suited for residential applications due to their small size and capacity to provide power and heat to individual dwellings. Larger-scale applications, such as industrial facilities or district heating, may benefit from SOFCs' reliability and efficiency. Secondly, the availability of many fuel cell technologies increases adaptability to different fuel sources, including hydrogen, natural gas, and biogas. This adaptability is especially useful in areas where some fuels are more abundant or cost-effective. Thirdly, competition among different fuel cell technologies stimulates innovation and cost reduction, resulting in increased use of CHP systems. Overall, the diversity of fuel cell CHP systems enhances their applicability, efficiency, and economic viability across a spectrum of energy demands and environmental considerations. Only a few manufacturers produce commercial fuel cell CHP systems, and Japan is one of the leaders in this industry, having developed the world's first commercial fuel cell CHP systems. Table 2 summarizes the technical specifications and performance of each fuel cell technology (Dodds et al., 2015). The table shows that SOFC systems, whether for residential or commercial applications, have electrical efficiencies ranging from 45% to 60%, while total efficiencies are often in the 85% to 90% range. On the other hand, fuel processing for PEMFCs incurs larger losses, resulting in lower electrical efficiencies but better thermal efficiencies. The top-performing residential systems achieve an electrical efficiency of 39% and a total efficiency of 95%. European residential systems have not attained the same level of performance as the best models in Japan and Australia, with modern SOFCs and PEMFCs achieving electrical efficiencies ranging from 30% to 35% (Ekins & Shah, 2017).

According to the literature, fuel cell CHP systems deliver higher electrical efficiency than other CHP technologies (e.g., internal combustion engine CHP). However, the majority of existing CHP systems currently rely on natural gas (Hawkes et al., 2009). Hydrogen can replace natural gas as a cleaner and more environmentally friendly fuel for CHP systems when its production volume achieves on a large scale through green and sustainable methods (such as electrolysis using RESs). This transition to green

Table 2. An overview of each fuel cell's technical specifications and performance

Application	Unit	PEMFC	SOFC	PAFC	MCFC
		Res	Res/Com	Com	Com
Electrical capacity	(kW)	0.75-2	0.75-250	100-400	300+
Thermal capacity	(kW)	0.75-2	0.75-250	110-450	450+
Electrical efficiency[a]	(LHV)	35-39%	45-60%	42%	47%
Thermal efficiency[a]	(LHV)	55%	30-45%	48%	43%
Expected lifetime	('000 hrs)	60-80	20-90	80-130	20
	(years)	10	3-10	15-20[c]	10[c]
Degradation rate[b]	(per year)	1%	1-2.5%	0.5%	1.5%

[a]Rated specifications when new.[b]Loss of peak power and efficiency. [c]Requires an overhaul of the fuel cell stack half-way through the working lifetime. Res: residential. Com: commercial.

hydrogen can contribute to reducing GHG emissions and making CHP systems more beneficial for our planet in terms of sustainability and environmental impact.

Gas turbines (GTs) are another method for converting hydrogen into electrical energy. GTs represent a financially, environmentally and safety-reliable choice for power and heat production. They find extensive use across diverse sectors, powering airplanes, ships, trains, and electricity generators. In essence, the GT is a internal combustion engine that operates continuously and follows the Brayton cycle (Nazari et al., 2022). The GT consists of three primary parts: a compressor, a combustor, and a turbine, all mounted on the same shaft as the compressor. The features of the combustion process influence the effectiveness of cycles employing GTs. The kind of fuel utilized in the combustion engine is a critical element influencing both combustion efficiency and emissions. So far, fossil fuels such as coal, oil, and natural gas have dominated the combustors of these cycles. While GTs have significant advantages and appeal for power generation, their main disadvantage is the GHG emissions caused by the usage of fossil fuels. The future plans aim using alternative, ecologically benign fuels like biofuels and hydrogen in GTs to address this problem. Biofuels are considered a potential alternative to traditional fuels to mitigate emissions, but one of their significant drawbacks is their relatively low energy density. In contrast, hydrogen, which has a high energy density per unit mass, becomes a more attractive option in this setting. Furthermore, adding hydrogen to the fuel mixture increases reactivity, which is helpful in specific applications like GTs. Hydrogen can be blended with various fuels and employed within the GT's combustion. For example, (Meziane & Bentebbiche, 2019) conducted a numerical simulation to study the impact of introducing hydrogen into natural gas during combustion. They focused on a rich/quench/lean combustor configuration. Their findings revealed that a minor addition of hydrogen, only 10%, to the natural gas led to a substantial reduction in NO emissions by 14% and CO emissions by 60%. In another study, (Yoshimura et al., 2005) examined the impact of introducing hydrogen into natural gas within a swirl-stabilized model combustor operating under lean premixed conditions. Their findings suggested that the addition of hydrogen could increase the combustor's stability range while decreasing nitrogen oxide (NOx) emissions. Furthermore, the study found that including a pilot injection could improve the system's emission performance when hydrogen was added to natural gas. On the other hand, using pure hydrogen as the fuel for GT combustors results in 0% CO_2 emissions. However, hydrogen combustion occurs at high temperatures, resulting in NOx levels up to three times higher than those produced by natural gas combustion (Nazari et al., 2022). NOx emissions are recognized as air pollutants with potential health risks. Therefore, one of the primary goals of hydrogen GT development is to find ways to reduce or eliminate these emissions.

Hydrogen GTs can be integrated into various energy systems to generate power. (Ebaid et al., 2015) conducted a study focusing on a novel energy system configuration comprising photovoltaic (PV) cells, an electrolyzer, a storage tank, and a hydrogen GT. The research explored two distinct scenarios: case (a) considered a worst-case scenario with minimal sunshine hours and solar radiation annually, while case (b) represented average sunshine hours and solar radiation throughout the year. According to their assumptions, the system yielded a profit equal to 25% of the total initial cost. In cases (a) and (b), the electricity prices produced by the system were \$0.12/kWh and \$0.16/kWh, respectively. The payback periods for these cases were calculated as 13 and 15 years, respectively. In different study, (Cen et al., 2021) examined a novel system configuration comprising a GT fueled by biofuel, solar PV cells, an electrolyzer, and a hydrogen post-firing chamber. In this configuration, the solar-powered electrolyzer generates hydrogen for post-firing in the GT. The system's performance was compared with a combined cycle using biomass without hydrogen post-firing. The results revealed that, apart from the capacity increase, post-firing system with hydrogen led to a significant 22.7% reduction in CO_2 emissions under optimal conditions. Numerous studies are underway on this subject, and there are plans for a 34 MW power plant that will operate solely on hydrogen generated from RESs, scheduled to be operational in the near future (Custom, 2022). The collaborative demonstration project, called "H2GT-Lingen," is significant because it demonstrates the feasibility of industrial-scale hydrogen-based electricity generation, highlighting its pivotal role in achieving decarbonized energy production. The H2GT project will utilize KHI's NOx emission reduction technology, which was previously trialed in 2018 within a 1 MW hydrogen gas turbine co-generation system manufactured by KHI in Kobe, Japan. The German project aims to increase the use of this technology via KHI's 'L30A turbine,' which has the capacity to generate 34.38 MW of energy with an efficiency of 40.3% when operated with hydrogen fuel (Nature Research Custom Media & Kawasaki, 2024).

Thermal Energy

CHP refers to systems that generate power and heat simultaneously in a single integrated unit. The fact that the hydrogen production that feeds fuel cells working with CHP originates from RESs will minimize CO_2 emissions. Fuel cell CHP systems are more efficient than other low-scale CHP technologies. While generating power from the chemical energy of hydrogen with fuel cells, low-temperature heating or hot water can be obtained by recovering the heat released as a by-product. CHP systems generally include a generator and heat conversion module (Olabi et al., 2020). However, comprehensive CHP systems should include the following auxiliary units in addition to fuel cells: a) a fuel processor for obtaining hydrogen from fossil fuels; b) a hydrogen storage tank to meet instant hydrogen needs; c) a heat conversion module for hot water production; d) a converter and power regulator to provide AC current compatible with the existing grid; e) a gas boiler for heat demand during peak hours; and f) control mechanisms to ensure the safety of the entire system (Cheng et al., 2023; Staffell, 2015). The more efficient CHP systems, with energy efficiency ranging between 75 and 80%, depend on the fuel used, the technology, and the location chosen to reduce energy losses during transmission (Razeghi et al., 2023). Companies such as Ceramic Fuel Cells, Panasonic, Toshiba, EneosCellTech, Kyocera, Aisin Seiki, and JxEneos are available trademarks for micro fuel cell CHP systems. Japan is a world leader in developing the technology of fuel cell CHP systems. The current high initial investment cost of fuel cell CHP systems, due to the limited manufacturing volume, can be decreased by around 25% by doubling output, with additional savings gained after commercialization (Ellamla et al., 2015).

Heating systems must be equipped with clean and sustainable technologies in order to reduce GHG emissions from buildings by 80-95%. Decarbonization of heating systems (CHP, heat pump, waste incineration, etc.), as well as gas supply (hydrogen, biomethane, blended gases, etc.) and electrification (heat pump, resistance heater, electric boiler, etc.), are required for this approach. Heat pumps convert free or waste energy to a higher temperature through the reverse Carnot cycle and pump it into the building. Although heat pumps vary according to their operating principles, application areas and sizes, the most basic classification area is heat sources. In this regard, there are four types of heat pumps (Abbasi et al., 2021):

- Air-source heat pumps: The heat source is ambient air.
- Water-source heat pumps: The heat source is local water resources.
- Ground-source heat pump: The heat source is geothermal resources.

- Other sources: The heat source is waste-water, industrial exhaust heat, flue gas, and district cooling.

Heat pumps provide a more efficient process of obtaining high-grade heat than pure electric heating or direct fuel burning. Heat pumps have different technological forms, such as thermoelectric cooler/heater (Peltier device), trans-critical R744 (CO_2 cycle), vapor compression cycle (VCC) or absorption/sorption chiller (Kwan et al., 2019). Coefficient of performance (COP) expresses the efficiency of a heat pump and is obtained by dividing the amount of heat produced to the amount of energy required by the compressor (Liu et al., 2017). Integrating fuel cell stacks with heat pumps is the optimum combination for meeting increasing heat demand in cold regions and throughout the winter months while minimizing energy use (Kwan et al., 2019). (Capuano et al., 2023) reported an overall COP of more than 1.5 and a cost savings of up to 60% when using an air source heat pump and a high-temperature fuel cell for residential space heating. In another study, (Jin et al., 2019) found a 7.6% improvement in COP value with open cathode PEM fuel cell and heat pump cogeneration. (Ozcelep et al., 2023) conducted a study using a solar photovoltaic-hydrogen-heat pump system to meet the electricity needs for heating greenhouses in Turkey. 24 m^2 solar panel area for 1000 m^2 floor area is sufficient for the energy demand with the proposed system. There are also other studies in the literature that use hydrogen in cooling systems, and these studies are particularly relevant in food processing and supply chains (Hassan et al., 2023; Segura et al., 2023).

In order to decarbonize current energy systems, especially high-heating industrial processes, integration of RESs in these systems through hydrogen is possible, but the creation of the distribution infrastructure of hydrogen makes this challenging. One cost-effective solution is to blend hydrogen with natural gas in existing gas pipes. Hydrogen injected directly into natural gas plays an intermediary role in converting surplus power into a grid-compatible gas and interconnecting the electricity and gas grid. The variables to consider while mixing hydrogen with natural gas are divided into three: safety, operational, and economic (Erdener et al., 2023). Since the physical properties of natural gas-hydrogen blending, such as the Wobbe index (W), higher heating value (HHV), specific gravity, flame speed, and flammability or explosion limit, vary, hydrogen must be mixed in definite proportions to ensure safe working conditions. If the W index, which measures how gas quality affects pipeline capacity, is high, it will result in overheating, while a low value will result in flame instability. Because hydrogen's HHV (12.1 MJ/m^3) is one-third that of methane (38.3 MJ/m^3), a higher volumetric flow rate is necessary to accomplish the same energy transfer as methane alone, which may need the modifications to current equipment. Specific gravity is also decisive parameter for the total gas flow rate. Good flame management is necessary, as the 10% hydrogen in the gas blending can increase the flame speed by 5–10%. Furthermore, the lower explosion limit (LEL) of the gas blending with 10% hydrogen falls from 5% to 4.6%, which is the value of pure hydrogen and pure methane. The penetration of small molecules of hydrogen into the microstructure of pipeline steel harms the steel's mechanical strength over time, causing hydrogen embrittlement and gas leaks (Chae et al., 2022).

Direct reduction of iron ore with hydrogen instead of coal in blast furnaces will prevent annual global emissions of 2.3 Gt of CO_2 from the iron and steel industries (Bhaskar et al., 2020). Green hydrogen can be used as fuel to heat the lime kiln to 1450 °C in cement, another industry that requires high heat like steel, by adapting existing equipment for such use (Oliveira et al., 2021). Using hydrogen to provide thermal energy for clinker production will make the cement industry, which has a large share in CO_2 emissions, more eco-friendly (Juangsa et al., 2022). Green hydrogen as fuel instead of natural gas is also an option in the glass industry to supply thermal energy demands that vary depending on different products for glass melting (Zier et al., 2021). The heat supplied by the combustion of green hydrogen and hydrogen-based steam converts bauxite into alumina in aluminum processing (Neuwirth et al., 2022).

SOLUTIONS AND RECOMMENDATIONS

Although there are various colors for hydrogen production, the resulting hydrogen is a colorless, odorless, and tasteless gas. At this point, it may be fair to classify hydrogen production according to whether it depends on renewable or non-renewable energy sources. The green hydrogen production through water electrolysis requires a significant investment in renewable energy infrastructure, which can be costly and time-consuming. The water supply required for electrolysis is not equal; this is a disadvantage for regions with limited water resources. Furthermore, the water vapor produced by excess hydrogen release may humidify the stratospheric ozone layer, causing it to deplete. Another issue with hydrogen production is that the concentration of hydroxyl (-OH) radicals, which react with other chemicals in the air to eliminate them, drops as they react with hydrogen. However, studies anticipate that these adverse effects of hydrogen will not contribute significantly to global warming (Jang et al., 2022). Scaling up production to meet growing demand poses logistical challenges, and the efficiency of electrolysis technology must continue to improve to make the process economically viable. Hydrogen safety remains a concern, as it is highly flammable, requiring stringent safety measures in its production, transportation, and storage (Öztürk Aydın & Bayrakçeken Yurtcan, 2023). A robust and efficient hydrogen distribution network is crucial to make it widely accessible. Addressing these challenges requires collaborative efforts among governments, industries, and researchers to drive innovation, improve technologies, and establish a sustainable hydrogen ecosystem that can play a pivotal role in our clean energy future. Despite some problems, researchers predict that hydrogen is a cleaner and more advantageous fuel than fossil fuels, and studies in this context are accelerating. Solar energy stands out as an ideal choice for hydrogen synthesis from electrolysis because it is a constant source of hydrogen production that has existed for as long as the world has existed and is practically equal everywhere on Earth.

Several key strategies are essential across diverse technologies to efficiently use green hydrogen for electricity production. Fuel cell technology should undergo continuous improvement in terms of efficiency, durability, and cost-effectiveness. Government support and industry investments are vital for scaling up production and infrastructure, including hydrogen refueling stations for fuel cell vehicles. Additionally, optimizing CHP systems to maximize electrical and thermal output while minimizing waste heat is essential. Incentives should encourage the deployment of green hydrogen-based CHP systems in various settings. Researchers should concentrate on reducing NOx emissions linked to hydrogen use and enabling safe hydrogen blending with natural gas for gas turbines. Updating regulations and standards to accommodate hydrogen-rich fuels will facilitate their adoption. Transitioning to a hydrogen-based energy system necessitates comprehensive policy support and regulatory frameworks to address safety concerns, standards, and incentives.

FUTURE RESEARCH DIRECTIONS

Hydrogen technology offers a wide range of applications. Hydrogen is viewed not just as a clean and unique fuel to replace fossil fuels but also as a resource for electricity production. Hydrogen is crucial for smart grids, and research into the future integration of hydrogen into smart grids has begun. Intersections in RESs negatively affect networks, causing voltage fluctuations, flashes, and frequency changes. This situation may reduce the quality of the electricity service provided to the consumer. In the relationship of hydrogen with smart grids, in cases where RESs are abundant, it is recommended to store the excess energy as hydrogen and use it intensively. However, hydrogen storage technology has not yet reached the desired level, and studies on storage methods are still in progress (Lin et al., 2020).

Green hydrogen produced through renewable energy-driven electrolysis will serve a versatile tool for grid resilience and stability. It will act as a buffer against the intermittency of RESs, storing excess energy during periods of surplus and releasing it when demand surges, thereby ensuring a reliable power supply. Moreover, green hydrogen will empower localized energy systems and microgrids, fostering energy self-sufficiency and reducing transmission losses. This versatile energy carrier will extend its influence to the transportation sector, facilitating the electrification of vehicles through fuel cells and powering hydrogen-based transportation modes. In industrial settings, green hydrogen will be a critical feedstock and an on-site source of electricity generation, enhancing energy efficiency and environmental sustainability. Regional hydrogen hubs will emerge as significant players, streamlining production, distribution, and utilization of hydrogen for electricity generation, thus creating a more integrated and eco-friendly energy ecosystem. Essentially, green hydrogen represents a paradigm shift in electrical energy production, offering a comprehensive solution to grid challenges, carbon emissions reduction, decentralized energy generation, and sustainable transportation. Its potential impact extends across sectors and industries, promising a cleaner, more resilient, and more efficient energy future.

CONCLUSION

Green hydrogen's ability to serve as a carbon-neutral energy carrier contributes to achieving climate goals by minimizing the release of greenhouse gases. The electrochemical conversion of green hydrogen into electrical energy, particularly within fuel cells, represents a groundbreaking advancement with inherent advantages for electricity and heat production. As an enabler of clean electricity and heat, it empowers industries and communities to transition away from fossil fuels, marking a crucial step toward a sustainable and environmentally responsible energy landscape. The production of green hydrogen involves the electrolysis of water using renewable electricity, such as solar or wind power. This clean and sustainable method ensures that hydrogen generation is entirely carbon-free.

Green hydrogen plays a significant role in clean electricity generation. When used in fuel cells, it undergoes a highly efficient electrochemical process, converting hydrogen directly into electricity with minimal emissions. This capability is instrumental in many applications, from propelling zero-emission electric vehicles to providing backup power to critical facilities. As the integration of renewable energy sources like wind and solar continues to grow, green hydrogen offers a means of storing surplus energy during periods of excess and releasing it when needed, thus addressing the intermittency challenge and enhancing grid stability. In thermal energy, green hydrogen holds immense potential. Its combustion in specialized burners or gas turbines provides a high-temperature heat, making it invaluable for industrial processes such as steel and cement manufacturing. Beyond industry, green hydrogen's clean combustion can heat residential and commercial buildings, replacing natural gas and dramatically lowering greenhouse gas emissions in the heating sector. Consequently, green hydrogen production and its utilization in electrical and thermal energy applications represent a pivotal shift towards a cleaner, more sustainable energy ecosystem by decarbonizing various sectors.

The existing literature acknowledges some limitations of green hydrogen while also emphasizing its significant benefits. Challenges such as high initial costs, the need for significant renewable energy input, and the need for advanced storage solutions make it difficult to expand the use of green hydrogen. Ongoing research aims to solve these restrictions, paving the way for greater acceptance and integration of green hydrogen technologies into mainstream energy systems. Government support and industrial investments are substantial in accelerating the dissemination of green hydrogen technologies. As we move towards a hydrogen-based energy environment, future research should prioritize innovations in technology, materials, and policies. In this sense, advancements in electrolysis efficiency, hydrogen storage and safety technologies, and the development of international standards will be critical in realizing the full potential of green hydrogen for electricity and thermal energy applications.

REFERENCES

Abbasi, M. H., Abdullah, B., Ahmad, M. W., Rostami, A., & Cullen, J. (2021). Heat transition in the European building sector: Overview of the heat decarbonisation practices through heat pump technology. *Sustainable Energy Technologies and Assessments*, *48*, 101630. doi:10.1016/j.seta.2021.101630

Ajanovic, A., Sayer, M., & Haas, R. (2022). The economics and the environmental benignity of different colors of hydrogen. *International Journal of Hydrogen Energy*, *47*(57), 24136–24154. doi:10.1016/j.ijhydene.2022.02.094

Arcos, J. M. M., & Santos, D. M. F. (2023). The Hydrogen Color Spectrum: Techno-Economic Analysis of the Available Technologies for Hydrogen Production. *Gases*, *3*(1), 25–46. doi:10.3390/gases3010002

Bhaskar, A., Assadi, M., & Somehsaraei, H. N. (2020). Decarbonization of the Iron and Steel Industry with Direct Reduction of Iron Ore with Green Hydrogen. *Energies*, *13*(3), 758. doi:10.3390/en13030758

Boulmrharj, S., Khaidar, M., Bakhouya, M., Ouladsine, R., Siniti, M., & Zine-dine, K. (2020). Performance Assessment of a Hybrid System with Hydrogen Storage and Fuel Cell for Cogeneration in Buildings. *Sustainability (Basel)*, *12*(12), 4832. doi:10.3390/su12124832

Capuano, M., Sorrentino, M., & Agelin-Chaab, M. (2023). Design and analysis of a hybrid space heating system based on HT-PEM fuel cell and an air source heat pump with a novel heat recovery strategy. *Applied Thermal Engineering*, *231*, 120947. doi:10.1016/j.applthermaleng.2023.120947

Cen, S. H., Li, K., Liu, Q. W., & Jiang, Y. L. (2021). Solar energy-based hydrogen production and post-firing in a biomass fueled gas turbine for power generation enhancement and carbon dioxide emission reduction. *Energy Conversion and Management*, *233*, 113941. doi:10.1016/j.enconman.2021.113941

Chae, M. J., Kim, J. H., Moon, B., Park, S., & Lee, Y. S. (2022). The present condition and outlook for hydrogen-natural gas blending technology. *Korean Journal of Chemical Engineering*, *39*(2), 251–262. doi:10.1007/s11814-021-0960-8

Chauhan, P. R., Kaushik, S. C., & Tyagi, S. K. (2022). A review on thermal performance enhancement of green cooling system using different adsorbent/refrigerant pairs. *Energy Conversion and Management-X, 14*. https://doi.org/ARTN 10022510.1016/j.ecmx.2022.100225

Chen, Z., Yiliang, X., Hongxia, Z., Yujie, G., & Xiongwen, Z. (2023). Optimal design and performance assessment for a solar powered electricity, heating and hydrogen integrated energy system. *Energy, 262*, 125453. doi:10.1016/j.energy.2022.125453

Cheng, L. X., Guo, Z. X., & Xia, G. D. (2023). A Review on Research and Technology Development of Green Hydrogen Energy Systems with Thermal Management and Heat Recovery. *Heat Transfer Engineering*. Advance online publication. doi:10.1080/01457632.2023.2191435

Corbo, P., Migliardini, F., & Veneri, O. (2011). Fuel Cells for Automotive Applications. In P. Corbo, F. Migliardini, & O. Veneri (Eds.), *Hydrogen Fuel Cells for Road Vehicles* (pp. 71–102). Springer. doi:10.1007/978-0-85729-136-3_3

Custom, N. R. (2022). *Hydrogen gas turbine offers promise of clean electricity*. Retrieved 29/09/2023 from https://www.nature.com/articles/d42473-022-00211-0

Daş, E., Ungan, H., Bayrakçeken Yurtcan, A., & Fıçıcılar, B. (2020). Other possible fuels and possible use of blended fuels in fuel cells. In R. G. Akay & A. Bayrakçeken Yurtcan (Eds.), *Direct Liquid Fuel Cells Fundamentals, Advances and Future* (pp. 249–273). Academic Press. doi:10.1016/B978-0-12-818624-4.00012-1

Dodds, P. E., Staffell, L., Hawkes, A. D., Li, F., Grunewald, P., McDowall, W., & Ekins, P. (2015). Hydrogen and fuel cell technologies for heating: A review. *International Journal of Hydrogen Energy, 40*(5), 2065–2083. doi:10.1016/j.ijhydene.2014.11.059

Dou, Y., Sun, L., Ren, J., & Dong, L. (2017). Opportunities and Future Challenges in Hydrogen Economy for Sustainable Development. In A. Scipioni, A. Manzardo, & J. Ren (Eds.), *Hydrogen Economy Supply Chain, Life Cycle Analysis and Energy Transition for Sustainability* (pp. 277–305). Academic Press. doi:10.1016/B978-0-12-811132-1.00010-9

Ebaid, M. S. Y., Hammad, M., & Alghamdi, T. (2015). THERMO economic analysis OF PV and hydrogen gas turbine hybrid power plant of 100 MW power output. *International Journal of Hydrogen Energy, 40*(36), 12120–12143. doi:10.1016/j.ijhydene.2015.07.077

EG&G Technical Services. (2004). Technology Overview. In Fuel Cell Handbook (7th ed., pp. 1-32). U.S. Department of Energy Office of Fossil Energy National Energy Technology Laboratory.

Ekins, P., & Shah, N. (2017). Extended Summary. In I. Staffell & P. E. Dodds (Eds.), The role of hydrogen and fuel cells in future energy systems. H2FC SUPERGEN.

Ellamla, H. R., Staffell, I., Bujlo, P., Pollet, B. G., & Pasupathi, S. (2015). Current status of fuel cell based combined heat and power systems for residential sector. *Journal of Power Sources, 293*, 312–328. doi:10.1016/j.jpowsour.2015.05.050

Erdener, B. C., Sergi, B., Guerra, O. J., Chueca, A. L., Pambour, K., Brancucci, C., & Hodge, B. M. (2023). A review of technical and regulatory limits for hydrogen blending in natural gas pipelines. *International Journal of Hydrogen Energy*, *48*(14), 5595–5617. doi:10.1016/j.ijhydene.2022.10.254

Fichtner, M., & Idrissova, F. (2009). Fundamental properties of hydrogen. In M. Ball & M. Wietschel (Eds.), *The Hydrogen Economy Opportunities and Challenges* (pp. 271–276). Cambridge University Press. doi:10.1017/CBO9780511635359.010

Głuchy, D., Kurz, D., & Trzmiel, G. (2014). The use of fuel cells to store energy from renewable sources. *Przegląd Elektrotechniczny, R. 90*(4), 81-84. doi:10.12915/pe.2014.04.18

Hassan, M., El-Sharkawy, I. I., & Harby, K. (2023). Study of an innovative combined absorption-adsorption cooling system employing the same evaporator and condenser. *Case Studies in Thermal Engineering*, *42*, 102690. doi:10.1016/j.csite.2022.102690

Hawkes, A., Staffell, I., Brett, D., & Brandon, N. (2009). Fuel cells for micro-combined heat and power generation. *Energy & Environmental Science*, *2*(7), 729–744. doi:10.1039/b902222h

Hočevar, S., & Summers, W. (2008). Hydrogen Production. In A. Leon (Ed.), Hydrogen Technology Mobile and Portable Applications (pp. 15-79). Springer. doi:10.1007/978-3-540-69925-5_2

Hordeski, M. F. (2008). *Alternative Fuels-The Future of Hydrogen*. Fairmont Press.

Hwang, J., Maharjan, K., & Cho, H. J. (2023). A review of hydrogen utilization in power generation and transportation sectors: Achievements and future challenges. *International Journal of Hydrogen Energy*, *48*(74), 28629–28648. doi:10.1016/j.ijhydene.2023.04.024

Incer-Valverde, J., Korayem, A., Tsatsaronis, G., & Morosuk, T. (2023). "Colors" of hydrogen: Definitions and carbon intensity. *Energy Conversion and Management*, *291*, 117294. doi:10.1016/j.enconman.2023.117294

Jang, D., Kim, J., Kim, D., Han, W. B., & Kang, S. (2022). Techno-economic analysis and Monte Carlo simulation of green hydrogen production technology through various water electrolysis technologies. *Energy Conversion and Management*, *258*, 115499. doi:10.1016/j.enconman.2022.115499

Jin, Y. H., Sun, L., & Shen, J. (2019). Thermal economic analysis of hybrid open-cathode hydrogen fuel cell and heat pump cogeneration. *International Journal of Hydrogen Energy*, *44*(56), 29692–29699. doi:10.1016/j.ijhydene.2019.03.098

Juangsa, F. B., Cezeliano, A. S., Darmanto, P. S., & Aziz, M. (2022). Thermodynamic analysis of hydrogen utilization as alternative fuel in cement production. *South African Journal of Chemical Engineering*, *42*, 23–31. doi:10.1016/j.sajce.2022.07.003

Karayel, G. K., Javani, N., & Dincer, I. (2023). Hydropower for green hydrogen production in Turkey. *International Journal of Hydrogen Energy*, *48*(60), 22806–22817. doi:10.1016/j.ijhydene.2022.04.084

Karayel, K., Javani, N., & Dincer, I. (2022). Green hydrogen production potential for Turkey with solar energy. *International Journal of Hydrogen Energy*, *47*(45), 19354–19364. doi:10.1016/j.ijhydene.2021.10.240

Kumar, S. S., & Lim, H. (2022). An overview of water electrolysis technologies for green hydrogen production. *Energy Reports*, *8*, 13793–13813. doi:10.1016/j.egyr.2022.10.127

Kwan, T. H., Wu, X. F., & Yao, Q. H. (2019). Performance comparison of several heat pump technologies for fuel cell micro-CHP integration using a multi-objective optimisation approach. *Applied Thermal Engineering*, *160*, 114002. doi:10.1016/j.applthermaleng.2019.114002

Lacko, R., Drobnic, B., Mori, M., Sekavcnik, M., & Vidmar, M. (2014). Stand-alone renewable combined heat and power system with hydrogen technologies for household application. *Energy*, *77*, 164–170. doi:10.1016/j.energy.2014.05.110

Leon, A. (2008). Introduction. In A. Leon (Ed.), Hydrogen Technology Mobile and Portable Applications (pp. 11-13). Springer. doi:10.1007/978-3-540-69925-5_1

Lin, R. H., Zhao, Y. Y., & Wu, B. D. (2020). Toward a hydrogen society: Hydrogen and smart grid integration. *International Journal of Hydrogen Energy*, *45*(39), 20164–20175. doi:10.1016/j.ijhydene.2020.01.047

Liu, Z., Deng, Z., Davis, S., & Ciais, P. (2023). Monitoring global carbon emissions in 2022. *Nature Reviews. Earth & Environment*, *4*(4), 205–206. doi:10.1038/s43017-023-00406-z PMID:37065615

Liu, Z. J., Xu, W., Zhai, X., Qian, C., & Chen, X. (2017). Feasibility and performance study of the hybrid ground-source heat pump system for one office building in Chinese heating dominated areas. *Renewable Energy*, *101*, 1131–1140. doi:10.1016/j.renene.2016.10.006

Longoria, G., Lynch, M., & Curtis, J. (2021). Green hydrogen for heating and its impact on the power system. *International Journal of Hydrogen Energy*, *46*(53), 26725–26740. doi:10.1016/j.ijhydene.2021.05.171

Lucia, U. (2014). Overview on fuel cells. *Renewable & Sustainable Energy Reviews*, *30*, 164–169. doi:10.1016/j.rser.2013.09.025

Meziane, S., & Bentebbiche, A. (2019). Numerical study of blended fuel natural gas-hydrogen combustion in rich/quench/lean combustor of a micro gas turbine. *International Journal of Hydrogen Energy*, *44*(29), 15610–15621. doi:10.1016/j.ijhydene.2019.04.128

Nami, H., Rizvandi, O. B., Chatzichristodoulou, C., Hendriksen, P. V., & Frandsen, H. L. (2022). Techno-economic analysis of current and emerging electrolysis technologies for green hydrogen production. *Energy Conversion and Management*, *269*, 116162. doi:10.1016/j.enconman.2022.116162

Nature Research Custom Media & Kawasaki. (2024). *Hydrogen gas turbine offers promise of clean electricity*. Nature Portfolio. Retrieved 14.01.2024 from https://www.nature.com/articles/d42473-022-00211-0

Nazari, M. A., Alavi, M. F., Salem, M., & Assad, M. E. (2022). Utilization of hydrogen in gas turbines: A comprehensive review. *The International Journal of Low Carbon Technologies*, *17*, 513–519. doi:10.1093/ijlct/ctac025

Neuwirth, M., Fleiter, T., Manz, P., & Hofmann, R. (2022). The future potential hydrogen demand in energy-intensive industries-a site-specific approach applied to Germany. *Energy Conversion and Management*, *252*, 115052. doi:10.1016/j.enconman.2021.115052

Olabi, A. G., Wilberforce, T., Sayed, E. T., Elsaid, K., & Abdelkareem, M. A. (2020). Prospects of Fuel Cell Combined Heat and Power Systems. *Energies*, *13*(16), 4104. doi:10.3390/en13164104

Oliveira, A. M., Beswick, R. R., & Yan, Y. S. (2021). A green hydrogen economy for a renewable energy society. *Current Opinion in Chemical Engineering*, *33*, 100701. doi:10.1016/j.coche.2021.100701

Ozcelep, Y., Bekdas, G., & Apak, S. (2023). Meeting the electricity demand for the heating of greenhouses with hydrogen: Solar photovoltaic-hydrogen-heat pump system application in Turkey. *International Journal of Hydrogen Energy*, *48*(7), 2510–2517. doi:10.1016/j.ijhydene.2022.10.125

Öztürk Aydın, A., & Bayrakçeken Yurtcan, A. (2023). Hydrogen Safety Issues: The Inherent Challenges of Hydrogen and Possible Precautions. In R. A. Felseghi (Ed.), *Hydrogen Fuel Cell Technology for Mobile Applications* (pp. 207–237). IGI Global. doi:10.4018/978-1-6684-6721-3.ch009

Panic, I., Cuculic, A., & Celic, J. (2022). Color-Coded Hydrogen: Production and Storage in Maritime Sector. *Journal of Marine Science and Engineering*, *10*(12), 1995. doi:10.3390/jmse10121995

Pasupathi, S., Gomez, J. C. C., Su, H., Reddy, H., Bujlo, P., & Sita, C. (2016). *Recent Advances in High Temperature PEM Fuel Cells*. Academic Press. doi:10.1016/C2015-0-06917-9

Pimm, A. J., Cockerill, T. T., & Gale, W. F. (2023). Reducing industrial hydrogen demand through preheating with very high temperature heat pumps. *Applied Energy*, *347*, 121464. doi:10.1016/j.apenergy.2023.121464

Razeghi, M., Hajinezhad, A., Naseri, A., Noorollahi, Y., & Moosavian, S. F. (2023). An overview of renewable energy technologies for the simultaneous production of high-performance power and heat. *Future Energy*, *2*(2), 1–11. doi:10.55670/fpll.fuen.2.2.1

Sasaki, K., Li, H.-W., Hayashi, A., Yamabe, J., Ogura, T., & Lyth, S. M. (2016). *Hydrogen Energy Engineering A Japanese Perspective*. Springer. doi:10.1007/978-4-431-56042-5

Savu, L. M., Amalia, M. A., Remus, S. E., Vasile, A. M., & Tabita, N. D. (2021). Possibilities to combine the hydrogen with heat pumps in heating systems of buildings. *Annals of 'Constantin Brancusi' University of Targu-Jiu. Engineering Series*, (4), 36–41.

Segura, F., Vivas, F. J., Andujar, J. M., & Martinez, M. (2023). Hydrogen-powered refrigeration system for environmentally friendly transport and delivery in the food supply chain. *Applied Energy*, *338*, 120945. doi:10.1016/j.apenergy.2023.120945

Sherif, S. A., Goswami, Y., Stefanakos, E., & Steinfeld, A. (2014). Handbook of Hydrogen Energy (S. A. Sherif, Y. Goswami, E. Stefanakos, & A. Steinfeld, Eds. 1 ed.). CRC Press.

Staffell, I. (2015). Zero carbon infinite COP heat from fuel cell CHP. *Applied Energy*, *147*, 373–385. doi:10.1016/j.apenergy.2015.02.089

Staffell, I., Scamman, D., Abad, A. V., Balcombe, P., Dodds, P. E., Ekins, P., Shah, N., & Ward, K. R. (2019). The role of hydrogen and fuel cells in the global energy system. *Energy & Environmental Science*, *12*(2), 463–491. doi:10.1039/C8EE01157E

Veneri, O. (2011). Hydrogen as Future Energy Carrier. In P. Corbo, F. Migliardini, & O. Veneri (Eds.), *Hydrogen Fuel Cells for Road Vehicles* (pp. 33–70). Springer. doi:10.1007/978-0-85729-136-3_2

Winter, C. J. (2005). Into the hydrogen energy economy - milestones. *International Journal of Hydrogen Energy, 30*(7), 681–685. doi:10.1016/j.ijhydene.2004.12.011

Yoshimura, T., McDonell, V., & Samuelsen, S. (2005). Evaluation of hydrogen addition to natural gas on the stability and emissions behavior of a model gas turbine combustor. *Proceedings of the ASME Turbo Expo 2005,* 2, 573-581. 10.1115/GT2005-68785

Yu, M. L., Wang, K., & Vredenburg, H. (2021). Insights into low-carbon hydrogen production methods: Green, blue and aqua hydrogen. *International Journal of Hydrogen Energy*, *46*(41), 21261–21273. doi:10.1016/j.ijhydene.2021.04.016

Yu, S., Fan, Y., Shi, Z., Li, J., Zhao, X., Zhang, T., & Chang, Z. (2023). Hydrogen-based combined heat and power systems: A review of technologies and challenges. *International Journal of Hydrogen Energy*, *48*(89), 34906–34929. Advance online publication. doi:10.1016/j.ijhydene.2023.05.187

Zhang, J. Z., Li, J., Li, Y., & Zhao, Y. (2014). Introduction to Basic Properties of Hydrogen. In *Hydrogen Generation, Storage and Utilization* (pp. 1–3). Wiley. doi:10.1002/9781118875193.ch1

Zier, M., Stenzel, P., Kotzur, L., & Stolten, D. (2021). A review of decarbonization options for the glass industry. *Energy Conversion and Management-X, 10.* https://doi.org/ARTN 10008310.1016/j.ecmx.2021.100083

Zun, M. T., Ahmad, M. S., Fayaz, H., Selvaraj, J., Ahmed, W., Wang, Y. R., Ben Khedher, N., Silitonga, A. S., Elfasakhany, A., Kalam, M. A., & Rashid, B. (2023). Towards techno-economics of green hydrogen as a primary combustion fuel for recreational vehicle vapor absorption refrigeration system. *Sustainable Energy Technologies and Assessments*, *56.* https://doi.org/ARTN 10300710.1016/j.seta.2022.103007

Züttel, A., Borgschulte, A., Schlapbach, L., Chorkendorff, I., & Suda, S. (2008). Properties of Hydrogen. In A. Züttel, A. Borgschulte, & L. Schlapbach (Eds.), *Hydrogen as a Future Energy Carrier* (pp. 71–147). Wiley. doi:10.1002/9783527622894.ch4

KEY TERMS AND DEFINITIONS

Coefficient of Performance (COP): The COP value, which determines the performance of a heat pump or cooling system, is the heating or cooling rate achieved in response to the work done.

Combined Heat and Power (CHP) System: In these systems, also called cogeneration, heat and electrical energy are produced simultaneously, where heat production is the limiting factor for energy efficiency.

Electrolysis: The electrolysis process is the decomposition of water by giving hydrogen and oxygen gas through externally supplied electrical energy.

Fuel Cell: Fuel cells are a clean energy conversion technology that can generate electrical energy directly from the chemical energy of the fuel.

Gas Turbine (GT): Gas turbines, which work according to the Brayton cycle principle, produce power by rotating a turbine due to the expansion of compressed, high-heated air.

Green Hydrogen: It is the hydrogen obtained from electrolysis powered by renewable energy sources accepted as the principal production route.

Greenhouse Gas (GHG): A greenhouse gas is characterized by its ability to absorb infrared radiation, capturing and re-emitting heat energy from the Earth's surface.

Heat Pump: It is equipment that transfers heat from a low-temperature environment to a high-temperature environment with externally supplied electrical energy.

Natural Gas Blending: It is blending clean hydrogen with natural gas in some proportions in the existing pipe system to reduce greenhouse gas emissions.

Wobbe Index: It is an index that compares the combustion energy outputs of gas fuels with different compositions.

Chapter 6
Energy-Efficient Grid Interactive Building Optimization With Green Energy for Net-Zero Emission

Dharmbir Prasad
https://orcid.org/0000-0002-9010-9717
Asansol Engineering College, India

Rudra Pratap Singh
https://orcid.org/0000-0001-7352-855X
Asansol Engineering College, India

Gaurav Gupta
Asansol Engineering College, India

Manish Kumar
https://orcid.org/0009-0004-6408-8328
Asansol Engineering College, India

Jatin Anand
Asansol Engineering College, India

Ranadip Roy
https://orcid.org/0000-0003-2111-2581
Sanaka Educational Trust's Group of Institutions, Durgapur, India

ABSTRACT

In view of increasing energy demands and the pressing need for sustainable solutions, grid-interactive building infrastructure has emerged as a viable and efficient option to meet the energy needs of communities and buildings. The following introduction explores the concept of grid-interactive building infrastructure, with a focus on a specific case study in Kashinagar, Odisha, India. Currently, the electric needs of the undertaken site are met through the conventional power grid, resulting in significant operating costs amounting to $6,039 per year. A proposed system is being introduced to address this challenge, incorporating 496 kW and 36 kW of solar and wind energy generation capacity, respectively. By integrating renewable energy sources, this solution aims to reduce operating costs to a negative value of -$11,432 per year. In addition to the economic and environmental benefits, the studied grid-interactive building infrastructure suggests an independent and resilient energy system.

DOI: 10.4018/979-8-3693-1014-4.ch006

INTRODUCTION

Decarbonisation and decentralization are pivotal cornerstones for meeting the enormous increase in energy demand caused by growing urbanization, enabling a global low-carbon economy. Due to greenhouse gas emissions, the world is experiencing considerable and accelerating climate change. As part of its August 2022 Nationally Determined Contribution (NDC) goals to the United Nations Framework Convention on Climate Change (UNFCCC), India committed to two of the four targets at Conference of the Parties (COP26), 26^{th}session. These two goals are to reduce the intensity of its emissions by 45% below 2005 levels by 2030 and to expand the share of non-fossil power capacity to 50% by 2030 (Kacker & Srivastava, 2023). According to an International Energy Agency (IEA) report, infrastructure and building construction account for one-third of worldwide energy consumption and generate around 15% of global CO_2 emissions. With a population of 1.4 billion people and counting, India's housing industry will need to create 400-600 million square meters of developed space per year until 2030 to meet escalating demand (IEA, 2022). As the demand for buildings and infrastructure grows, so does the demand for energy surged in the form of power consumption for cooling, heating, lighting, cooking and many other requirements.

Paper layout: Remaining portion of the paper has been described as follows: energy potential assessments of the site, problem statements, literature survey, mathematical modeling, study methodology, hybrid system modeling, component costing and financial analysis, simulation results, and at end conclusion.

Energy Potential Assessments of the Site

In the Gajapati district in the Indian state of Odisha, Kashinagar serves as both the town and the administrative center for a community development block of the same name. There were 9684 people living in Kashinagar as of the 2011 India Census. The state DISCOs are in charge of medium voltage power distribution and transmission throughout the several districts of Odisha. The tehsil office, community health center, block office, notified area council office, revenue inspector office, etc. are only a few of the prominent buildings there with high energy requirements. These enterprises will receive sustainable power from the hybrid renewable energy. The grid-interactive building infrastructure is extremely important for the local community as well as the larger energy environment. The site has advantages from its strategic position, plentiful sunshine, and favorable wind conditions, giving it a good place for deploying renewable energy based building solutions. Furthermore, this system's adoption is consistent with national and international obligations to combat climate change and reduce greenhouse gas emissions. Kashinagar's adoption of grid-interactive building infrastructure contributes to India's ambitious renewable energy targets and supports global efforts towards a low-carbon future.

Table 1. Geographical co-ordinates of the undertaken location

Country	Location	State	Latitude (°N)	Longitude (°E)
India	Kashinagar	Odisha	18° 51.5'N	83° 52.8'E

Problem Statements

The graphical illustration of undertaken objectives along with the HOMER software tool based approach has been given in Figure 1.

Recent Studies and Research Motivation

To expedite progress, future strategies must prioritize sustainable and energy-efficient infrastructure. Government initiatives alone cannot achieve carbon neutrality and energy efficiency; building owners and occupants must also be motivated to adopt green energy practices. The Energy Conservation Building Code, introduced in 2007 and revised in 2017, aims to cut energy consumption by 50% by 2030. The inception of the Smart Cities Mission (SCM) in 2015 presents another potential avenue. Recent SCM reports reveal projects benefiting marginalized communities and promoting sustainable growth (Vyas et al., 2014). Earlier initiatives by the Bureau of Energy Efficiency (BEE) reportedly saved 300 million tons of CO_2 across Indian regions (Kacker & Srivastava, 2023). Numerous researchers have recently contributed to eco-friendly building infrastructure solutions. The following key studies as indicated in following broad areas motivate the current research:

(i). Smart Energy System Optimization and Sustainability: Noteworthy studies in this domain includes the optimization of distribution and emission reduction in smart energy systems (Byun et al., 2011), an optimization approach for carbon reduction in energy systems (Morvaj et al., 2016), integration of intelligent control systems for zero-energy buildings (Zheng & HienWong, 2016), cooperative power control in green buildings (Dagdougui et al., 2016), innovative energy management models (Kirmani et al., 2018), energy optimization for smart microgrids (Ullah et al., 2021), study emphasizes decarbonization, digitization (Wu et al., 2022), power network assessment with zero-energy buildings (Zhang et al., 2022), net-zero management for city commercial buildings (Liu et al., 2022), building information modelling (BIM) based decision-making for building upgrades (Chen et al.,

Figure 1. Optimization objectives and different approach for optimal solution

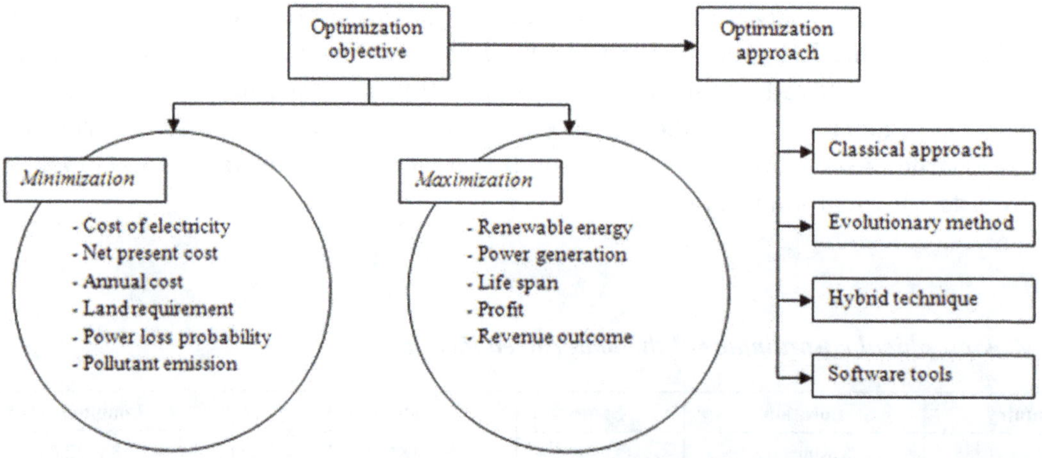

2022), maximizing renewable energy in the UK (Sadhukhan et al., 2022) and data-driven methods enhance building efficiency (Mousavi et al., 2023).

(ii). Energy Efficiency and Demand Management: A study offering innovative demand management models combining solar, wind, and coal power to reduce emissions, using regression models is presented (Kirmani et al., 2018). Another research focuses on lowering energy consumption in commercial buildings via energy-efficient appliances and photovoltaic systems, aiming for 90% efficiency and a 6-year payback time (Qamar et al., 2022).

(iii). Renewable Energy Integration and Carbon Emission Reduction: In (Mousavi et al., 2023), sustainable towns with almost zero energy use; the study focuses on positive energy buildings (PEBs). It investigates data-driven methods like machine learning to enhance building efficiency, occupant comfort and renewable energy provision. Wu et al., (2022) emphasizes on decarbonisation, digitization and decentralization of buildings energy feeding system. The US's net-zero emissions and carbon-free electricity grid goal is explored, addressing implementation challenges and offering guidance (Arent et al., 2022). Another study focused on optimizing profit and reducing costs and emissions in energy management. It efficiently manages energy price changes while promoting renewable energy use (Wang et al., 2023).

(iv). Internet of things (IoT) and Technology Integration for Energy Management: One study investigates combining IoT and business intelligence to advance cognitive building development. It delves into challenges and proposes solutions (Moudgil et al., 2023). Another research enhances smart home energy management through real-time building energy monitoring enabled by the internet. This involves integrating IoT for energy monitoring to improve performance and cost-effective renewable energy utilization (AlFaris et al., 2017).

(v). Energy Management in Specific Contexts: Need for energy-efficient sports facilities, particularly in hot city regions, highlighting a research gap are presented in (Elnour et al., 2022). Further, the HOMER software based optimization methods for hybrid RE systems demonstrates residential buildings application is reported in (Thirunavukkarasu et al., 2023). Different studies relating to smart and energy-efficient buildings are presented in (Wang et al., 2023), (Pan et al., 2023), (Liu et al., 2023), (Ngo et al., 2023). Corresponding, recent studies have been summarized in Table 2.

These studies highlight strategies for energy efficiency, renewable integration and sustainable growth. The research underscores the significance of interdisciplinary approaches, data-driven methodologies, and the integration of renewable energy sources for achieving

MATHEMATICAL MODELING

Considering economically sound and sustainable energy solutions for the design parameters, the electrical energy generated by a PV panel is expressed in terms of temperature and reference temperature as given in (1) (Kacker & Srivastava, 2023), (Mariano-Hernández et al., 2021),

$$P_{PV} = G_t \eta_{PV\_ref}\left[1-\beta\left(\left\{T_{amb}+\left(\frac{NOCT-20}{800}\right)G_t\right\}-T_{ref}\right)\right] \tag{1}$$

Table 2. Summary of observations in recent studies

Reference	Renewable Source	Objective	Solution Technique	Findings	Location
AlFaris et al., 2017	• Solar Panel	• Energy performance optimization using intelligent home technology may lead to net zero energy dwellings. • Technologies may effectively control energy use.	• Intelligent Home Energy Management System.	• International energy conservation code. • Home Energy Management System.	• Saudi Arabia
Dagdougui et al., 2016	• PV Module. • Wind Turbine.	• To show efficacy through simulation. • Optimize internal power regulation in a network of green generation, and loads while taking uncertainties into account.	• Integrating energy storage, generation, and loads	• Optimize energy management throughout the network of green buildings.	-
Huang et al., 2021	• Solar	• Application of net-zero energy community (NZEC) concepts.	• Supplying electricity, heat for domestic water, and heating/cooling.	• System modeling to address technical challenges and enables simulation-based assessment of building-to-grid integration controls.	• Florida, USA
Kirmani et al., 2018	• PV system. • Wind turbine. • Hybrid grid connectivity system.	• To determine if it is feasible to combine solar and wind power systems with currently operating coal-based facilities.	• Creating hybrid energy systems that combine photovoltaic systems from the sun and the wind with sophisticated control methods and grid integration plans.	• Underlines the necessity for sophisticated control strategies to improve grid integration and system efficiency while emphasizing the importance of hybrid energy systems.	• Ramapuram, Chennai
Liu et al., 2022	• Rooftop solar PV. • Offshore wind power. • Static pumped hydro storage.	• To establish net-zero energy management techniques for urban commercial buildings.	• Utilizing the hybrid renewable energy and storage technologies is being implemented.	• Green energy framework for decarbonizing the construction and transportation sectors may dramatically cut carbon emissions.	• Hong-Kong, China.
Nosratabadi et al., 2022	• Photovoltaic powerplant.	• To evaluate the effectiveness of demand-side management (DSM).	• Using a probabilistic model based on historical data to account for the random nature of solar radiation in the multi-energy microgrid.	• Conversion of solar radiation is taken into account in the mathematical model of the PV unit.	-
Pan et al., 2023	• Wind • Solar • Battery storage system	• Simulation for performance-driven design for new building and retrofit design. • Integrated simulation using data measurements. • Building demand response.	• Building energy modelling.	• Reviewed building energy simulation.	-
Sadhukhan, et al., 2022	• Wind • Solar • Hydro • Geothermal • Nuclear-gas with CCUS	• Transitioning to an electricity system with net-zero greenhouse gas emissions.	• Optimizes capacities of renewable bioenergy, nuclear, CCUS and storage systems • NLP to minimize cost	• Minimum costs for these maximum electricity mixes are 110, 100, and 190 EURO/ MWh.	• UK regions, South West, Greater London, and South East,
Tariq, et al 2021	• Solar • Wind • Diesel Generator • Battery System	• Optimization to achieve goals like preserving water quality while lowering net-present-cost, energy costs, and environmental pollutants.	• Estimation of energy consumption of water treatment plants and power demand • Economic study of the hybrid energy system	• Values obtained with HOMER for the NPC and COE	• Yucatan, Mexico

where, G_t: incidence solar radiation, $\eta P_{V\_ref:}$ efficiency under the reference temperature ($Tr_{ef=}$ 20°C), β: temperature coefficient and $Tam_{b:}$ ambient temperature.

This power is then supplied as a function of the wind speed at the wind turbines height. The change in wind speed with elevation above ground level is calculated using (2) (Kacker & Srivastava, 2023), (Mariano-Hernández et al., 2021),

$$v_w(t) = v_{an}(t)\left\{\frac{\ln\left(\frac{H_{hb}}{H_{an}}\right)}{\ln\left(\frac{H_{an}}{H_0}\right)}\right\} \tag{2}$$

where, $v_w(t)$: wind speeds at hub height (m/s), $v_{an}(t)$: wind speeds at the anemometer elevation (m/s), H_{hb}: hub height (m), H_{an}: anemometer height (m) and H_0: surface roughness length (m).

Low-loss power conversion between AC and DC components is carried out via converters. The conversion efficiency is set at 95% for the AC to DC stage as well as the DC to AC stage. To account for allowed overloads, a converter's rating is more than the maximum power flow in both directions, as calculated by (3) (Kacker & Srivastava, 2023), (Mariano-Hernández et al., 2021),

$$\left.\begin{aligned} \eta_{dc/ac}^{con}\{P_b(t)+P_{pv}(t)\} &\le N_{con}.PSC_r \\ \eta_{ac/dc}^{con}\{P_w(t)\} &\le N_{con}.PSC_r \end{aligned}\right\} \tag{3}$$

where, $\eta_{dc/ac}^{con}$: dc-ac conversion efficiency, $\eta_{ac/dc}^{con}$: ac-dc conversion efficiency, $P_b(t)$: battery power, $P_{pv}(t)$: PV power, N_{con}: number of converters, PSC_r: rating of power converters.

The net zero emissions by 2050 Scenario is an IEA normative scenario that illustrates a roadmap for the global energy industry to reach net zero CO2 emissions by 2050. The CO2 emission (kgCO2/kWh) is computed using (4) (Kacker & Srivastava, 2023), (Mariano-Hernández et al., 2021),

$$CO_2 \text{ emissions} = \sum_{i=1}^{\text{number of power sources}} (CO_2 \text{ emission factor} \times \text{capacity of power source}) \tag{4}$$

Converting between one-time charges and yearly costs is made easier by using the annual real interest rate. The discount factor and annualized costs from present costs are calculated by the HOMER tool using the annual real interest rate, which is given by (5) (Kacker & Srivastava, 2023), (Mariano-Hernández et al., 2021).

$$i = \frac{i' - f}{i + f} \tag{5}$$

where, f: expected inflation rate, i' : nominal interest rate, i: annual real interest rate (%).

Total of NPC value represents the cost of system life cycle as using equation (6) (Kacker & Srivastava, 2023), (Mariano-Hernández et al., 2021), which, indicates the summation of the cash flow of the t-year over the factor and the initial capital cost. The costs contain capital cost, fuel cost, operation cost, replacement cost, maintenance cost, etc. Income contains electricity selling and salvage value after life cycle.

$$\left.\begin{aligned} NPC &= CF_0 + \frac{CF_1}{(1+i)^1} + \frac{CF_2}{(1+i)^2} + \frac{CF_3}{(1+i)^3} + \dots \frac{CF_N}{(1+i)^N} \\ &= CF_0 + \sum_{t=1}^{N} \frac{CF_t}{(1+i)^t} \end{aligned}\right\} \quad (6)$$

where, CF_0: initial capital cost ($), CF_t: cash flow of t-year ($), t: number of year, i: annual real interest rate (%), N: project life time (year).

The CRF ratio is used to calculate the annuity's present value over the course of the project. It could be stated as follows in (7) (Kacker & Srivastava, 2023), (Mariano-Hernández et al., 2021),

$$CRF = \frac{i(1+i)^t}{(1+i)^t - 1} \quad (7)$$

where, i: annual real interest rate (%), t: number of years.

The levelized cost of energy, which may be calculated using (8) (Kacker & Srivastava, 2023), (Mariano-Hernández et al., 2021), is the price necessary to create one unit of energy.

$$LCOE = \frac{\text{Systems' life cycle total cost}}{\text{Systems' total annual energy generation}} \quad (8)$$

By dividing the TAC by the total annualized useable electric energy production, the levelized COE (in $ per kWh) is determined. The TAC is the NPC's annualized value (in $ per year), which may be found in (9) (Kacker & Srivastava, 2023), (Mariano-Hernández et al., 2021).

$$TAC = NPC \times CRF(i,N) \quad (9)$$

STUDY METHODOLOGY

Simulation Techniques

The HOMER Pro is a powerful energy simulation tool necessary for the study of energy-efficient grid-interactive building infrastructure optimization with green energy sources. It plays an important role in the design and evaluation of the optimal building energy system. The tool helps determine the best way to use energy by simulating different scenarios and using optimization techniques as shown in Fig. 2.

It allows researchers to assess the potential for energy savings and environmental benefits, considering variables such as building characteristics, energy needs, and green energy potential.

Renewable Energy Generation and Load Profiles

(i). Solar GHI Resource at Ste: A chart of solar GHI resource data illustrates the amount of solar radiation received by a surface horizontal to the ground as given in Table 3. It is important for assessing the potential and designing the appropriate solar energy system for the site. Monthly average GHI data of the selected location over 22 years is obtained from the NASA POWER database. Figure 3 shows the monthly average GHI data for Kashinagar. The minimum GHI value (*i.e.,*3.600 kWh/m^2/day)occurred in August; whereas, the maximum value (*i.e.,* 6.300 kWh/m^2/day) appeared in April.

(ii). Temperature Resource at Site: A data chart of temperature presented in Table 3 shows its variations at undertaken site over time for assessing the performance and efficiency of renewable energy systems. Figure 4 depicts the monthly average temperature data in the selected location. The minimum monthly average air temperature value (*i.e.,* 20.57°C) at the site occurred in December; whereas, the maximum monthly value (*i.e.,*32.24°C) reached in May.

(iii). Wind Resource at Site: A wind resource data (as presented in Table 3) shows the wind speed and direction variations at a site over time. It is important for assessing the potential of wind energy at the site and designing an appropriate wind energy system for the site. Figure 5 shows the proposed site's monthly mean wind speed data over 30 years at 50 m height. The maximum monthly mean wind speeds occurred in June with a value of 5.010 m/s.

(iv). Power Demand Profile: Electric load refers to the amount of electrical power consumed by a system or device. The daily, seasonal, and yearly profile of electric load describes how the amount of electrical power consumed varies over different time periods. The daily profile of electric load typically shows peak demand during the daytime hours when people are awake and active, while, lower demand during the night when people are asleep. This daily profile is also affected by daily routines and patterns,

Table 3. Monthly average atmospheric data

Month	Clearness Index	Daily Radiation (kWh/m^2/day)	Daily Temperature (°C)	Wind Speed Average (m/s)
January	0.628	4.810	21.690	2.850
February	0.642	5.530	24.880	3.110
March	0.620	6.030	28.980	3.590
April	0.598	6.300	31.640	4.550
May	0.555	6.030	32.240	4.770
June	0.389	4.250	29.500	5.010
July	0.335	3.640	27.400	4.920
August	0.339	3.600	26.840	4.600
September	0.414	4.120	26.500	3.800
October	0.509	4.530	25.230	3.560
November	0.576	4.520	22.560	3.710
December	0.618	4.530	20.570	3.120

Figure 2. Flow chart for grid interactive intelligent building load management

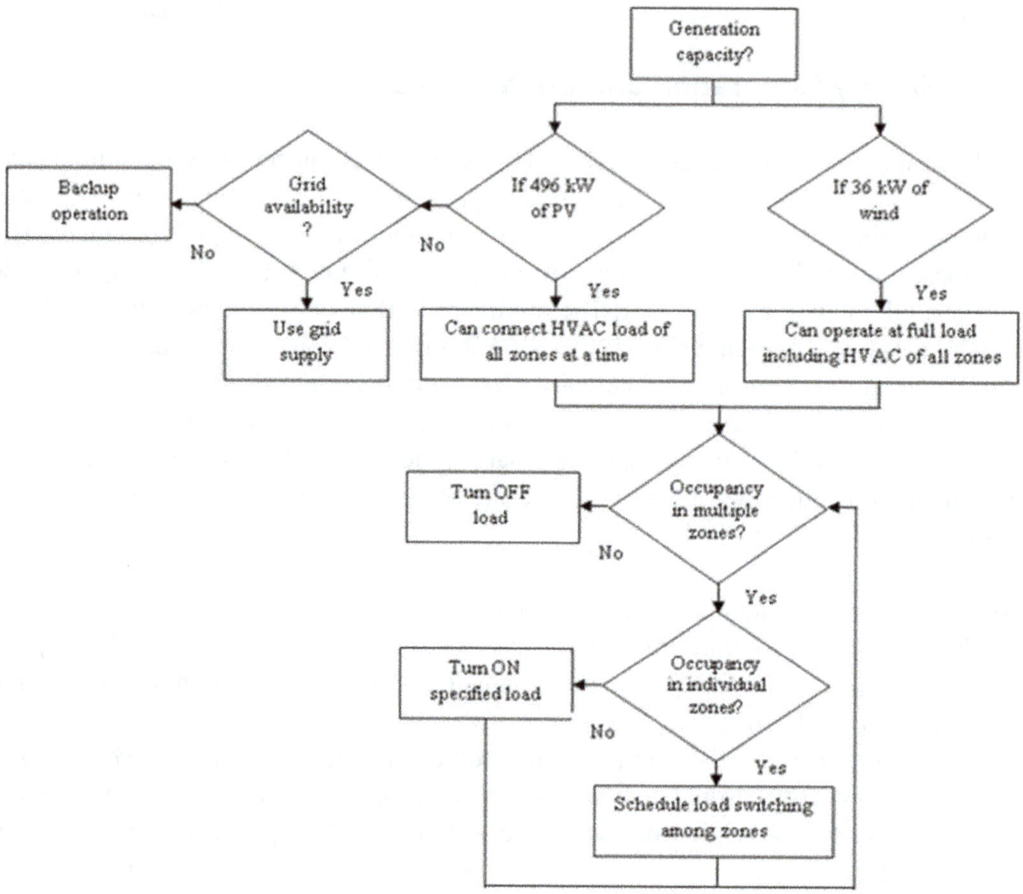

Figure 3.Graphical representation of solar GHI resource data

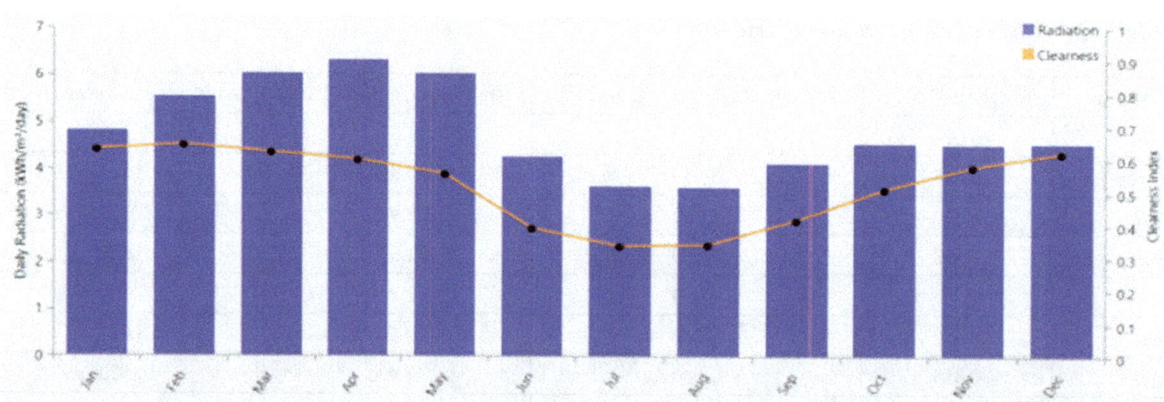

such as work schedules and meal times. Seasonal variations in electric load can be seen when different weather patterns occur throughout the year, which can affect heating and cooling needs. For example, during the summer months, the demand for electricity often increases as people use more air conditioning.

Figure 4. Graphical representation of temperature resource data

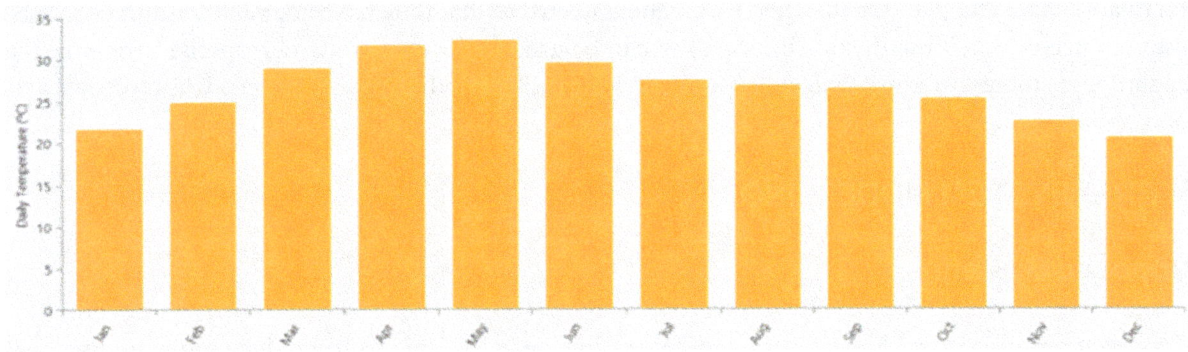

Figure 5. Graphical representation of wind resource data

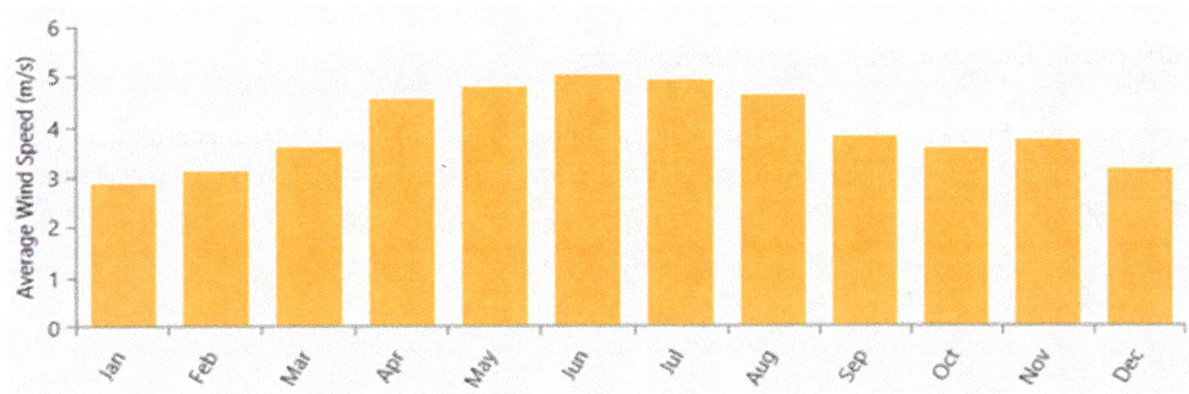

Figure 6. Daily power demand profile

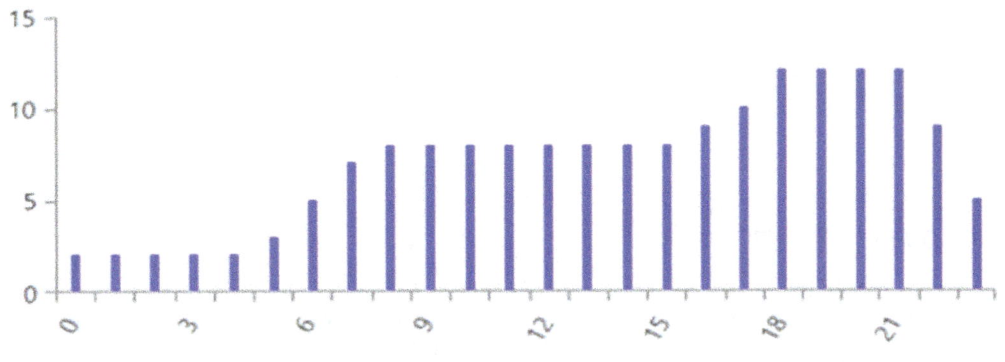

The yearly profile of electric load shows the overall pattern of electricity demand throughout the year. This profile is affected by many factors, including population growth, changes in weather patterns, and economic conditions. The yearly profile also reflects long-term changes in energy consumption patterns, such as the increasing use of renewable energy sources and the adoption of energy-efficient technologies.

Figures 6-8 depicts the daily, seasonal, and yearly profiles, respectively. Utilities can use this information to anticipate and plan for changes in demand and ensure that they have adequate resources to meet customer needs. Additionally, this information can be used to develop demand response programs that encourage customers to adjust their energy usage during peak periods to reduce strain on the electrical grid.

HYBRID SYSTEM MODELLING

Proposed System

It has been proposed to add 36 kW of wind generation capacity. This would reduce the operating costs to -$11,432/yr. Here, the investment has a payback of 10.6 years and an internal rate of return (IRR) of 6.73%. However, previous system was consisting of grid and PV (rated capacity: 496 kW) generation with operating costs of -$10,232 per year.

Schematic Diagram of Integrated System

In this analysis, Fig. 9 presents a schematic diagram illustrating the hybrid system configuration. The diagram showcases the integration of different components to form an efficient hybrid power generation system.

Figure 7. Seasonal power demand profile

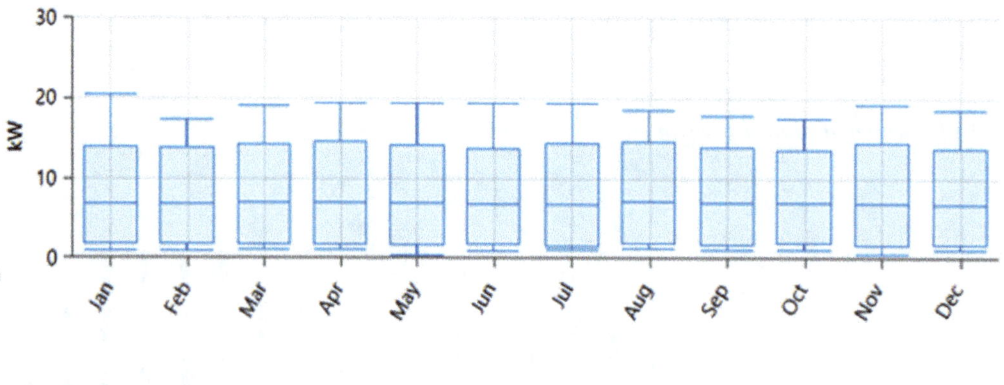

Figure 8. Yearly power demand profile

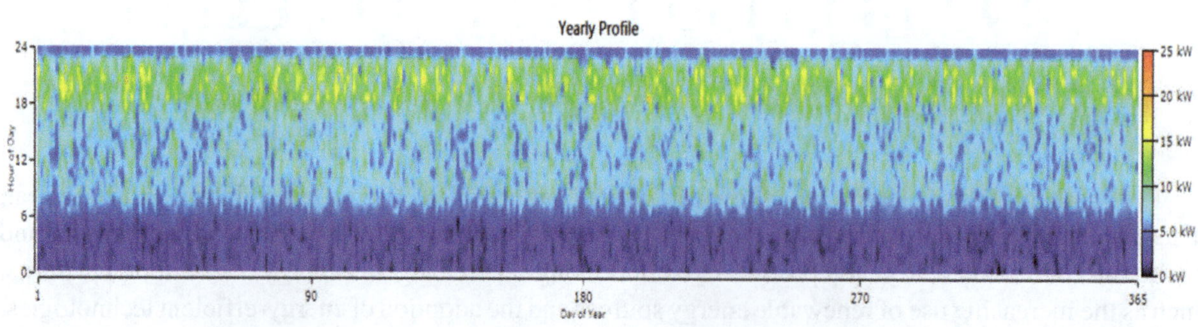

Figure 9. Schematic diagram of the proposed model

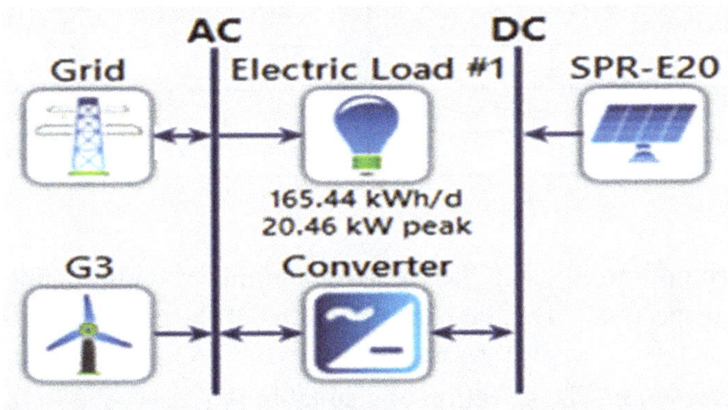

(i). PV System Modeling: The efficient interconnection of crystalline silicon solar cells within a PV module is of paramount importance to ensure the long-term performance and durability of the system, enabling it to consistently operate for its expected lifespan of up to 25 years. Proper interconnection techniques play a critical role in maintaining the structural integrity and electrical efficiency of the PV module. For the purposes of simulation in this study, the SunPower E20-327 model PV module was selected. This module is equipped with high-quality mono-crystalline silicon solar cells, known for their superior efficiency and reliability. The LR6-60PB model boasts a rated power output of 305 Wp and an impressive efficiency rating of 18.70%, ensuring optimal conversion of sunlight into electrical energy. With a projected working life of 25 years, the SunPower E20-327 PV module demonstrates the capability to endure the rigors of long-term operation in various environmental conditions. Additionally, a derating factor of 88% is considered, accounting for any potential performance degradation over time due to factors like aging, module mismatch, soiling, and other operational aspects. The detailed specifications are given in Table 4.

(ii). Power Conversion System Modeling: The converter is a crucial component of the renewable energy hybrid system as it is responsible for converting the DC power generated by the renewable sources into AC power for use in the grid. With a lifetime of 15 years and an efficiency of 95%, the capacity

Table 4. Technical data of the PV panels

Parameter	Value
Panel Type	Flat plate
PV slope	36°
Azimuth	0°
Derating factor	88%
Temperature effects on power	-0.38%/°C
Nominal operating cell temperature	45°C
Efficiency of PV arrays at STC	20.40%
Lifetime	25 years
Capacity	1000 kW

Table 5. Technical data of the selected converter

Parameter	Value
Efficiency	95%
Lifetime	15 years
Capacity	0-1000 kW

of the converter has been optimized using the HOMER optimizer, with the upper and lower limits set at 1000 kW and 0 kW, respectively. Detailed specifications of the converter used in this study can be found in Table 5.

(iii). Wind Turbine Modeling: The selection of a suitable WT is crucial in the design of a wind energy system. For this study, 12 individual Generic 3 kW WT was deemed a favorable option due to its notable economic advantage and commendable performance. The assumed capital, replacement, and O&M costs of each WT are $18,000, $18,000, and $180 per year, respectively. To optimize the capacity of the WT, the HOMER optimizer has been employed. In practical applications, the variation of wind speed with height and terrain is typically modeled by an exponential law model. This model allows for the calculation of wind speeds at different heights, such as the anemometer and hub heights of WTs, by applying the exponential law. The detailed specifications of the WT are presented in Table 6.

(iv). Grid Interaction Modeling: The electrical grid supplies power when renewable energy sources cannot meet the demand, and it consumes excess renewable energy when available. The utility charges for grid power have two components: the price to buy electricity from the grid and the price when selling back excess electricity. This research estimates a fixed rate, equivalent to $0.10 per kWh, based on the LCOE. After analyzing different sell-back price values, the study selects a suitable fixed sell-back price of $0.05 per kWh. This helps optimize the integration of renewable energy into the grid and encourages consumers to return excess power, promoting a more sustainable and balanced energy system. The findings offer valuable insights for policymakers and utilities to create a fair and effective framework for grid interactions with renewable energy. The grid has been used because it can function in both grid-connected and islanded modes, and also as it can support frequency and offer damping. The use of grid is to supply the energy produced, to the buildings and also for the sellout of extra energy produced.

Table 6. Technical and cost data of the selected wind turbine

Parameter	Value
Rated power	3 kW
Hub height	17 m
Capital cost	18,000 $/unit
Replacement cost	18,000 $/unit
O&M cost	180 $/year/unit
Lifetime	20 years
Quantity	0-20

COMPONENT COSTING AND FINANCIAL ANALYSIS

System Component Cost

Table 7 presents the categorization of costs for the hybrid energy system, including capital cost, replacement cost, and O&M expenses. In this analysis, the capital and replacement costs are considered as the same to simplify the analysis process. The capital cost of the PV system is estimated at 1300 $/kW, encompassing balance of system components, transportation, installation engineering, and other associated expenses. By providing a breakdown of costs into capital, replacement, and O&M categories, this analysis enables a comprehensive evaluation of the financial aspects associated with the system. The inclusion of installation and transportation costs for the relevant components ensures a thorough assessment of the total capital investment required for the system implementation. This detailed cost analysis is crucial for accurately estimating the economic viability and feasibility of the hybrid energy system, enabling informed decision-making and planning for the deployment of sustainable and cost-effective energy solutions.

Rate Chart of RE Components

The statistics pertaining to the key rates and parameters considered in this analysis. The discount rate utilized for evaluating the economic viability of the project is set at 8%. This rate represents the desired rate of return or the opportunity cost of investing capital in the project. It is used to calculate the net present value (NPV) and determine the economic attractiveness of the hybrid energy system. The inflation rate, denoted as 2%, reflects the expected annual increase in prices and accounts for the impact of inflation on project costs and revenues over time. It is an important factor to consider in economic analysis, as it affects the future cash flows and the purchasing power of the investment.The annual capacity shortage, set at 0%, indicates that the system is designed to meet the entire load demand without any capacity shortages. This constraint ensures a reliable power supply throughout the project's lifetime. Lastly, the project is evaluated over a 25year lifetime, considering the long-term sustainability and operational performance of the hybrid energy system.

Table 7. Summary of cost for system integration components

Description	PV system	Wind turbine	Power converter
Capital cost ($/kW)	1,300	18,000	300
Replacement cost ($/kW)	1,300	18,000	300
O&M cost ($/kW/year)	20	180	10

SIMULATION RESULTS AND DISCUSSIONS

Optimization Results for Techno-Economic Aspects

The optimization results of the proposed grid-interactive building infrastructure in Kashinagar, Odisha reveal significant improvements in energy management and cost-effectiveness. Through the integration of renewable energy sources and the efficient utilization of the power grid, the system achieves impressive optimization outcomes (as portrayed in Fig. 10). The introduction of 496 kW of PV capacity and 36 kW of wind generation capacity results in a substantial reduction in operating cost. The projected operating costs of -$11,432 per year indicate a complete elimination of energy expenses and a surplus of energy generation. This optimization not only leads to financial savings but also promotes sustainable energy practices and reduces dependence on the traditional power grid. The proposed system demonstrates its ability to achieve substantial cost savings, rapid payback, and high financial returns. By optimizing energy generation, consumption, and utilization, the grid-interactive building infrastructure contributes to sustainable energy practices, financial benefits, and a greener future for the community.

Electrical Energy Outcomes

Based on the simulation parameters and the results obtained from the HOMER analysis, as presented in Fig. 11-12, it is determined that the total annual power delivery of the system is recorded as 805,794 kWh per year. Additionally, the analysis reveals that the system generates annual carbon dioxide (CO_2) emissions totaling 13,891 kg. The results obtained from the analysis demonstrate that the system exhibits a consistent and reliable capacity throughout the year, successfully meeting the annual load demand. This indicates that the system is designed to effectively cater to the electricity requirements of the desired application without experiencing any capacity shortages. By providing these quantitative dataon, power delivery, and CO_2 emissions, the HOMER analysis offers valuable insights into the operational performance and environmental impact of the system. These findings contribute to the overall understanding and evaluation of the system's efficiency, effectiveness, and sustainability, supporting informed decision-making in the design and implementation of hybrid energy solutions.

(i). Solar Energy Generation: The SunPower E20-327 PV module presents a comprehensive analysis of its performance and energy generation characteristics. With a rated capacity of 496 kW, the PV module exhibits a mean output of 86.7 kW, resulting in a capacity factor of 17.5%. This factor indicates the module's efficiency in converting solar energy into electricity and serves as a crucial metric for evaluating its effectiveness. It achieves a mean daily output of 2,080 kWh, contributing to a total annual

Figure 10. Optimization result summary snapshots

Architecture								Cost			
				SPR-E20 (kW)	G3	Grid (kW)	Converter (kW)	Dispatch	NPC ($)	COE ($)	Operating cost ($/yr)	Initial capital ($)
				496	12	999,999	93.0	CC	-$104,203	-$0.0216	-$11,432	$43,589
				498		999,999	93.0	CC	-$103,076	-$0.0225	-$10,232	$29,195
					32	999,999		CC	$68,375	$0.0523	$2,319	$38,400
						999,999		CC	$78,064	$0.100	$6,039	$0.00

Figure 11. Electrical system analysis

Production	kWh/yr	%
SunPower E20-327	759,245	94.2
Generic 3 kW	24,569	3.05
Grid Purchases	21,979	2.73
Total	805,794	100

Consumption	kWh/yr	%
AC Primary Load	60,386	16.2
DC Primary Load	0	0
Deferrable Load	0	0
Grid Sales	312,449	83.8
Total	372,835	100

Quantity	kWh/yr	%
Excess Electricity	415,786	51.6
Unmet Electric Load	0	0
Capacity Shortage	0	0

Quantity	Value	Units
Renewable Fraction	94.1	%
Max. Renew. Penetration	516	%

Figure 12. Graphical illustration for monthly electric production

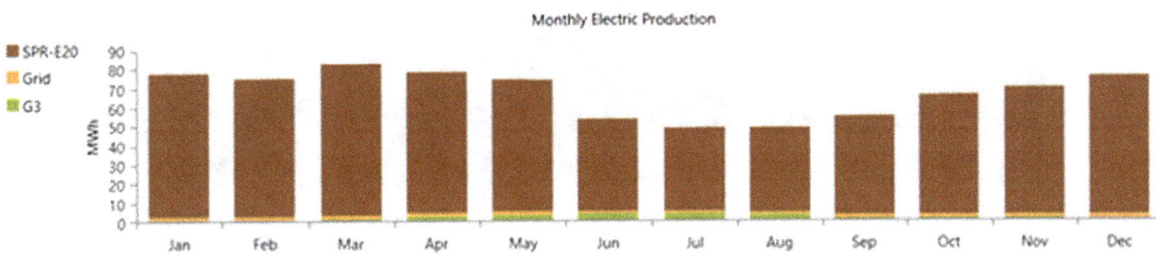

Figure 13. Pollutant emissions summary

Quantity	Value	Units
Carbon Dioxide	13,891	kg/yr
Carbon Monoxide	0	kg/yr
Unburned Hydrocarbons	0	kg/yr
Particulate Matter	0	kg/yr
Sulfur Dioxide	60.2	kg/yr
Nitrogen Oxides	29.5	kg/yr

production of 759,245 kWh. The impressive performance is further highlighted by its ability to operate over 4,352 hours per year, ensuring consistent energy generation throughout the day and across seasons. The LCOE is a crucial economic indicator, estimated to be at a low value of $0.000157 per kWh. This LCOE value affirms the economic viability and competitiveness of the PV module as a cost-effective renewable energy solution as presented in Figure 14.

(ii). Wind Energy Generation: A generic 3 kW wind turbine illustrates a comprehensive analysis of its performance and energy generation characteristics. With a total rated capacity of 36.0 kW, the wind turbine exhibits a mean output of 2.80 kW, resulting in a capacity factor of 7.79%. This factor reflects the turbine's efficiency in harnessing wind energy and converting it into electricity, providing essential insights

Figure 14. Electricity generation from solar panel

Quantity	Value	Units
Rated Capacity	496	kW
Mean Output	86.7	kW
Mean Output	2,080	kWh/d
Capacity Factor	17.5	%
Total Production	759,245	kWh/yr

Quantity	Value	Units
Minimum Output	0	kW
Maximum Output	480	kW
PV Penetration	1,257	%
Hours of Operation	4,352	hrs/yr
Levelized Cost	0.000157	$/kWh

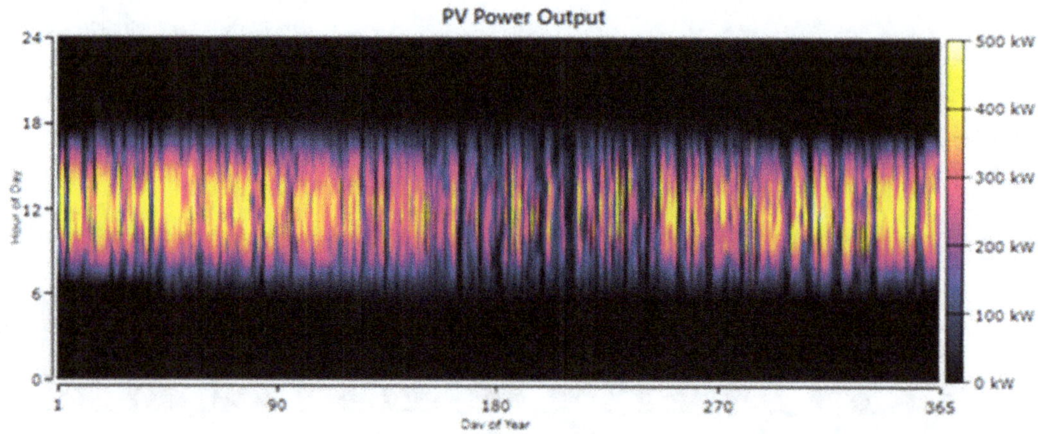

into its effectiveness. It achieves a total annual production of 24,569 kWh, indicating its contribution to sustainable energy generation. The turbine's ability to operate for 5,940 hours per year demonstrates its reliability and consistent energy production, even in varying wind conditions. The LCOE is a crucial economic metric, and the report estimates it to be $0.0575 per kWh as given in Figure 15. This LCOE value indicates the cost-effectiveness of the turbine as a viable renewable energy option. The annual energy purchased from the grid is 21,979 kWh and the annual energy sold to the grid is 312,449 kWh. Its grid interactive results have been presented in Figure 16 and Table 8.

(iii). System Converter Operation: The system converter provides a comprehensive analysis of its performance and energy conversion characteristics. With a capacity of 93.0 kW, the system converter exhibits a mean output of 37.2 kW, resulting in a capacity factor of 40.1%. This capacity factor reflects the efficiency of the converter in converting energy from one form to another, indicating its effectiveness in energy conversion processes. It operates for 4,352 hours per year, resulting in an energy output of 326,286 kWh annually (as portrayed in Figures 17-18). It receives an energy input of 343,459 kWh per year, leading to losses of 17,173 kWh annually. The analysis of energy losses is crucial as it identifies inefficiencies and potential areas for improvement in the conversion process. It demonstrates its flexibility by showing a minimum output of 0 kW and a maximum output of 93.0 kW, indicating its capability to adapt to different energy demand levels and effectively handle varying energy inputs.

(iv). Cost Analysis of the Proposed System: Figure 19 shows the cost summary of a system consisting of a 496 kW solar panel, 36 kW Wind Turbine. The system is optimized for dispatch and has an NPC of $104,203, a COE of $0.0216/kWh, an annual operating cost of $11,432 and an initial capital cost of $43,589.

Figure 15. Electricity generation from wind mills

Quantity	Value	Units
Total Rated Capacity	36.0	kW
Mean Output	2.80	kW
Capacity Factor	7.79	%
Total Production	24,569	kWh/yr

Quantity	Value	Units
Minimum Output	0	kW
Maximum Output	36.0	kW
Wind Penetration	40.7	%
Hours of Operation	5,940	hrs/yr
Levelized Cost	0.0575	$/kWh

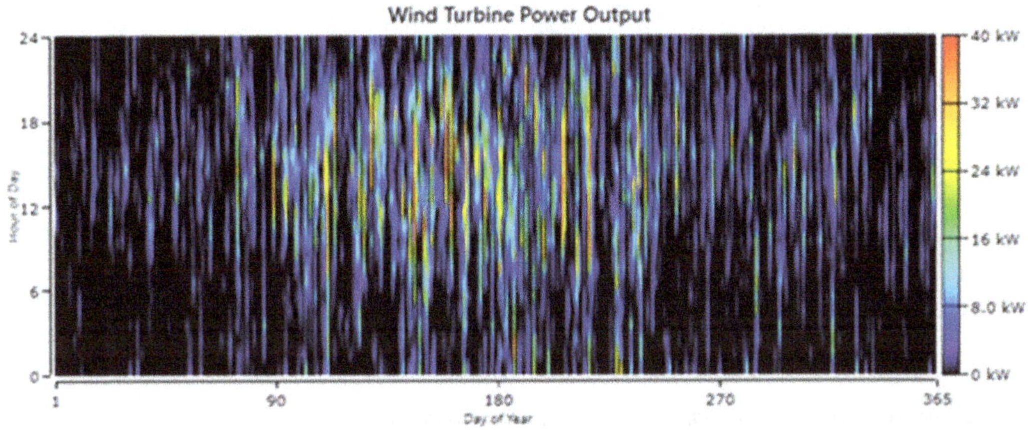

Figure 16. Electricity exchange operation with grid

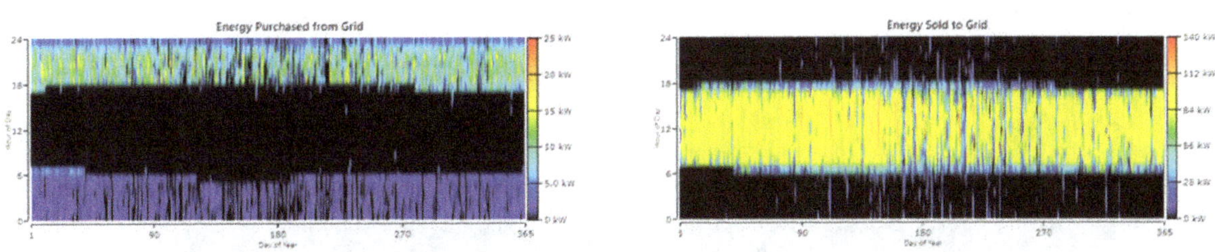

The cash flow diagram represents the financial performance system at Kashinagar. The diagram shows the flow of cash over a 25-year period, starting from year 0, which is the installation year of the system. Figure 20 shows that the initial capital cost of $43,589 is incurred in year 0 for the installation of the system.

CONCLUSION

The proposed grid-interactive building infrastructure at Kashinagar showcases impressive optimization results, both in terms of financial viability and environmental impact. By incorporating 496 kW of PV capacity and 36 kW of wind generation capacity, the system reduces operating costs to a negative value of -$11,432 per year. This significant cost reduction not only contributes to the financial well-being of

Table 8. Results for energy exchange operation with grid

Month	Energy Purchased (kWh)	Energy Sold (kWh)	Net Energy Purchased (kWh)	Peak Load (kW)	Energy Charge ($)	Demand Charge ($)	Total Charge ($)
January	2,435	25,399	-22,964	20.2	-1,026	NR	-1,026
February	1,869	25,094	-23,225	17.2	-1,068	NR	-1,068
March	1,945	27,693	-25,747	18.2	-1,190	NR	-1,190
April	1,577	29,118	-27,542	18.6	-1,298	NR	-1,298
May	1,467	30,028	-28,561	17.0	-1,355	NR	-1,355
June	1,382	26,408	-25,026	19.4	-1,182	NR	-1,182
July	1,485	25,398	-23,914	18.7	-1,121	NR	-1,121
August	1,611	23,766	-22,155	17.0	-1,027	NR	-1,027
September	1,790	22,608	-20,818	17.2	-951.42	NR	-951.42
October	2,058	25,853	-23,795	17.0	-1,087	NR	-1,087
November	2,038	25,295	-23,257	18.3	-1,061	NR	-1,061
December	2,323	25,789	-23,466	18.5	-1,057	NR	-1,057
Annual	21,979	312,449	-290,469	20.2	-13,425	NR	-13,425

*NR: It means not report to the literature.

Figure 17. Energy conversion performance of system converter

Quantity	Inverter	Rectifier	Units
Capacity	93.0	93.0	kW
Mean Output	37.2	0	kW
Minimum Output	0	0	kW
Maximum Output	93.0	0	kW
Capacity Factor	40.1	0	%

Quantity	Inverter	Rectifier	Units
Hours of Operation	4,352	0	hrs/yr
Energy Out	326,286	0	kWh/yr
Energy In	343,459	0	kWh/yr
Losses	17,173	0	kWh/yr

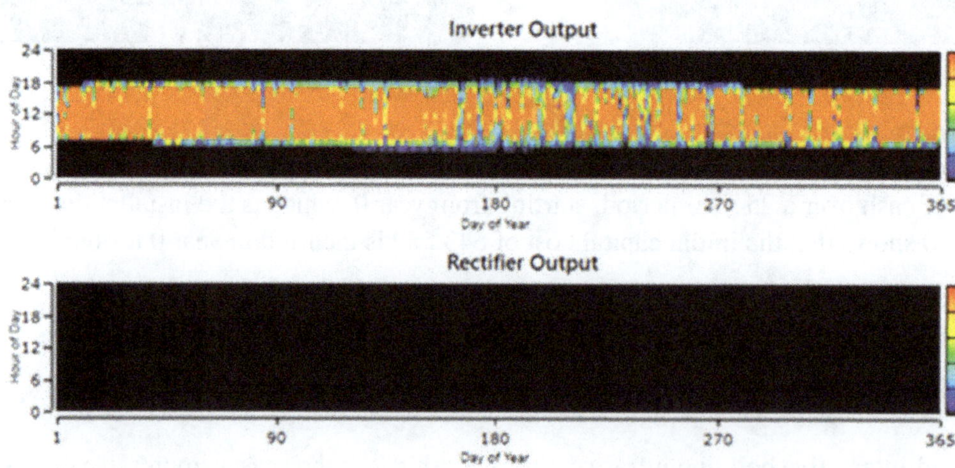

the community but also fosters energy independence and resilience. The financial analysis highlights the attractive return on investment (ROI) and the rapid payback period of 2.37 years, indicating the economic

Figure 18. Graphical representation for cumulative cash flow over project lifetime

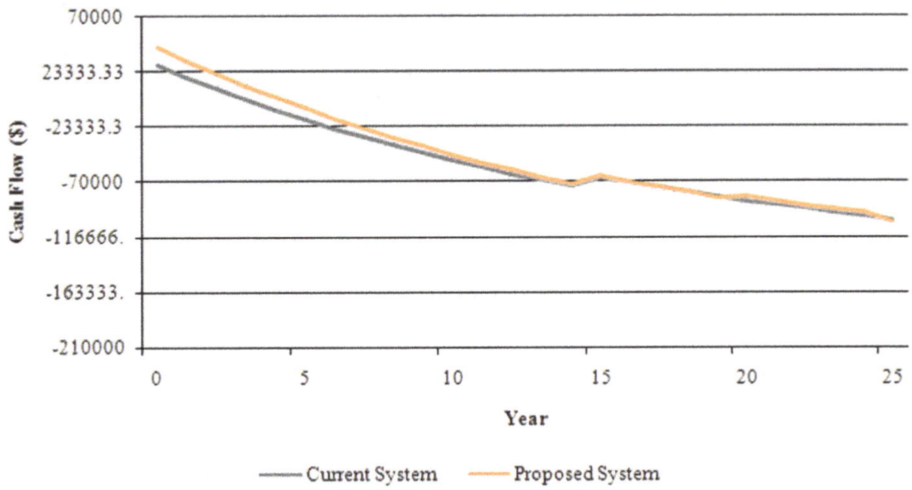

Figure 19. Cost analysis of the proposed system

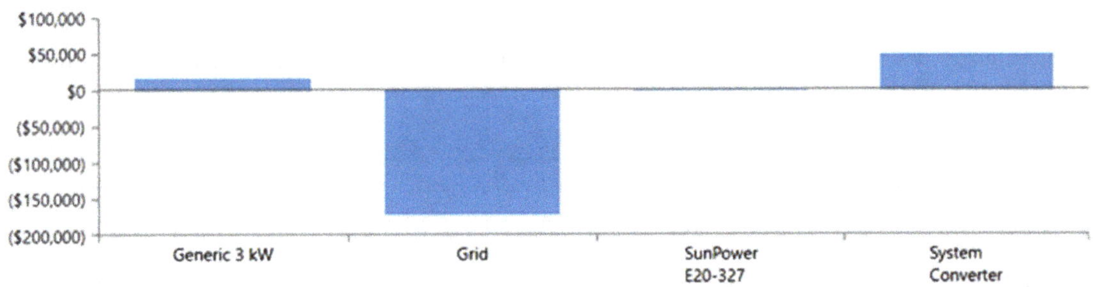

Figure 20. Cash flow diagram of the studied system

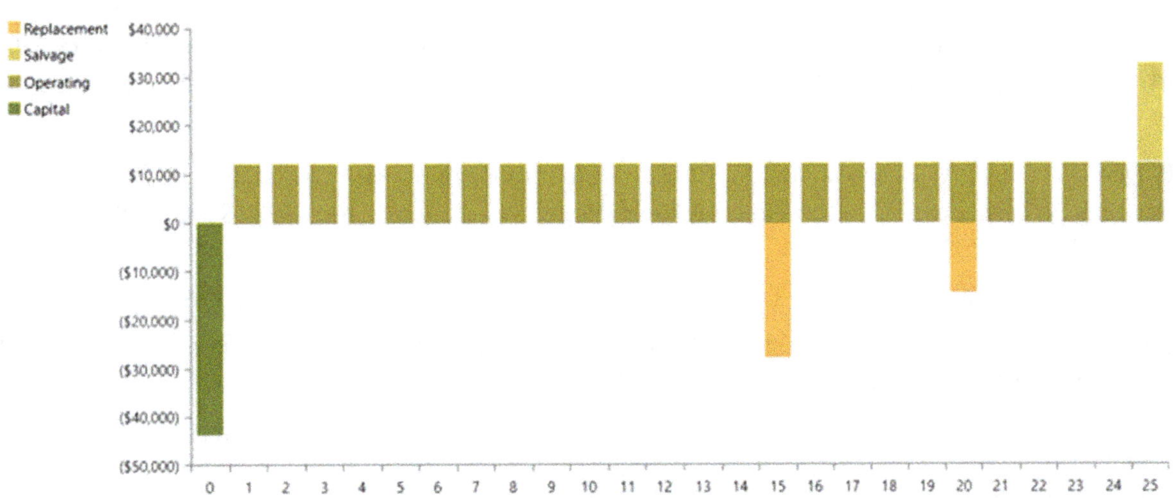

viability of the proposed system. With an IRR of 42.0% and a NPV of $182,267 over the project's 25-year lifetime, the grid-interactive building infrastructure in Kashinagar demonstrates its long-term financial benefits. Moreover, the integration of renewable energy sources reduces greenhouse gas emissions and minimizes the environmental footprint of the community. The SunPower E20-327 PV system and the wind turbine system contribute to a significant reduction.

REFERENCES

AlFaris, F., Juaidi, A., & Manzano-Agugliaro, F. (2017). Intelligent homes' technologies to optimize the energy performance for the net zero energy home. *Energy and Building*, *153*, 262–274. doi:10.1016/j.enbuild.2017.07.089

Arent, D. J., Green, P., Abdullah, Z., Barnes, T., Bauer, S., Bernstein, A., Berry, D., Berry, J., Burrell, T., Carpenter, B., Cochran, J., Cortright, R., Curry-Nkansah, M., Denholm, P., Gevorian, V., Himmel, M., Livingood, B., Keyser, M., King, J., ... Turchi, C. (2022). Challenges and opportunities in decarbonizing the US energy system. *Renewable & Sustainable Energy Reviews*, *169*, 112939. doi:10.1016/j.rser.2022.112939

Byun, J., Hong, I., Kang, B., & Park, S. (2011). A smart energy distribution and management system for renewable energy distribution and context-aware services based on user patterns and load forecasting. *IEEE Transactions on Consumer Electronics*, *57*(2), 436–444. doi:10.1109/TCE.2011.5955177

Chen, Z. S., Yang, L. P., Rodríguez, R. M., Zhu, Z., Pedrycz, W., & Skibniewski, M. J. (2022). BIM-aided large-scale group decision support: Optimization of the retrofit strategy for existing buildings. *Applied Soft Computing*, *131*, 109790. doi:10.1016/j.asoc.2022.109790

Dagdougui, H., Ouammi, A., Dessaint, L., & Sacile, R. (2016). Global energy management system for cooperative networked residential green buildings. *IET Renewable Power Generation*, *10*(8), 1237–1244. doi:10.1049/iet-rpg.2015.0282

Elnour, M., Fadli, F., Himeur, Y., Petri, I., Rezgui, Y., Meskin, N., & Ahmad, A. M. (2022). Performance and energy optimization of building automation and management systems: Towards smart sustainable carbon-neutral sports facilities. *Renewable & Sustainable Energy Reviews*, *162*, 112401. doi:10.1016/j.rser.2022.112401

ElSayed, M., Foda, A., & Mohamed, M. (2022). Autonomous drone charging station planning through solar energy harnessing for zero-emission operations. *Sustainable Cities and Society*, *86*, 104122. doi:10.1016/j.scs.2022.104122

Huang, S., Wang, J., Fu, Y., Zuo, W., Hinkelman, K., Kaiser, R. M., He, D., & Vrabie, D. (2021). An open-source virtual testbed for a real Net-Zero Energy Community. *Sustainable Cities and Society*, *75*, 103255. doi:10.1016/j.scs.2021.103255

IEA. (2022). *World energy outlook 2022*. IEA.

Kacker, R., & Srivastava, N. (2023). Elements and Possible Constraints for a Smooth Energy Transition in India. *Ecology. Economy and Society–the INSEE Journal*, *6*(1), 5–11.

Kirmani, S., Jamil, M., & Akhtar, I. (2018). Economic feasibility of hybrid energy generation with reduced carbon emission. *IET Renewable Power Generation*, *12*(8), 934–942. doi:10.1049/iet-rpg.2017.0288

Liu, J., Zhou, Y., Yang, H., & Wu, H. (2022). Net-zero energy management and optimization of commercial building sectors with hybrid renewable energy systems integrated with energy storage of pumped hydro and hydrogen taxis. *Applied Energy*, *321*, 119312. doi:10.1016/j.apenergy.2022.119312

Liu, Z., Chen, Y., Yang, X., & Yan, J. (2023). Power to Heat: Opportunity of Flexibility Services Provided by Building Energy Systems. *Advances in Applied Energy*, *11*, 100149. doi:10.1016/j.adapen.2023.100149

Mariano-Hernández, D., Hernández-Callejo, L., Zorita-Lamadrid, A., Duque-Pérez, O., & García, F. S. (2021). A review of strategies for building energy management system: Model predictive control, demand side management, optimization, and fault detect & diagnosis. *Journal of Building Engineering*, *33*, 101692. doi:10.1016/j.jobe.2020.101692

Morvaj, B., Evins, R., & Carmeliet, J. (2016). Optimization framework for distributed energy systems with integrated electrical grid constraints. *Applied Energy*, *171*, 296–313. doi:10.1016/j.apenergy.2016.03.090

Moudgil, V., Hewage, K., Hussain, S. A., & Sadiq, R. (2023). Integration of IoT in building energy infrastructure: A critical review on challenges and solutions. *Renewable & Sustainable Energy Reviews*, *174*, 113121. doi:10.1016/j.rser.2022.113121

Mousavi, S., Marroquín, M. G. V., Hajiaghaei-Keshteli, M., & Smith, N. R. (2023). Data-driven prediction and optimization toward net-zero and positive-energy buildings: A systematic review. *Building and Environment*, *242*, 110578. doi:10.1016/j.buildenv.2023.110578

Ngo, N. T., Truong, N. S., Truong, T. T. H., Pham, A. D., & Huynh, N. T. (2023). Implementing a Web-based Optimized Artificial Intelligence (AI)System with Metaheuristic Optimization for Improving Building Energy Performance. *Journal of Asian Architecture and Building Engineering*.

Nosratabadi, S. M., Moshizi, H. N., & Guerrero, J. M. (2022). Strategy for demand side management effectiveness assessment via a stochastic risk-based bidding approach in a multi-energy microgrid containing combined cooling, heat and power and photovoltaic units. *IET Renewable Power Generation*, *16*(10), 2036–2058. doi:10.1049/rpg2.12482

Pan, R., Liu, D., Yang, Y., & Yang, J. (2024). Network based impedance analysis of grid forming based MMC-HVDC with wind farm integration. *Electric Power Systems Research*, *229*, 110120. doi:10.1016/j.epsr.2024.110120

Pan, Y., Zhu, M., Lv, Y., Yang, Y., Liang, Y., Yin, R., Yang, Y., Jia, X., Wang, X., Zeng, F., Huang, S., Hou, D., Xu, L., Yin, R., & Yuan, X. (2023). Building energy simulation and its application for building performance optimization: A review of methods, tools, and case studies. *Advances in Applied Energy*, *10*, 100135. doi:10.1016/j.adapen.2023.100135

Panda, D. K., & Das, S. (2021). Smart grid architecture model (SGAM) for control, optimization and data analytics of future power networks with more renewable energy. *Journal of Cleaner Production*, *301*, 126877. doi:10.1016/j.jclepro.2021.126877

Qamar, A., Iqbal, J., Saher, S., Shah, A. A., & Basit, A. (2022). Design of optimized energy system based on active energy-saving technologies in very low-energy smart buildings. *Transactions on Emerging Telecommunications Technologies*, *33*(2), e3691. doi:10.1002/ett.3691

Qi, H., Huang, X., & Sheeraz, M. (2023). Green financing for renewable energy development: Driving the attainment of zero-emission targets. *Renewable Energy*, *213*, 30–37. doi:10.1016/j.renene.2023.05.111

Sadhukhan, J., Sen, S., Randriamahefasoa, T. M. S., & Gadkari, S. (2022). Energy system optimization for net-zero electricity. *Digital Chemical Engineering*, *3*, 100026. doi:10.1016/j.dche.2022.100026

Shandiz, S. C., Rismanchi, B., & Foliente, G. (2021). Energy master planning for net-zero emission communities: State of the art and research challenges. *Renewable & Sustainable Energy Reviews*, *137*, 110600. doi:10.1016/j.rser.2020.110600

Su, S. B., He, C., Shu, Y., Chen, Q. L., & Zhang, B. J. (2023). Total site modeling and optimization for petrochemical low-carbon retrofits using multiple CO2 emission reduction methods. *Journal of Cleaner Production*, *383*, 135450. doi:10.1016/j.jclepro.2022.135450

Tariq, R., Cetina-Quiñones, A. J., Cardoso-Fernández, V., Daniela-Abigail, H. L., Soberanis, M. E., Bassam, A., & De Lille, M. V. (2021). Artificial intelligence assisted technoeconomic optimization scenarios of hybrid energy systems for water management of an isolated community. *Sustainable Energy Technologies and Assessments*, *48*, 101561. doi:10.1016/j.seta.2021.101561

Thirunavukkarasu, M., Sawle, Y., & Lala, H. (2023). A comprehensive review on optimization of hybrid renewable energy systems using various optimization techniques. *Renewable & Sustainable Energy Reviews*, *176*, 113192. doi:10.1016/j.rser.2023.113192

Ullah, K., Hafeez, G., Khan, I., Jan, S., & Javaid, N. (2021). A multi-objective energy optimization in smart grid with high penetration of renewable energy sources. *Applied Energy*, *299*, 117104. doi:10.1016/j.apenergy.2021.117104

Vyas, S., Ahmed, S., & Parashar, A. (2014). BEE (Bureau of energy efficiency) and Green Buildings. *International Journal of Research*, *1*(3), 23–32.

Wang, D., Almojil, S. F., Ahmed, A. N., Chaturvedi, R., & Almohana, A. I. (2023). An intelligent design and environmental consideration of a green-building system utilizing biomass and solar having a bidirectional interaction with the grid to achieve a sustainable future. *Sustainable Energy Technologies and Assessments*, *57*, 103287. doi:10.1016/j.seta.2023.103287

Wang, Y., Liu, Z., Wang, J., Du, B., Qin, Y., Liu, X. & Liu, L. (2023). A Stackelberg game-based approach to transaction optimization for distributed integrated energy system. Energy, 128475.

Wu, Y., Wu, Y., Guerrero, J. M., & Vasquez, J. C. (2022). Decentralized transactive energy community in edge grid with positive buildings and interactive electric vehicles. *International Journal of Electrical Power & Energy Systems*, *135*, 107510. doi:10.1016/j.ijepes.2021.107510

Xu, Y., Yan, C., Liu, H., Wang, J., Yang, Z., & Jiang, Y. (2020). Smart energy systems: A critical review on design and operation optimization. *Sustainable Cities and Society*, *62*, 102369. doi:10.1016/j.scs.2020.102369

Zhang, W., Yan, C., Xu, Y., Fang, J., & Pan, Y. (2022). A critical review of the performance evaluation and optimization of grid interactions between zero-energy buildings and power grids. *Sustainable Cities and Society*, *86*, 104123. doi:10.1016/j.scs.2022.104123

Zheng, Z., LixiongWang, & HienWong, N. (2016). Intelligent control system integration and optimization for zero energy buildings (ZEBs)to mitigate urban heat island (UHI). *Procedia Engineering*, *169*, 100–107. doi:10.1016/j.proeng.2016.10.012

Zhu, Q., Leibowicz, B. D., Busby, J. W., Shidore, S., Adelman, D. E., & Olmstead, S. M. (2022). Enhancing policy realism in energy system optimization models: Politically feasible decarbonization pathways for the United States. *Energy Policy*, *161*, 112754. doi:10.1016/j.enpol.2021.112754

NOMENCLATURE

AI: Artificial intelligence
BEE: Bureau of Energy Efficiency
BIM: Building information modelling
COP26: Conference of the Parties, 26th Session
IEA: International Energy Agency
IoT: Internet of things
IRR: Internal rate of return
NDC: Nationally Determined Contribution
NPV: Net present value
PEB: Positive energy building
SCM: Smart Cities Mission
UHI: Urban heat island
UNFCCC: United Nations Framework Convention on Climate Change
ZEB: Zero energy building

Chapter 7
Usage of Green Hydrogen in the Steel Industry

G. Prasad
https://orcid.org/0000-0002-5709-9182
Chandigarh University, India

ABSTRACT

Since independence, steel has been the backbone of the Indian economy. Water, hydrocarbons (such as methane), and other organic substances all include hydrogen, which is abundant in nature. As a result, India is geopolitical reliance on coal, natural gas, and petroleum products as energy sources would be reduced as a result of the shift to hydrogen. In the long run, the transition would be beneficial since it might play a big role in other sectors such as transportation, process industry, agriculture, and long-term energy storage, making economies of scale easier to attain. The capital expenses are tough for the steel sector to absorb because it is traded on a global scale. If India wishes to maintain its position as one of the world's leading steel producers, the government should give incentives for hydrogen adoption through laws. This will draw the attention of independent hydrogen producers and lower the capital load on each steel plant, urging India to move toward emission-free steel production.

INTRODUCTION

The created hydrogen could be put to use in a variety of businesses, including the chemical industry, the steel industry, and the cement industry. Hydrogen that is created using renewable energy has the potential to decarbonize significant greenhouse gas emitters such as steel. Green hydrogen is produced by using renewable energy. Hydrogen is an essential component in fertilizers and refineries; hence, there is a possibility that green hydrogen could assist these firms in lowering their overall emissions. In addition, it might be utilized to reduce emissions in the steel manufacturing process by substituting coal as both a source of energy and a reducing agent when it is needed. Green hydrogen may have even greater potential for decarbonization as a zero-carbon energy source in key sectors that are otherwise difficult to decarbonize. These sectors include the production of iron and steel, cement, and heavy-duty transportation. If green hydrogen can demonstrate that it is cost-competitive with black carbon for existing indus-

DOI: 10.4018/979-8-3693-1014-4.ch007

trial uses, then it may have even greater potential for decarbonization. At this point in time, the cement and steel sectors are ready to transition away from coal as a source of energy and instead utilize green hydrogen. Value addition should be the primary emphasis of industry, and this can be accomplished by operating metal firms that mine items using green hydrogen as a source of energy. Hassan et al (2024) analysed the Green hydrogen as a pathway to a sustainable energy future which is sustainable. Hassan et al (2024) conducted a comprehensive Review of International Renewable Energy Growth and suggested green hydrogen as a main source. Schmidhalter et al (2024) studied the green hydrogen levelized cost assessment from wind energy in Argentina with dispatch constraints. Althiyabi (2023) examined a progressive regulatory framework to enable innovation in Neom's Green Hydrogen ecosystem. Zainal et al (2024) studied the recent advancement and assessment of green hydrogen production technologies. Kouchaki et al (2024) explained the role of hydrogen in a net-zero emission economy under alternative policy scenarios. Mio et al (2024) suggested that sustainability analysis of hydrogen production processes is the core research area. Chakrabarty et al (2024) conducted research on Hydrogen production by electrolysis as a sustainable pathway.

At this time, green hydrogen is being utilized as an alternative to coal as a source of energy in the steel industry. In the first place, hydrogen can be utilized in either of two different ways. To generate heat or power, it can be burned or fed into a fuel cell. Both of these methods are viable options. The majority of the gas that is now being used as a chemical in industrial settings is either dark or gray in color. Grey gas is produced through the reformation of steam methane, in contrast to brown gas, which is produced through the gasification of coal or lignite. There is no environmental benefit associated with either of these approaches. The steam methane reformation process, which also includes carbon capture and storage, results in the production of blue hydrogen, a more environmentally friendly form of hydrogen. In contrast, green hydrogen has been acknowledged as a sustainable energy source that has the potential to assist the global community in reaching the goal of achieving net-zero emissions. Through the utilization of renewable energy sources to power water electrolysis, it is possible to almost entirely eradicate this emission. In an electrolyzer that is powered by renewable energy sources such as wind and solar, water is divided into hydrogen and oxygen, which results in the production of green hydrogen. One of the primary reasons for the low utilization of this particular type of hydrogen is the high cost of production. Environmentally friendly hydrogen has the potential to be utilized in the manufacturing sector, as well as in conjunction with fuel cells to provide power to any device that is powered by electricity, including electronic devices and electric cars.

In Figure 1, the pure green hydrogen is depicted as the primary source of energy utilization. Fuel cells that run on hydrogen do not need to be recharged and will not run out of available electricity. Electric vehicles are not the only vehicles that are capable of operating on green hydrogen. Hybrid trucks that are environmentally friendly are also being developed. Additionally, Airbus is already working on developing aircraft that are fueled by hydrogen. The use of hydrogen in homes as a substitute for natural gas for the purposes of cooking and heating is possible. It is anticipated that by the year 2050, the majority of homes throughout the world will be powered by hydrogen heating.

BACKGROUND OF THE RESEARCH:

Choi & Kang (2023) studied on Greenhouse gas reduction and economic cost of technologies using green hydrogen in the steel industry with the merits and demerits. Griffin & Hammond (2021) analysed on

Figure 1. Hydrogen as a green

prospects for 'green steel' making in a net-zero economy, a UK perspective. Gusev et al (2023) proposed production of hydrogen and carbon in the petrochemical industry by cracking of hydrocarbons in the process of heat utilization in steel production. Guidi (2022) developed an economic evaluation of the technologies to power a hydrogen steel production process under a CO2 emission limit. Holappa (2020) studies a general vision for reduction of energy consumption and CO2 emissions from the steel industry. Ledari et al (2023) studied on Greening steel industry by hydrogen, Lessons learned for the developing world. Marocco et al (2023) assessed the role of hydrogen to produce high-temperature heat in the steel industry. Patnaik et al (2023) studied on reducing CO2 emissions in the iron industry with green hydrogen. Rechberger et al (2020) developed a green hydrogen-based direct reduction for low-carbon steelmaking. Souza et al (2022) studied on green steel at its crossroads: Hybrid hydrogen-based reduction of iron ores. Swennenhuis et al (2022) carried a research on towards a CO2-neutral steel industry: Justice aspects of CO2 capture and storage, biomass-and green hydrogen-based emission reductions. Additional goals for the year 2030 that are included in the "Breakthrough Agenda" include making clean energy "the most affordable and dependable option for all countries," making zero-emission road cars "the new normal," and making near-zero-emission steel the "preferred choice in global markets." Despite the fact that there are more than 250 gigawatts of green hydrogen projects currently under development around the world, as well as a seemingly endless stream of announcements to use vast quantities of the zero-carbon gas to decarbonize everything from trucks and heating to steel and aviation, the simple truth is that renewable hydrogen is scarce and expensive at the present time.Green hydrogen will be more affordable than any fossil fuel alternative during a third stage, which will take place between 2021 and 2050, as stated in the strategy paper. This will make it possible for green hydrogen to be utilized in the transportation industry, aviation, power plants, and steel manufacturing. To put it another way, consumers and businesses who use products that are manufactured with green hydrogen, like as steel or fertilizer, will be eligible for reduced tax rates or tax relief. In this way, the additional expense of utilizing a more expensive H_2 would be substantially reduced. There are additional green H_2 quotas for existing large hydrogen consumers, which are not defined but include ammonia and chemical factories as well as oil refineries. Additionally, "sectors other than industry (such as aviation)" are included in this category. The latter would be controversial due to the fact that hydrogen is not generally considered to be significantly more energy-intensive by volume than other fuels that could be utilized directly for aviation. Although it would be prohibitively expensive in comparison to bio-based sustainable aviation fuel, the gas might be utilized as a crucial ingredient in so-called "e-fuel," which is synthetic kerosene jet fuel manufactured by combining green hydrogen with collected carbon dioxide. This possibility exists despite the fact that the gas would be used. Green hydrogen has the potential to assist in the decarbonization of businesses such as shipping and transportation, where it may be utilized as a fuel, as well as industrial industries such as steel and chemicals, where it can be utilized both as a raw material and as a fuel. It is possible to use it to store renewable energy and to utilize it as a replacement for fossil fuels in the generation of power. Green hydrogen, in addition to ammonia, has the potential to be utilized in gas turbines for the purpose of making adjustments to the supply and demand of electricity. One example is the steel and iron industry, which is one of the most polluting businesses in the world and is responsible for seven percent of the world's greenhouse gas emissions.Steel, which is used in everything from bridges to cars, is anticipated to create 35% of India's carbon dioxide emissions from fossil fuel consumption and industry by the year 2050, according to a research conducted by The Energy and Resources Institute (TERI). Steel is utilized in everything from manufacturing to transportation. Both a reducing agent and a fuel, hydrogen is utilized in the steel industry during the production process. It was claimed in a report that was published in 2020 by the International Energy Agency that significant technological advancements in the manufacturing of iron and steel are necessary in order to make the process environmentally friendly and carbon neutral. The use of green hydrogen is one example of such an invention. Through the utilization of green hydrogen-based steel, India may be able to circumvent carbon border levies. As a result of the fact that the majority of the steel that is currently in use is produced in a blast furnace, which is a giant cylinder that is used for the purpose of dumping coal, iron ore, and other materials, steelmakers would require new equipment. Green hydrogen-based steel requires a shaft furnace, which is a separate technology. This is necessary in order to produce the steel.

To implement the usage of green hydrogen in the steel industry effectively, a comprehensive and multi-faceted methodology needs to be proposed. This methodology would encompass several key stages, from hydrogen production to its application in steelmaking processes. Here's an outline of a proposed methodology:

1. Hydrogen Production
   a. Renewable Energy Source Identification and Development
      - Identify and develop renewable energy sources (solar, wind, hydro, etc.) suitable for the location.
      - Ensure a steady and reliable supply of renewable energy to power the electrolysis process.

   b. Electrolysis for Hydrogen Production
      - Set up electrolyzers for water splitting to produce hydrogen.
      - Optimize electrolyzer technology for efficiency, durability, and scalability.

   c. Hydrogen Storage and Transportation
      - Develop safe and efficient hydrogen storage solutions (compression, liquefaction, etc.).
      - Establish a transportation network for hydrogen, considering pipelines, tankers, or other methods as appropriate.

2. Steel Production Process Integration
    a. Direct Reduction of Iron Ore (DRI)
        - Replace conventional natural gas-based DRI with hydrogen-based DRI.
        - Adapt DRI plants to handle hydrogen, ensuring safety and efficiency.
    b. Electric Arc Furnace (EAF) Adaptation
        - Utilize electric arc furnaces for melting DRI to produce steel.
        - Optimize EAF operations to work efficiently with hydrogen-reduced iron.
3. Infrastructure and Supply Chain Development
    a. Site Selection and Facility Upgrades
        - Choose suitable locations for hydrogen-based steel plants, considering proximity to renewable energy sources and markets.
        - Upgrade existing steelmaking facilities or build new ones designed for hydrogen use.
    b. Supply Chain Optimization
        - Develop a robust supply chain for the procurement of raw materials and distribution of green steel.
        - Coordinate with suppliers and customers to ensure a smooth transition.
4. Economic and Environmental Assessment
    a. Cost-Benefit Analysis
        - Evaluate the economic viability of the hydrogen-based steel production process.
        - Consider initial investment, operational costs, and potential market premiums for green steel.
    b. Environmental Impact Assessment
        - Conduct a comprehensive assessment of the environmental impacts, including CO2 emissions reduction.
        - Analyze the life cycle of the process to ensure overall sustainability.
5. Policy Framework and Incentives
    a. Regulatory Support
        - Work with government bodies to ensure regulatory support and compliance.
        - Advocate for policies that incentivize the adoption of green hydrogen in steelmaking.
    b. Financial Incentives and Funding
        - Secure funding and incentives for initial investments and R&D.
        - Explore options like carbon credits, subsidies, and green bonds.
6. Research and Development
    a. Continuous Improvement
        - Invest in ongoing R&D to improve the efficiency and scalability of hydrogen-based steel production.
        - Stay updated with technological advancements and integrate them into the process.
    b. Pilot Projects and Scaling Up
        - Start with pilot projects to test and refine the process.
        - Plan for gradual scaling up to commercial levels, learning from pilot project experiences.
7. Stakeholder Engagement and Education
    a. Collaboration with Industry Partners
        - Collaborate with other steel companies, hydrogen producers, and technology providers.
        - Form consortia or partnerships for shared learning and resource pooling.

b. Community and Consumer Engagement
    - Engage with local communities and consumers to raise awareness about green steel.
    - Market green steel effectively to create demand and consumer acceptance.

By following this proposed methodology, the steel industry can transition towards using green hydrogen in a structured and efficient manner, balancing economic viability with environmental sustainability.

CHALLENGES

Making the transition to green hydrogen economically and commercially viable is the facet of the transition that presents the greatest challenge. Steel produced using green hydrogen is between fifty and one hundred three percent more expensive than steel produced with conventional coal. In order for India to successfully complete the transition to green steel and green hydrogen, it is necessary for the country to establish some kind of regulatory structure or encouragement. However, the government of India has not yet developed a policy framework for the deployment of environmentally friendly hydrogen. Despite the fact that many countries, such as India, have established national hydrogen programs, there are no plans in place for how green hydrogen would be commercialized on a large scale. The deployment of green hydrogen is now at the experimental stage. Based on the findings of the research, it is suggested that a multi-country alliance be established in order to provide financial and scientific support for the development of green hydrogen technologies. Making the transition to green steel as quickly as feasible should be a collaborative effort between businesses and the government. Companies such as Reliance Industries and JSW Steel, in addition to experts from think tanks in India, have joined forces to form the India Hydrogen Alliance in order to design the green hydrogen roadmap for the country. The utilization of green hydrogen in green steel is another area that needs more research to be conducted in India. At the moment, just the deployment of green steel is the center of attention; research to develop creative ways to use green hydrogen is not being prioritized. In addition, India has the potential to gradually transition to green steel by blending green and grey hydrogen. In order to make the conversion to steelmaking that does not use fossil fuels overnight, the cost will be prohibitively exorbitant. The lowest cost of production for green steel in the year 2030 would still be 22% more than the cost of producing steel based on coal. An examination of the ways in which low-carbon hydrogen for direct reduction and electric arc furnaces that are powered by renewable electricity can be of assistance to the Indian steel industry in the production of green steel is presented in the following brief. In doing so, it highlights the potential of green hydrogen technologies and the practicality of these technologies over the long term in India. In addition to this, it investigates the extent of the emission reductions that may be accomplished in the steel sector through the utilization of hydrogen with zero or low carbon emissions, and it also provides suggestions for the advancement of the technology.4% of all carbon dioxide emissions in Europe are attributed to the steel industry, which is also responsible for 22% of all carbon emissions from industrial sources. The following are some of the decarbonization options available:

- Improving the efficiency of contemporary manufacturing processes.
- A recycling program for steel It is projected that carbon capture and storage, often known as CCS, will play a role in the process of decarbonizing the steel industry. Steel will always be required for industries that require high-quality steel, such as the automotive industry, because impurities like

copper accumulate with time. This means that fresh steel will always be required. Carbon capture and storage (CCS) can be directly utilized in steel mills or in the creation of hydrogen. In recent times, there has been a high level of interest in hydrogen as a potential method for decarbonizing industry.

There are two distinct ways in which hydrogen can be utilized in the manufacturing of steel:

- H2-BF is an auxiliary reducing agent that can be utilized in the BF-BOF process. Hydrogen can be used in this capacity.
- It is possible to use hydrogen as the sole reducing agent in a process that is referred to as the direct reduction of iron.

For the purpose of ensuring that any reductions in emissions are not simply equated to "climate-neutral" or "green" steel, it is imperative that they be rigorously monitored. In general, the claim made by the corporation to produce climate-neutral steel by the year 2022 is questionable and requires additional research. There is a difference between hydrogen BF steel and hydrogen direct reduction of iron ore, which can produce carbon-neutral steel with very minimum emissions. They should not be confused with one another. Within the coal-rich nation, there is a certain amount of unease over the industry of making "green steel" through the utilization of hydrogen and renewable energy. It is possible that this will occur if the cost of producing environmentally friendly hydrogen continues to decrease. And in the event that the global community makes the decision to levy a price for every tonne of carbon dioxide that is released. After that, it would be able to overcome the most significant obstacle, which is the cost, drawing in steel producers to participate. It is currently the case that practically all steel is manufactured using iron oxide and "met" coal. Through the use of ovens that are heated to 1100 degrees Celsius, water and other impurities are extracted from the coal. In the end, it results in the production of coke, which is a source of pure carbon that is utilized in the production of steel.

The novelty of using green hydrogen in the steel industry lies in its potential to revolutionize one of the most carbon-intensive sectors of the global economy. This approach represents a significant shift from traditional steelmaking processes, introducing a new, more sustainable method. The key aspects of novelty in using green hydrogen in steelmaking include:

- Carbon-Neutral Steel Production
- Integration with Renewable Energy
- Technological Innovation
- Economic Transformation of the Steel Industry
- Enabling Circular Economy
- Policy and Regulatory Frameworks
- Global Environmental Impact

In essence, the novelty of using green hydrogen in the steel industry is not just in its potential to reduce carbon emissions, but also in how it integrates with broader energy systems, drives technological innovation, transforms industry economics, supports circular economy principles, influences policy making, and contributes to global environmental sustainability.

INTERNATIONAL STATUS

A number of other European nations, including Germany and Italy, have undertaken efforts to manufacture steel by employing green hydrogen, which is generated through the electrolysis of renewable energy sources. Because of the proliferation of cities and the fact that steel is essential for the production of a wide variety of goods, including automobiles, bridges, highways, and buildings, the consumption of steel will increase at an exponential rate. A consultancy estimates that the steel industry is responsible for eight percent of the world's carbon dioxide emissions, and that fourteen percent of the value of all steel companies could be lost if they do not decarbonize.

The nations that make use of hydrogen are graphically represented in Figure 2. Steel production that makes use of energy sources that do not produce carbon dioxide is currently technically feasible, but it is an expensive option. It is possible that this will change if certain conditions are met: the prices of producing green hydrogen would need to decrease, and the costs of carbon emissions would need to be paid for. On the other hand, the steel industry will be required to participate at some point in the future. Indeed, according to projections made by the International Energy Agency, by the year 2050, green hydrogen steel will constitute a percentage of the total steel output that is lower than 10,000 percent. Steel production relies on coal as a raw material. In the early years, the cost of green hydrogen as an energy source would add between twenty and thirty percent to the cost of steel production. This is an even more convincing argument when taken into consideration.The storage facility will provide the power for the pilot plant that is located nearby and is responsible for the production of green steel. Storage of hydrogen that does not require the use of fossil fuels will be an additional step toward the production of green steel. In addition to claiming to have opened the world's first pilot plant dedicated to the production of fossil-free steel the previous year, the hydrogen storage cavern is continuing to lead the development

Figure 2. Showing the production of hydrogen all around the world
https://www.rechargenews.com

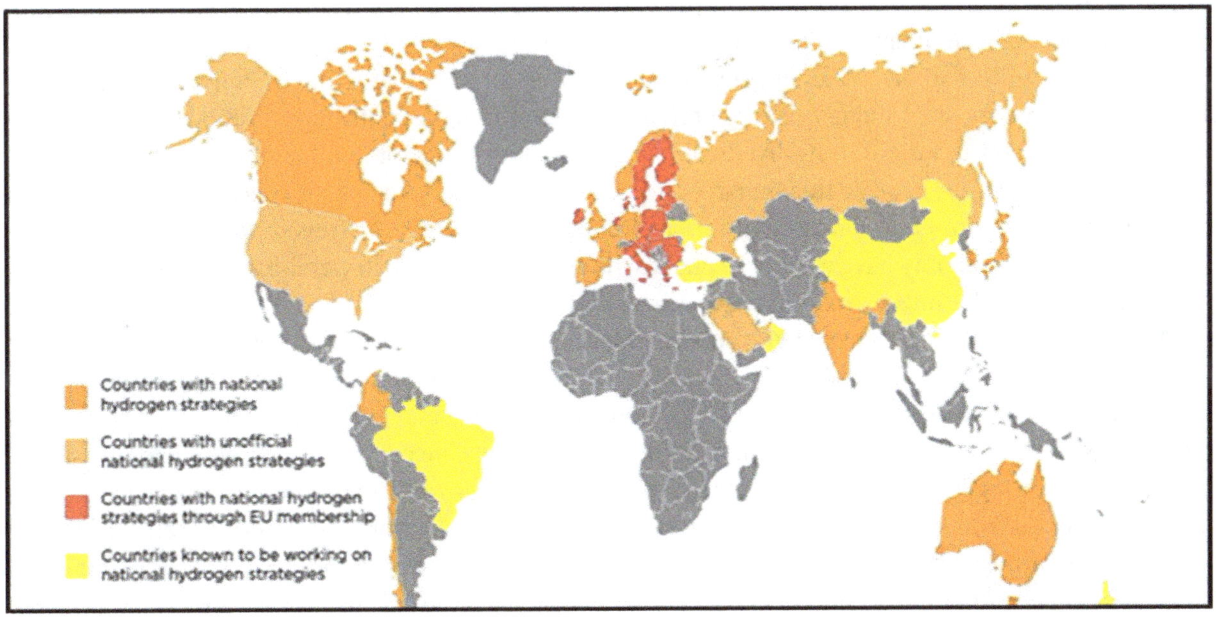

of efficient fossil-free steel production by constructing a pilot storage facility for large-scale fossil-free hydrogen. It is anticipated that the cavern will be operational between the years 2021 and 2030.According to the International Energy Agency (IEA), the steel industry is one of the most challenging to control. It is responsible for seven percent of the total carbon dioxide emissions that are caused by fossil fuels around the world. According to the findings of a study that was published in the previous year, the agency estimated that by the middle of the century, renewable hydrogen will power less than 10 percent of all steel that is produced around the world.

On the other hand, hydrogen-to-metals initiatives are becoming increasingly popular, particularly in Europe. Industry heavyweights like as Salzgitter in Germany and Arcelor-Mittal in India have proposed trial projects in Germany and Spain, respectively.Permanent structural shifts in the global steel price are here to stay. The first is, of course, the expenditures that governments all over the world, including India, make on national infrastructure. It is only relatively recently that money has started to flow, even in India. Increases in investment on infrastructure are expected to continue. The consumption of steel around the world is growing at a pace of 49%, mostly due to regions such as India, as well as the United States of America and other countries. Compared to the previous year, even Europe has seen a 14–15% increase. As a consequence of this, the developed world and emerging countries from other countries besides China will take the initiative. Secondly, there has been a movement in cost structures, and this shift will continue to occur. When compared to the price of steel three years ago, the cost of coking coal is still significantly greater. Because of this, the expenses of the inputs are quite substantial. As a result of China's reduction in output in specific regions, the prices of a significant number of the consumables that the steel sector purchases from China are gradually going up. China's exporting strategy has been less aggressive in recent years compared to what it was a few years ago. The global steel markets will become significantly more disciplined as a consequence of this.A hydrogen infrastructure as well as a gas infrastructure are currently in the process of being developed. Therefore, steel plants such as ours that are interested in switching from coal to gas and hydrogen have the infrastructure that has been developed by the government, and the supply of gas and hydrogen is promised to be provided at costs that are affordable. Everything is coming together, and as a consequence, the United Kingdom and the Netherlands are both working on building plans to transition to a greener future with the support of the government. These plans are being developed in conjunction with the government. Additionally, consumers are willing to pay a higher price for steel that is more in line with environmental standards. The existing primary steel industry's large reliance on coal [which contributes around 75% of the industry's energy consumption], long-lived capital assets, and the sector's vulnerability to international trade and competitiveness make this transition to near-zero emissions hard. In spite of the fact that it is being pushed as a zero-emissions option for steel manufacture, green hydrogen that is produced from renewable energy will only have a small impact to the reduction of industry emissions between now and 2050, according to the International Energy Agency (IEA).

Annealing processes and hot-dip galvanizing lines are two examples of the ways in which hydrogen is being utilized in the steel production industry with the assistance of an industrial gas processor. If hydrogen were used instead of natural gas, there would be a significant reduction in the amount of carbon dioxide emissions that are linked with electric arc furnace steel. The project, which will also investigate strategies for storing hydrogen as well as the oxygen that is produced as a by-product of electrolysis for use in the steelmaking process, has not been provided with a budget or a schedule. When it comes to lowering the carbon footprint of one of the industries that consumes the most energy and is the most challenging to decarbonize, the incorporation of green hydrogen into the manufacture of steel and other metals is considered to be of primary importance. According to the Sustainable Development Scenario, the emissions that are produced by the steel industry is required to be decreased by 90 percent by the year 2070. In light of the fact that this scenario makes the assumption that the global energy sector would reach net-zero emissions by the year 2070, carbon removal technologies would be necessary in order to compensate for the remaining 10% of emissions from the steel sector. The International Energy Agency (IEA) acknowledges that it will be challenging for the steel industry to meet the emissions reductions that are required under the Sustainable Development Scenario. This is because a significant portion of the technology that is required is still in the process of being developed, and governments will need to find ways to ensure that greener steel producers, who will most likely face higher costs, do not lose out in a global market that is becoming increasingly commoditized. A Sustainable Development Scenario (SDS) that is in line with the goals of the Paris Agreement states that by the middle of the century, less than ten percent of the steel that is produced around the world will use renewable hydrogen as its energy source. This is stated in the new report that was released by the agency, which is titled Iron and Steel Technology Roadmap: Towards More Sustainable Steelmaking. Even under these circumstances, coal will continue to supply more than half of the energy that the steel industry requires in the year 2050. Additionally, approximately one third of this energy will be utilized through carbon capture and storage.

RESULTS AND DISCUSSION

The use of green hydrogen in the steel industry is a topic of growing interest, particularly as industries worldwide seek to reduce their carbon footprint and comply with increasingly stringent environmental regulations. The steel industry is one of the largest industrial sources of CO2 emissions, traditionally relying on carbon-intensive processes. Green hydrogen, produced via electrolysis powered by renewable energy sources, offers a promising alternative to conventional fossil fuels. Here's a breakdown of the results and discussion surrounding the usage of green hydrogen in the steel industry:

1. Reduction in CO2 Emissions: The primary advantage of using green hydrogen in steel production is the significant reduction in CO2 emissions. Traditional blast furnaces use coking coal, which releases large amounts of CO2. Replacing coal with hydrogen can drastically cut these emissions.
2. Energy Efficiency: Early studies and pilot projects have shown that hydrogen-based steel production can be more energy-efficient compared to traditional methods, especially when the hydrogen is produced using surplus renewable energy.
3. Quality of Steel: Preliminary results indicate that the quality of steel produced with green hydrogen is comparable to, or even better than, steel produced with conventional methods. This is crucial for its acceptance in various industries.
4. Economic Viability: The cost of green hydrogen production is currently a significant barrier. However, with the advancement in electrolysis technology and an increase in renewable energy capacity, it's expected that the costs will decrease, making it economically viable.
5. Scalability: Initial pilot projects have been successful on a small scale, but scaling up to meet the demands of major steel producers remains a challenge. This includes not only the production of hydrogen but also the necessary infrastructure for storage and transport.

6. Technological Challenges: While the results are promising, there are still significant technological hurdles to overcome, particularly in terms of the efficiency and durability of electrolyzers, as well as the development of hydrogen-based steelmaking furnaces at an industrial scale.
7. Renewable Energy Dependency: The environmental benefits of green hydrogen are contingent on the availability of renewable energy sources. This necessitates substantial growth in renewable energy infrastructure.
8. Investment and Policy Support: Achieving the transition to hydrogen-based steel production will require significant investment, both in research and development and in the infrastructure required. Government policies and incentives will play a crucial role in facilitating this shift.
9. Market Dynamics: The market acceptance of green steel (steel produced using green hydrogen) will depend on various factors, including cost, regulatory pressures, and consumer demand, especially from industries that are highly sensitive to their carbon footprint.
10. Global Implications: The shift to green hydrogen in steel production could have significant global implications, including changes in the global trade of steel, impacts on coal-exporting countries, and the potential for new leaders in the steel industry emerging from countries with abundant renewable energy resources.

CONCLUSION

In conclusion, the use of green hydrogen in the steel industry presents a promising path towards decarbonization. While the technology is in its nascent stages and faces challenges in terms of cost and scalability, ongoing research and investment, along with supportive policies, could make it a key component in the global effort to combat climate change. In order to provide monetary opportunities. In industries that are difficult to regulate, fossil fuel-based technologies should be phased out. it is necessary to establish quotas and rules for hydrogen, environmentally friendly products, and "basic materials" such as steel and green fertilizer. The production of green hydrogen, which is a fuel that is both 100% sustainable and adaptable, can be stored and transported for use on demand, and is produced from wind power that is both affordable and competitive. It is possible to manufacture green hydrogen anywhere, and it can be utilized in industries that are challenging to decarbonize, such as the aviation and shipping industries, as well as heavy industries, such as the iron and steel industry, the chemical industry, and the glass industry. Green hydrogen can be produced anywhere and used in industries that are very difficult to decarbonize, such as aviation and shipping, as well as heavy industry, such as iron and steel, chemicals, and glass. This presents a massive opportunity for the transition to green energy by driving the transformation of the energy system. Green hydrogen can be produced anywhere. Hydrogen that is produced by wind can be utilized to lessen the amount of emissions produced by nations and industries that are working to lessen their overall global carbon footprint.

REFERENCES

Althiyabi, A. (2023). Developing a Progressive Regulatory Framework to Enable Innovation in Neom's Green Hydrogen Ecosystem. *Migration Letters : An International Journal of Migration Studies*, *20*(8), 1102–1123.

Chakrabarty, P., Alam, K. C. A., Paul, S. K., & Saha, S. C. (2024). Hydrogen production by electrolysis: A sustainable pathway. In *Hydrogen energy conversion and management* (pp. 81–102). Elsevier. doi:10.1016/B978-0-443-15329-7.00013-2

Choi, W., & Kang, S. (2023). Greenhouse gas reduction and economic cost of technologies using green hydrogen in the steel industry. *Journal of Environmental Management*, *335*, 117569. doi:10.1016/j.jenvman.2023.117569 PMID:36842354

Griffin, P. W., & Hammond, G. P. (2021). The prospects for 'green steel'making in a net-zero economy: A UK perspective. *Global Transitions*, *3*, 72–86. doi:10.1016/j.glt.2021.03.001

Guidi, F. (2022). An economic evaluation of the technologies to power a hydrogen steel production process under a CO2 emission limit. *Energy*, *2004*, 2965.

Gusev, A. L., Jabbarov, T. G., Mamedov, S. G., Malikov, R., Hajibalaev, N. M., Abdullaeva, S. D., & Abbasov, N. M. (2023). Production of hydrogen and carbon in the petrochemical industry by cracking of hydrocarbons in the process of heat utilization in steel production. *International Journal of Hydrogen Energy*, *48*(40), 14954–14963. doi:10.1016/j.ijhydene.2022.12.341

Hassan, Q., Algburi, S., Sameen, A. Z., Salman, H. M., & Jaszczur, M. (2024). Green hydrogen: A pathway to a sustainable energy future. *International Journal of Hydrogen Energy*, *50*, 310–333. doi:10.1016/j.ijhydene.2023.08.321

Hassan, Q., Algburi, S., Sameen, A. Z., Tariq, J., Al-Jiboory, A. K., Salman, H. M., ... & Jaszczur, M. (2024). *A Comprehensive Review of International Renewable Energy Growth. Energy and Built Environment*. Academic Press.

Holappa, L. (2020). A general vision for reduction of energy consumption and CO2 emissions from the steel industry. *Metals*, *10*(9), 1117. doi:10.3390/met10091117

Kouchaki-Penchah, H., Bahn, O., Bashiri, H., Bedard, S., Bernier, E., Elliot, T., Hammache, A., Vaillancourt, K., & Levasseur, A. (2024). The role of hydrogen in a net-zero emission economy under alternative policy scenarios. *International Journal of Hydrogen Energy*, *49*, 173–187. doi:10.1016/j.ijhydene.2023.07.196

Ledari, M. B., Khajehpour, H., Akbarnavasi, H., & Edalati, S. (2023). Greening steel industry by hydrogen: Lessons learned for the developing world. *International Journal of Hydrogen Energy*.

Marocco, P., Gandiglio, M., Audisio, D., & Santarelli, M. (2023). Assessment of the role of hydrogen to produce high-temperature heat in the steel industry. *Journal of Cleaner Production*, *388*, 135969. doi:10.1016/j.jclepro.2023.135969

Mio, A., Barbera, E., Pavan, A. M., Bertucco, A., & Fermeglia, M. (2024). Sustainability analysis of hydrogen production processes. *International Journal of Hydrogen Energy*, *54*, 540–553. doi:10.1016/j.ijhydene.2023.06.122

Patnaik, D., Pattanaik, A. K., Bagal, D. K., & Rath, A. (2023). Reducing CO2 emissions in the iron industry with green hydrogen. *International Journal of Hydrogen Energy*, *48*(61), 23449–23458. doi:10.1016/j.ijhydene.2023.03.099

Rechberger, K., Spanlang, A., Sasiain Conde, A., Wolfmeir, H., & Harris, C. (2020). Green hydrogen-based direct reduction for low-carbon steelmaking. *Steel Research International, 91*(11), 2000110.

Schmidhalter, I., Mussati, M. C., Mussati, S. F., Oliva, D. G., Fuentes, M., & Aguirre, P. A. (2024). Green hydrogen levelized cost assessment from wind energy in Argentina with dispatch constraints. *International Journal of Hydrogen Energy, 53*, 1083–1096. doi:10.1016/j.ijhydene.2023.12.052

Souza Filho, I. R., Springer, H., Ma, Y., Mahajan, A., da Silva, C. C., Kulse, M., & Raabe, D. (2022). Green steel at its crossroads: Hybrid hydrogen-based reduction of iron ores. *Journal of Cleaner Production, 340*, 130805. doi:10.1016/j.jclepro.2022.130805

Swennenhuis, F., de Gooyert, V., & de Coninck, H. (2022). Towards a CO2-neutral steel industry: Justice aspects of CO2 capture and storage, biomass-and green hydrogen-based emission reductions. *Energy Research & Social Science, 88*, 102598. doi:10.1016/j.erss.2022.102598

Zainal, B. S., Ker, P. J., Mohamed, H., Ong, H. C., Fattah, I. M. R., Rahman, S. A., Nghiem, L. D., & Mahlia, T. I. (2024). Recent advancement and assessment of green hydrogen production technologies. *Renewable & Sustainable Energy Reviews, 189*, 113941. doi:10.1016/j.rser.2023.113941

Chapter 8
Safe Green Hydrogen:
Production and Storage as an Emerging Source of Energy for Sustainable Future

Shuaibu Chiroma Hassan
https://orcid.org/0000-0002-8779-9462
Isa Kaita College of Education, Nigeria

Sanjeev Kumar
https://orcid.org/0000-0002-7375-7341
Lovely Professional University, India

ABSTRACT

Energy has been a vital resource in our daily lives and, simultaneously, one of the worldwide main problems. However, energy is an essential resource in almost everything in our daily lives, ranging from transportation, cooking, warming residential places, electronics, factories, schools, banks, etc. The most popular forms of energy used are from biofuels, which pollute the environment and cause some hazards to humans and plants, in the states of coal, oil and gas, nuclear, and so on. Due to these concerns, a new form of energy is needed that is environmentally friendly, cost-effective, and more sustainable. With the various types of research carried out across the globe, different forms of energy were identified and tested: solar, wind, reusable batteries, etc. Various research institutions, organizations, multinational organizations, and companies are embarking on research to find meaningful solutions to the problems. None has been shown to hold much promise like the green hydrogen (GH). Against this backdrop, this research will be conducted to show how much potential it holds.

INTRODUCTION

The primary purpose of this study is to present the prospects and potentials of Green Hydrogen as a critical substitute to bio-fuels (fossil fuels), to show the progress made on/in the storage and transportation of the Green Hydrogen; and the trend on future of the Green Hydrogen.Energy has been a vital commodity for producing goods and services in transportation and family uses – making it an essential

DOI: 10.4018/979-8-3693-1014-4.ch008

ingredient for human living across the globe. However, its supply, transportation, processing, storage, and consumption create many environmental and human problems due to its associated crises (Huseynli, 2022). Notably, the production, processing, and utilization of various forms of energy, coal, fossil fuels, nuclear, etc., cause much environmental harm – leading to global warming (Ásványi et al., 2017). There is a growing concern worldwide over using bio-fuels that affect the environment through Green Gas Emissions (GGE) .Energy has been among the main problems worldwide due to its complexities, like pollution, high revenues, politics, and so on (Ball &Wietschel, 2009). The IEA (International Energy Agency) has stressed the need for natural gas as an energy source in the global markets. However, it emphasized the need to address the emission concern from gas supplies (Okonkwo, 2023). Faith Birol, the IEA Executive Director, states that natural gas is still needed in the global transition to clean energy sources, further states (Huppmann& Egging, 2014). Hydrogen has been a researched source to provide a novel way of utilizing it to provide energy sustainably.

Green Hydrogen is identified as a crucial ingredient in meeting the Greenhouse Gas (GHG) Emissions Goals by the United Nations in the Paris Agreement (New-Hydrogen, 2023). Many types of research covered Green Hydrogen worldwide; however, specific gaps were needed (Alzoubi, 2021). For instance, Hassan (2022) has covered safe and green energy from family, tourism, and hospitality dimensions, with some breakthroughs made in the middle of 2022 and prior, and therefore should have covered the recent development created in 2023. Due to this gap, there is a need for urgent and up-to-date research to bridge the hole for the current period. Meanwhile, (Hosseini& Wahid, 2016) looked at it from only hospitality and tourism perspectives, highlighted little on Green Hydrogen, and covered only 2022, hence the need for this chapter.

Hydrogen, especially the Green one, has shown a promising prospect – being a clean energy of the future. Hydrogen, as an energy carrier, holds a prominent advantage of a high Gravimetric Energy Density, with a net-zero emission and high abundance (Dougherty et al., 2009). Among the myriad uses of Hydrogen is that it plays vital roles in formulating fertilizers, producing steel, and refining oil in refineries. However, the majority of the Hydrogen produced in the world is from natural sources, like coal or gas (Grey Hydrogen), which leads to the emission of up to 900 million metric tons of CO_2 into the atmosphere, as reported by the International Energy Agency (IAE) (Nowak, 1994). With the ongoing research, it is evident that Hydrogen, especially Green Hydrogen, has been identified as the prominent potential toward a smooth transition to clean, safe, and sustainable energy (Eyring et al., 2007). Various methods and techniques innovated by scientists will be highlighted, as well as their potential to be environmentally friendly.

Therefore, Green Hydrogen production from cleaner energy sources is the ideal solution. Green Hydrogen is produced by the process of *Electrolysis*, which splits water into hydrogen (H) and oxygen (O_2) molecules (Mbah et al., 2010). The research will show how Hydrogen is produced (industrial production) and the source of the materials for its production. Similarly, it will present the level reached in storage and production. In the past, there were difficulties and risks associated with the storage and transportation of Hydrogen(Millet &Grigoriev, 2013). This research will present how global experts and scientists from different research institutions resolve the challenges.

HYDROGEN

Hydrogen is a gas and fuel as well as a 'feedstock' material, which has the advantage of burning to produce energy without the emission of carbon (CO_2) (Durakovic et al., 2023). Hydrogen does not occur freely in nature but is the cleanest and most abundant element in our universe, which we cannot live without (New-Hydrogen, 2023; Pilkington, 2022). However, this gas is produced with energy-intensive methods that detract from the sustainability benefits of Greenhouse Gas-free fuels (Sansaniwal et al., 2017). It is also the key ingredient in producing fertilizers, refining oils, glass, and steel making in pharmaceutical industries (New-Hydrogen, 2023). Due to these potentials, it is now seen as a beacon of hope in clean energy transition and sustainability. Hence, most research is centered on this promising gas.

New-Hydrogen (2023) states that the 'Gold Standard' for Green Hydrogen production is through Electrolysis – by using Electrolysis with wind or solar electricity to split the H_2O into O and $H_2$with just less than 1000^0C to produce the cheapest Green Hydrogen. However, the unfortunate thing is that it is costly as the electricity for Green Hydrogen production accounts for up to 73% of the total costs(Chi & Yu, 2018). Notably, the combustion reaction of Hydrogen produces energy comparable to burning fossil fuels. As a result, this enables the application of Hydrogen as a source of energy or power. Similarly, it is also used in fuel cells like internal combustion energies and batteries. In addition, it could also be added to gas-based fuels for cooking and heating – to release the amount of CO_2 they release when burning(Adamou et al., 2023). When Hydrogen is burnt, it combines with O_2 to produce H_2O; when manufactured, the process could make less or a lot of CO_2 – depending on the manufacturing mode and when the feedstock is a fossil fuel (Salvi& Subramanian, 2015). The disruptive breakthrough made by the *New-Hydrogen* is that novel *Molten Catalytic Liquid* is reduced in one chamber, oxidized in another compartment, and continuously recycled and reused – with only inputs such as heat and water (New-Hydrogen, 2023). It is another step in solving some hydrogen production and usage challenges (Heiman& Solomon, 2007).

There are different types of Hydrogen, which come in various colors - namely Green Hydrogen, Grey Hydrogen, Blue Hydrogen, white Hydrogen, and Turquois Hydrogen (Ajanovic et al., 2022)with its type of Method of production and characteristics as well as cost implications. WEF (2023) reveals that Green Hydrogen could be flexible, low carbon, and efficient, which can be produced in multiple ways, and is an essential means that shapes the current global energy architecture for achieving Net-Zero Carbon emission(Garcia &Oliva, 2023). White Hydrogen usually refers to Hydrogen found in the earth's crust previously believed to be too sparse to extract. It is also energy-intensive because it is combined with the other gasses that must be separated (Ma et al., 2023). These are the main types of Hydrogen in use with potential today.

In other words, Hydrogen could be categorized in various ways – by carbon emission intensity across the whole life-cycle or by production method. Therefore, Hydrogen's development has become a global priority and is incorporated into China's National Development planning (Meng et al., 2021). Similarly, the global energy crises and climate change have been aggravated by Russia's Special Military Operation in Ukraine and the current extreme weather affecting Europe (WEF, 2023). It has necessitated the exploration of Hydrogen as the potential future global energy solution (Xu& Liu, 2019). Hydrogen will, therefore, play a vital role in accelerating the Energy transition towards a net zero emission, provided it is clean and at scale. As a result, its rapid development (production, transportation, and storage) has become a global priority issue (WEF, 2023). It has necessitated the exploration of Hydrogen as the potential future global energy solution.

Environmental Benefits

When H_2 is combusted in the presence of O_2 (Oxygen), the only product to be released is H_2O (water). As a result, both large chemical energy and its clean 'reactivity' therefore makes H_2an extremely appealing for uses as a fuel in engines an automobile: $H_{2(g)} + O_{2(g)}$ $H_2O_{(g)}$(Barron, 2023).

The production of green Hydrogen releases no greenhouse gases, contributing to a reduction in carbon emissions and helping combat climate change. Green Hydrogen can replace fossil fuels in various sectors, such as transportation and industry, improving air quality by eliminating harmful pollutants (Hosseini, 2023). It can be used as a clean energy source for industries like steel, cement, and chemicals, where decarburization is challenging. Hydrogen can store excess renewable energy and release it when needed, enhancing the stability and reliability of renewable energy systems. Hydrogen can be burned in gas turbines or used in fuel cells to generate electricity when renewable sources are unavailable(Zimmer & Hoffmann, 2023). Hydrogen fuel cell vehicles (FCVs) have the potential to revolutionize the transportation sector, especially for long-haul trucks, buses, and potentially even aircraft, offering zero-emission solutions with longer ranges and shorter refueling times(Dunn, 2002).

Methods of Producing Hydrogen

(Ogden et al., 2018) identifies a way of producing Hydrogen by utilizing natural gas as a raw material in combination with a 'special ceramic membrane', in this process, carbon captures and hydrogen productions are simultaneously achieved within one step, making the Method highly efficient. Moreover, the Method has energy-intensive ratings of 70-75 percent, potentially reaching 90 percent (Balat, 2008). Eventually, the end product is compressed Hydrogen with high purity. Meanwhile, the Ceramic Membrane Reactor separates the carbon dioxide (CO_2) more efficiently and conveniently, allowing the Greenhouse Gas (GHG) to be sequestered and transported conveniently (Kothari et al., 2008). This technology serves as a promising clean energy synthesis.

H_2O + Electricity $H_2 + 1/2O_2$

Hydrogen is produced in various ways:

I. Green Hydrogen – this is made through the Electrolysis of water (H2O) and energy derived from renewable like wind and solar.
II. Blue Hydrogen – this is made with SMR with CO2 emissions captured and stored with an appropriate Carbon Capture Sequestration (CCS) Technology.
III. Grey Hydrogen – made with SMR - also uses methane emissions, the byproducts of other industrial processes.
IV. Pink Hydrogen – this is made through water electrolysis whereby the energy is derived from Nuclear Power (Piikington, 2022). Pilkington further reveals that Gray and blue Hydrogen account for around 95 percent of the entire Hydrogen produced today.

Figure 1. Process of hydrogen production from water by electrolysis
(Kumar, 2023)

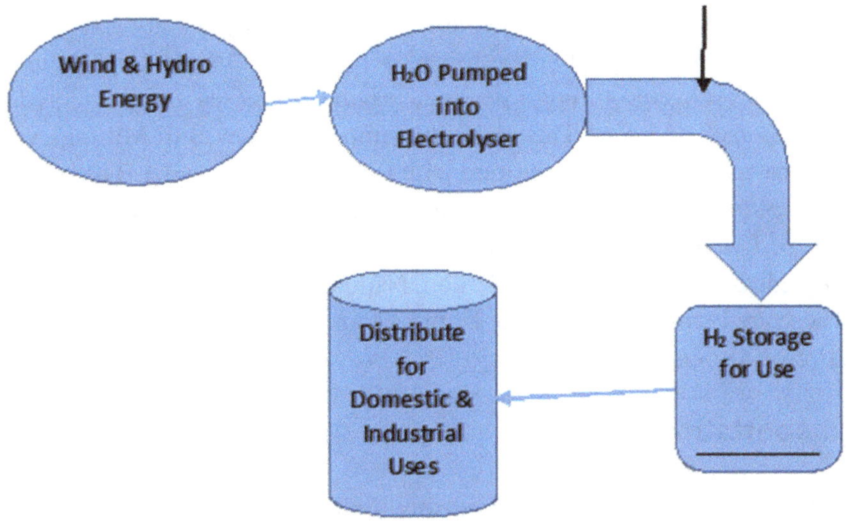

ELECTROLYSIS TO SPLIT WATER INTO HYDROGEN AND OXYGEN

Hydrogen Storage

Despite its advantages as a clean source of energy, high abundance, and high energy density, its storage and transport have become a bottleneck for its wide-scale application (Zhou, 2005). Gas is highly explosive – making very risky to store. As a result, regular maintenance of the highly pressurized storage system is necessary to prevent risks and hazardous situations. Currently, it is stored in high pressure of up to about 700 bars in fiber-reinforced composite tanks (FRC) as ideal storage equipment to detect strain sensors and emission to detect damages in the tanks. Therefore, due to its explosive nature, it needs a safe means of storage to avoid accidents (Crowl& Jo, 2007). Hydrogen (Dihydrogen), is a gas energy commodity at 37°C (room temperature), and gases are however occupying the most spaces of volumes as compared with other states of matter. Therefore, liquids are much easier to store (safely) than compressed gases. As a result, its energy density could be increased by storing the hydrogen gas under increased pressure at lower temperatures in a liquid forms (Barron, 2023).

India is currently working on a 'multi-pronged strategy' that simultaneously focuses on energy storage, Green Hydrogen, and offshore wind. The Government is also working on building energy storage (BESS) and pump-hydro (for pumping and storing Hydrogen). It is also realized that green Hydrogen will play a decisive role in India's Energy Transition (Mohammed et al., 2011). Various types of Hydrogen Storage systems or materials are being developed and investigated to address the challenge. For instance, Magnesium hydroxide (MgH_2) is considered one of the most promising storage due to its excellent reversibility, high capacity, low cost, and sufficient reserves (Industrial Chemistry & Materials, 2023). Previously, a team of researchers from Deakin University, Australia, discovered a novel method for Hydrogen Storage, whereby it is converted into powdered form, which revolutionizes the storage and transportation of Hydrogen and is also very cheap. This Method employs the Principle of Chemical Mechanics to initiate a chemical reaction that utilizes force instead of heat, light, or electrical potential differences. In another breakthrough, a new method was revealed by breaking the crude oil – notably petroleum, into petrol and gases (including Hydrogen) in refineries through Cryogenic Distillation. The absorbed gases could be recovered through heating, leaving the gases and powdered unchanged and allowing for immediate use and reuse. This technique employs a new Ball Milling system to store the gas in particular nonmaterial at room temperature with Boron nitride powder. The boron nitride powder could also be used to carry out some gas (H) separation and storage up to fifty times (Mohammed et al., 2011). It could also be *liqueed* (liquefied), as a gas or compressed to be stored, or stored in a process known as Metal Organic Materials (MOMs), Carbon, Nanostructures (CNS), or Metal Hydrides (MHs) (Barron, 2023). These are breakthroughs made from various researches across the developed world for hydrogen application as a source of clean energy.

Hydrogen Transportation

This new method provides industries with a solid way for gases with higher and more capacities than the current ones (Sharma &Ghoshal, 2015). Hydrogen transportation refers to using Hydrogen as a fuel source for various modes of transportation, including cars, buses, trucks, trains, ships, and even aircraft. It is a potential solution for reducing the transport sector's greenhouse gas emissions and air pollution. Hydrogen Fuel Cell Vehicles (FCVs) use Hydrogen to produce electricity through a chemical reaction in a fuel cell (Manoharan et al., 2019). The electricity powers an electric motor, propelling the vehicle. FCVs produce zero tailpipe emissions, as water vapor is the only byproduct of the hydrogen fuel cell reaction. FCVs typically have a more extended driving range than Battery Electric Vehicles (BEVs) on a single tank of Hydrogen(Shin et al., 2019). Hydrogen refueling takes a few minutes, similar to refueling a gasoline or diesel vehicle. Hydrogen refueling infrastructure needs to be improved, making it challenging for consumers to find refueling stations. FCVs and hydrogen production can be expensive, though costs are expected to decrease with scale. Ensuring the display of Hydrogen is green (using renewable energy) is essential to maximize environmental benefits. Several cities and transit agencies have adopted hydrogen fuel cell buses as a clean, quiet public transportation option. Hydrogen buses have longer ranges and shorter refueling times than battery electric buses, making them suitable for continuous, high-mileage routes(Mori & Hirose, 2009). Hydrogen is being explored for heavy-duty trucks and freight transportation, where long-range and quick refueling is crucial. Hydrogen fuel cells can power trains, especially in regions where electrification of rail lines is expensive or challenging. Hydrogen trains offer zero emissions and high efficiency. Hydrogen can be used to power ships, reducing emissions in the maritime industry. Hydrogen can be stored onboard in liquid or gaseous form. Hydrogen is being explored as a potential fuel for aircraft to reduce greenhouse gas emissions from aviation. Hydrogen-powered aircraft are in the early stages of development.

Challenges and Considerations: One of the main challenges faced in the transition towards the usage of Hydrogen is the cost of its conversion, and the processes identified that cause carbon emission during its production. Recently, some breakthroughs were made (Ren et al., 2020). A new form of catalyst known as Carbon Compound Nickel-iron-molybdenum-phosphate (NiFeMo-P-C) is anchored on the nickel foam, which significantly reduces the amount of electricity needed to generate both Oxygen and Hydrogen from water (H2O) – providing an efficient and clean way to produce H gas. Therefore, the catalyst lowers the energy needed for water electrolysis(Ball &Wietschel, 2009). O and H gases are split into H2O through Hydrogen Evolution Reaction (HER) and Oxygen Evolution Reaction (OER). In this process, the transition metal alloy or the mixture that contains at least one metal (NiFeMo) is used as the catalyst for the water electrolysis due to an incomplete filling of the electron orbital's in the transition metal atoms iron and nickel – making an ideal electron acceptor and donor in chemical reactions (Tsinghua University Press, 2023). Ideally, a catalyst (in water hydrolysis) does neither change nor degrade by the response; therefore, for the electrolyzes, this becomes so critical as the amount of Elecrocatalysts accounts for up to 50% of its total costs. Due to this, their lifetime and efficiency are essential to the future of the availability of Green Hydrogen and, thus, to a carbon-free economy (Bridgeland et al., 2022). The process reveals that Boron impurities produced during Hydrogen and Boron synthesis could enhance the catalyst's performance by expanding the Lattice structure. As a result, boron must be shielded inside the catalyst's Lattice structure from the Hydrogen produced on the surface of the stimulus. With these developments, the costs would be reduced as the amount of energy required to have the Hydrogen is diminished, and the efficiency of hydrogen production and density will be enhanced.

Therefore, building a robust hydrogen refueling infrastructure is a significant challenge, requiring substantial investment. Hydrogen production, storage, and transportation can be expensive, but costs will decrease as technology advances. Hydrogen must be produced using renewable energy sources to maximize the environmental benefits of hydrogen transportation. Hydrogen is highly flammable, and safety measures must be in place for production, storage, and transportation. Hydrogen production and distribution can have energy efficiency challenges, which must be addressed for a sustainable future. Hydrogen transportation has the potential to play a significant role in reducing emissions in the transport sector, especially for heavy-duty and long-range applications. However, its widespread adoption depends on addressing infrastructure challenges, cost competitiveness, and ensuring a green and sustainable hydrogen supply chain. To store hydrogen in liquefied form, it has to be kept in temperature below its boiling point (-253°C) – which needs an effective refrigeration system to keep it cooled. To achieve this, refrigeration system should be featured into vehicles which are now a challenge for vehicles (Barron, 2023). Otherwise, it will be risky due to its inflammability nature where fire outbreak can occur in case of any accidents which may affect lives, the vehicles carrying it, and the buildings in case it happens in the cities. Preventive measures need to be in place for safety purposes.

THE FUTURE PROSPECTS OF HYDROGEN

It is now evident that Hydrogen is the beacon of hope for the successful future energy transition and for the mobility sector due to its high density, with zero carbon emission, as well as high heat conversion (Rosen &Koohi-Fayegh, 2016). Hydrogen can significantly decarburize various sectors, including industry, power generation, and transportation, where electrification alone may not be feasible or efficient. Developing low-cost green hydrogen production methods, primarily through renewable energy-powered Electrolysis holds immense potential. It can lead to a significant reduction in carbon emissions. Hydrogen can serve as an energy carrier and storage medium, allowing excess renewable energy to be stored and transported when electricity demand is low, helping to balance the grid (Moriarty &Honnery, 2019). Hassan (2022) reported a H_2 storage breakthrough based on a team of researchers' findings from the University of Deakin (Australia) where they discovered a novel method of storing Hydrogen by converting it into powdered form for easier storage and transportation by chemical mechanics through a chemical reaction that utilizes force as opposed to light, heat or electrical potential differences. The method takes 10% of the energy which is used for producing gas.

Hydrogen fuel cell vehicles (FCVs) have the potential to revolutionize the transportation sector, especially for long-haul trucks, buses, and potentially even aircraft, offering zero-emission solutions with longer ranges and shorter refueling times. Hydrogen can be used in various industrial processes, such as steel and cement production, where high-temperature heat and direct electrification are challenging. Hydrogen is a critical feedstock for different chemical processes, including ammonia and methanol production. Green Hydrogen can replace fossil-based Hydrogen in these processes, reducing emissions (Junwen et al., 2021).

Some countries with abundant renewable energy resources are exploring the possibility of producing green Hydrogen for export, creating new economic opportunities and reducing global emissions. Ongoing research and development efforts focus on improving hydrogen production, storage, and transportation technologies, particularly cost reduction and efficiency improvements. Governments and private sector investments are being made to expand hydrogen infrastructure, including developing refueling stations, hydrogen pipelines, and storage facilities. Hydrogen can store excess energy and release it when needed, providing grid stability and supporting the integration of intermittent renewable energy sources. Governments worldwide are increasingly recognizing Hydrogen's potential and are implementing supportive policies, such as financial incentives and regulatory frameworks, to foster its growth. International collaboration is growing, with countries and organizations working together to establish common standards and best practices for hydrogen production, storage, and transportation. However, several challenges must be overcome to fully realize the prospects of Hydrogen, including cost competitiveness with fossil fuels, building a robust infrastructure, ensuring the sustainability of hydrogen production, and addressing safety concerns associated with its production and transportation.

CONCLUSION

The chapter looks at the prospects and development so far made on hydrogen, particularly the Green Hydrogen. Hydrogen, as a gas, does not occur freely in nature; however, its synthesis is cheaper through hydrolysis which makes it more sustainable. Many issues are raised ranging from its various types, methods of production and storage, safety and challenges issues associated with it and current breakthroughs made. Similarly, the chapter, though theoretical, draws from a case like India, on its development and prospects on hydrogen use for clean and sustainable energy. So far, it identifies some min challenges being faced on hydrogen application in the automobile industry as well as other domestic uses. In addition, the chapter highlights on the prospects of hydrogen for future uses to ensure sustainable energy consumption, as it does not produce any carbon gas that will pollute the environment, rather it only produces water in liquid or gas form. The main challenges are the way of storing the gas, transporting it and installing it on vehicles, due to its high flammability in case of accidents. The Chapter presents solutions to some of the challenges, though not all the challenges. Future researches can therefore look at how these challenges could be overcome for sustainable energy consumption.

REFERENCES

Adamou, P., Bellomi, S., Hafeez, S., Harkou, E., Al-Salem, S., Villa, A., Dimitratos, N., Manos, G., & Constantinou, A. (2023). Recent progress for hydrogen production from ammonia and hydrous hydrazine decomposition: A review on heterogeneous catalysts. *Catalysis Today*, *423*, 114022. doi:10.1016/j.cattod.2023.01.029

Ajanovic, A., Sayer, M., & Haas, R. (2022). The economics and the environmental benignity of different colors of hydrogen. *International Journal of Hydrogen Energy*, *47*(57), 24136–24154. doi:10.1016/j.ijhydene.2022.02.094

Alzoubi, A. (2021). Renewable Green hydrogen energy impact on sustainability performance. *International Journal of Computations, Information and Manufacturing, 1*(1).

Ásványi, K., Juhász-Dóra, K., Jászberényi, M., & Michalkó, G. (2017). Literature review of renewable energy in the tourism industry. *Journal of Environmental Management & Tourism*, *8*(2 (18)), 476.

Balat, M. (2008). Possible methods for hydrogen production. *Energy Sources. Part A, Recovery, Utilization, and Environmental Effects*, *31*(1), 39–50. doi:10.1080/15567030701468068

Ball, M., & Wietschel, M. (2009). The future of hydrogen–opportunities and challenges. *International Journal of Hydrogen Energy*, *34*(2), 615–627. doi:10.1016/j.ijhydene.2008.11.014

Barron, A. R. (2023). *2.10 Storage of Hydrogen for Use as a Fuel.* LibreTexts Chemistry.

Bridgeland, R., Chapman, A., McLellan, B., Sofronis, P., & Fujii, Y. (2022). Challenges toward achieving a successful hydrogen economy in the US: Potential end-use and infrastructure analysis to the year 2100. *Cleaner Production Letters*, *3*, 100012. doi:10.1016/j.clpl.2022.100012

Chi, J., & Yu, H. (2018). Water electrolysis based on renewable energy for hydrogen production. *Chinese Journal of Catalysis*, *39*(3), 390–394. doi:10.1016/S1872-2067(17)62949-8

Crowl, D. A., & Jo, Y.-D. (2007). The hazards and risks of hydrogen. *Journal of Loss Prevention in the Process Industries*, *20*(2), 158–164. doi:10.1016/j.jlp.2007.02.002

Dougherty, W., Kartha, S., Rajan, C., Lazarus, M., Bailie, A., Runkle, B., & Fencl, A. (2009). Greenhouse gas reduction benefits and costs of a large-scale transition to hydrogen in the USA. *Energy Policy*, *37*(1), 56–67. doi:10.1016/j.enpol.2008.06.039

Dunn, S. (2002). Hydrogen futures: Toward a sustainable energy system. *International Journal of Hydrogen Energy*, *27*(3), 235–264. doi:10.1016/S0360-3199(01)00131-8

Durakovic, G., del Granado, P. C., & Tomasgard, A. (2023). Are green and blue hydrogen competitive or complementary? Insights from a decarbonized European power system analysis. *Energy*, *282*, 128282. doi:10.1016/j.energy.2023.128282

Eyring, V., Corbett, J. J., Lee, D. S., &Winebrake, J. J. (2007). Brief summary of the impact of ship emissions on atmospheric composition, climate, and human health. *Document Submitted to the Health and Environment Sub-Group of the International Maritime Organization, 6.*

Garcia, M., & Oliva, S. (2023). Technical, economic, and CO2 emissions assessment of green hydrogen production from solar/wind energy: The case of Chile. *Energy*, *278*, 127981. doi:10.1016/j.energy.2023.127981

Hassan. (2022). *Toward a Green and Safe Energy: An Implication for Hospitality, Tourism and Family Usage*. Paper Presented at the 10th International Istanbul Scientific Research Congress.

Heiman, M. K., & Solomon, B. D. (2007). Fueling US transportation: The hydrogen economy and its alternatives. *Environment*, *49*(8), 10–25. doi:10.3200/ENVT.49.8.10-25

Hosseini, S. E. (2023). Hydrogen fuel, a game changer for the world's energy scenario. *International Journal of Green Energy*, 1–17.

Hosseini, S. E., & Wahid, M. A. (2016). Hydrogen production from renewable and sustainable energy resources: Promising green energy carrier for clean development. *Renewable & Sustainable Energy Reviews*, *57*, 850–866. doi:10.1016/j.rser.2015.12.112

Huppmann, D., & Egging, R. (2014). Market power, fuel substitution and infrastructure–A large-scale equilibrium model of global energy markets. *Energy*, *75*, 483–500. doi:10.1016/j.energy.2014.08.004

Huseynli, B. (2022). The Role of Renewable Energy in the Development of the Tourism Sector. *African Journal of Hospitality, Tourism and Leisure*, *11*(6), 1924–1936.

Junwen, C., Xiangfu, Q., Ga, G., Wenqiang, Z., & Bo, Y. (2021). Current status and prospects of hydrogen storage and transportation technology. *Acta Petrolei Sinica*, *37*(6), 1461.

Kothari, R., Buddhi, D., & Sawhney, R. (2008). Comparison of environmental and economic aspects of various hydrogen production methods. *Renewable & Sustainable Energy Reviews*, *12*(2), 553–563. doi:10.1016/j.rser.2006.07.012

Kumar, A. (2023). *Overview of Green Hydrogen Technology*. Presentation Slides for Lectures.

Ma, Z., Tian, T., Cui, Q., Shu, J., Zhao, J., & Wang, H. (2023). Rapid sizing of a hydrogen-battery storage for an offshore wind farm using convex programming. *International Journal of Hydrogen Energy*, *48*(58), 21946–21958. doi:10.1016/j.ijhydene.2023.03.037

Manoharan, Y., Hosseini, S. E., Butler, B., Alzhahrani, H., Senior, B. T. F., Ashuri, T., & Krohn, J. (2019). Hydrogen fuel cell vehicles; current status and future prospect. *Applied Sciences (Basel, Switzerland)*, *9*(11), 2296. doi:10.3390/app9112296

Mbah, J., Weaver, E., Srinivasan, S., Krakow, B., Wolan, J., Goswami, Y., & Stefanakos, E. (2010). Low voltage H2O electrolysis for enhanced hydrogen production. *Energy*, *35*(12), 5008–5012. doi:10.1016/j.energy.2010.08.021

Meng, X., Gu, A., Wu, X., Zhou, L., Zhou, J., Liu, B., & Mao, Z. (2021). Status quo of China hydrogen strategy in the field of transportation and international comparisons. *International Journal of Hydrogen Energy*, *46*(57), 28887–28899. doi:10.1016/j.ijhydene.2020.11.049

Millet, P., &Grigoriev, S. (2013). Water electrolysis technologies. *Renewable Hydrogen Technologies: Production, Purification, Storage, Applications and Safety, 19.*

Mohammed, M., Salmiaton, A., Azlina, W. W., Amran, M. M., Fakhru'l-Razi, A., & Taufiq-Yap, Y. (2011). Hydrogen rich gas from oil palm biomass as a potential source of renewable energy in Malaysia. *Renewable & Sustainable Energy Reviews*, *15*(2), 1258–1270. doi:10.1016/j.rser.2010.10.003

Mori, D., & Hirose, K. (2009). Recent challenges of hydrogen storage technologies for fuel cell vehicles. *International Journal of Hydrogen Energy*, *34*(10), 4569–4574. doi:10.1016/j.ijhydene.2008.07.115

Moriarty, P., & Honnery, D. (2019). Prospects for hydrogen as a transport fuel. *International Journal of Hydrogen Energy*, *44*(31), 16029–16037. doi:10.1016/j.ijhydene.2019.04.278

Nowak, D. J. (1994). Atmospheric carbon dioxide reduction by Chicago's urban forest. *Chicago's Urban Forest Ecosystem: Results of the Chicago Urban Forest Climate Project*, 83–94.

Ogden, J., Jaffe, A. M., Scheitrum, D., McDonald, Z., & Miller, M. (2018). Natural gas as a bridge to hydrogen transportation fuel: Insights from the literature. *Energy Policy*, *115*, 317–329. doi:10.1016/j.enpol.2017.12.049

Ren, X., Dong, L., Xu, D., & Hu, B. (2020). Challenges towards hydrogen economy in China. *International Journal of Hydrogen Energy*, *45*(59), 34326–34345. doi:10.1016/j.ijhydene.2020.01.163

Rosen, M. A., & Koohi-Fayegh, S. (2016). The prospects for hydrogen as an energy carrier: An overview of hydrogen energy and hydrogen energy systems. *Energy, Ecology & Environment*, *1*(1), 10–29. doi:10.1007/s40974-016-0005-z

Salvi, B. L., & Subramanian, K. (2015). Sustainable development of road transportation sector using hydrogen energy system. *Renewable & Sustainable Energy Reviews*, *51*, 1132–1155. doi:10.1016/j.rser.2015.07.030

Sansaniwal, S., Rosen, M., & Tyagi, S. (2017). Global challenges in the sustainable development of biomass gasification: An overview. *Renewable & Sustainable Energy Reviews*, *80*, 23–43. doi:10.1016/j.rser.2017.05.215

Sharma, S., & Ghoshal, S. K. (2015). Hydrogen the future transportation fuel: From production to applications. *Renewable & Sustainable Energy Reviews*, *43*, 1151–1158. doi:10.1016/j.rser.2014.11.093

Shin, J., Hwang, W.-S., & Choi, H. (2019). Can hydrogen fuel vehicles be a sustainable alternative on vehicle market?: Comparison of electric and hydrogen fuel cell vehicles. *Technological Forecasting and Social Change*, *143*, 239–248. doi:10.1016/j.techfore.2019.02.001

Xu, S., & Liu, J. (2019). Metal-based direct hydrogen generation as unconventional high density energy. *Frontiers in Energy*, *13*(1), 27–53. doi:10.1007/s11708-018-0603-x

Zhou, L. (2005). Progress and problems in hydrogen storage methods. *Renewable & Sustainable Energy Reviews*, *9*(4), 395–408. doi:10.1016/j.rser.2004.05.005

Zimmer, M., & Hoffmann, P. (2023). *The Green Industrial Revolution–Investment Pathways to Decarbonize the Industrial Sector in Europe*. Academic Press.

Chapter 9
Integrated Approach for Green Hydrogen Production and Sustainable Electrification of Railway Platform With Solar Energy Sources

Dharmbir Prasad
https://orcid.org/0000-0002-9010-9717
Asansol Engineering College, India

Ranadip Roy
https://orcid.org/0000-0003-2111-2581
Sanaka Educational Trust's Group of Institutions, Durgapur, India

Rudra Pratap Singh
https://orcid.org/0000-0001-7352-855X
Asansol Engineering College, India

Sudhangshu Chakraborty
Asansol Engineering College, India

Jatin Anand
Asansol Engineering College, India

ABSTRACT

This study provides a technical and financial analysis of the viability of producing green hydrogen and electrifying railway platforms at Hatia locations in India using solar energy and pumped hydro storage backup. The present study examines Hatia Railway Station's switch to green energy, highlighting the advantages and prospects it offers. Through the utilization of renewable energy sources like solar and wind power, the area may reduce greenhouse gas emissions, meet national climate objectives, and improve air quality. The solar potential of the area offers a strong basis for the generation of sustainable energy, reinforced by cutting-edge technology. Furthermore, creating green energy infrastructure has the potential to boost employment and economic growth, which would improve the community. The total energy produced by the PV panel is 2,266,982 kWh.

DOI: 10.4018/979-8-3693-1014-4.ch009

INTRODUCTION

The World is approaching a crucial turning point in the energy transition as a result of the quick depletion of fossil resources and the rising concern over climate change. In the middle of this paradigm change, hybrid renewable energy systems - especially those that use solar power technologies have become well known as effective ways to deal with the issues of energy sustainability. By combining these resources, the energy supply is more reliable and there is less chance of blackouts during inclement weather. Energy storage technologies that are included into hybrid systems also make it possible to store excess energy during times of high production. This allows the energy to be used during periods of low production, which improves system efficiency and minimizes waste. India and the majority of other major economies have committed to net zero goals. One of the main prerequisites for reducing emissions is the switch to green ammonia and hydrogen, particularly in the difficult-to-abate industries. An environmentally friendly substitute for fossil fuels is green hydrogen, which is created by electrolysis with the use of renewable energy sources like solar or wind power. Its effective and environmentally friendly application spans several industries:

Transportation: Fuel for Vehicles: Fuel cell electric vehicles (FCEVs), which provide extended ranges and quick refilling akin to gasoline/diesel cars, may be powered by green hydrogen.

Heavy duty transport: Hydrogen fuel cells may be used by trucks, buses, and trains to operate emissions-free, particularly on long-distance routes.

Maritime and aviation: By using hydrogen in fuel cells for ships and airplanes, emissions in these sectors which have historically proven difficult to decarbonize can be decreased.

Infrastructure: The installation of hydrogen filling stations will facilitate a broader adoption, particularly for fleets and public transit.

Applications in Industry:

- Hydrogen may be used as a clean fuel or reducing agent in industries with heavy emissions, such as cement and steel manufacture.
- For the production of industrial heat and electricity, hydrogen can be used as fuel in gas turbines or combined heat and power systems.
- Hydrogen is an essential feedstock that lowers carbon footprints in the manufacturing of ammonia, refining, and other chemical processes.
- When renewable energy output is low, excess energy can be used for electrolysis to produce hydrogen for later consumption.
- Hydrogen's ability to hold energy for extended periods of time is essential for counteracting seasonal fluctuations in the output of renewable energy.

Water Conservation and its Effects on the Environment:

- Electrolysis: Using renewable power, electrolysis splits water into hydrogen and oxygen to make green hydrogen.
- Water Recycling: In an effort to consume less freshwater, efforts are being made to recycle the water used in electrolysis.
- Environmental Benefits: Electrolysis with renewable energy lowers pollutants and greenhouse gas emissions.
- Local Impact: Green hydrogen generation may help communities and local ecosystems by lowering pollution and water use.

In order to ease the shift from fossil fuel-based feedstocks to green hydrogen and green ammonia as chemical form of energy for various industries and energy carriers, the Indian government has been considering a variety of regulatory initiatives. The Indian government has developed a policy on green hydrogen after giving it some thought (Raj et al., 2022). As part of the 'Hydrogen for Heritage' campaign, Indian Railways intends to run 35 hydrogen trains. India will join the group of nations actively investing in cutting-edge technologies to alter the transportation industry when the first of these hydrogen-fueled trains, which will travel between Jind and Sonipat, is launched the following year (ET EnergyWorld, 2023). In order to supply hydrogen for India's first hydrogen train for Indian Railways, Jind, Haryana-based GreenH Electrolysis, a joint venture between H2B2 Electrolysis Technologies and GR Promoter Group, announced today that it has signed a contract with Hyderabad-based Medha Servo Drives to construct a hydrogen production and refuelling station. There is currently no specialized state wide network of interconnected hydrogen pipelines for India's green hydrogen market. India should learn from Europe, which expanded on its already-existing gas network to create a network specifically for transporting hydrogen Table 1 (India Briefing, 2024).

The electrification of Hatia railway station's platform using renewable energy sources marks a progressive and eco-conscious endeavour. This study is designed to leverage sustainable and environmentally friendly energy solutions, specifically through solar power, to meet the electricity needs of the station's ticket room and other accessories. As a significant contributor to India's energy consumption, the state of Jharkhand's exploration of green energy alternatives takes center stage, reflecting a commitment to mitigating environmental impacts and advancing energy security. These solar panels will capture sunlight and convert it into electricity, which can be utilized to power various electrical equipment and lighting within the ticket room. Solar power offers a reliable, clean, and sustainable energy source that reduces the station's dependence on conventional grid electricity, while simultaneously contributing to a reduction in greenhouse gas emissions.

PLATFORM LOCATION FEASIBILITY MAPPING

The geographical feasibility of Hatia Railway Station is underscored by its strategic location within the larger context of Ranchi, Jharkhand. This assessment takes into account various geographical factors that contribute to its practicality and viability as a significant hub. Hatia Railway Station's accessibility is a notable factor in its geographical feasibility. It serves as a crucial railway junction that facilitates con-

Table 1. Green hydrogen future market scenario for sustainable application

Country	Demand of Hydrogen	Hydrogen Energy Application Area	Demand of Green H2 by 2030 (MMTPA)	Allotted Capital (US$)
India	6 MMTPA	Decrease carbon emissions and attain energy self sufficiency	12	2.1 billion
Australia	650 KTPA	Clean – blue / green	Not available	487 million
United States	10 MMTPA	Low carbon – blue/green/others	10	> 15 billion
European Union	8.2 MMTPA	Low carbon - green or blue	10	609 billion
South Korea	220 KTPA	Grey/blue/green	3.9	653 million
Germany	1.65 MMTPA	Carbon free – green or blue	2.7 – 3.3	15-25 billion
Japan	2 MMTPA	Blue	3	664 million

nectivity not only within Jharkhand but also to neighbouring states. This strategic positioning enhances the station's significance as a transit point for passengers and freight. Situated within the capital city of Ranchi, Hatia station is at the heart of administrative and economic activities. This proximity to key institutions and markets enhances the station's potential to cater to a diverse range of travellers, from commuters to business persons. The station's integration with various modes of transport, including buses and taxis, amplifies its accessibility. This connectivity simplifies the movement of travellers to and from the station, further establishing its geographical practicality. Its location mapping for energy access is shown in Figure 1.

The Hatia Railway Station situated within the larger context of Ranchi, Jharkhand benefits from a variety of energy resources available to fulfill its load demand. These resources contribute to the station's energy sustainability, operational efficiency, and environmental consciousness. Given Jharkhand's substantial solar potential, Hatia station can leverage solar energy as a renewable resource. Solar panels installed on the station's roof or adjacent spaces can harness sunlight to generate clean electricity. This can offset a portion of the station's energy demand, especially during daylight hours. Implementing energy storage solutions, such as batteries, can store excess energy generated during periods of low demand (e.g., solar energy during the day) and release it during peak hours. This helps manage load fluctuations, reduces strain on the grid, and ensures a stable power supply. Energy-efficient technologies, such as LED lighting, energy-efficient HVAC systems, and optimized station design, can collectively reduce the station's overall energy demand. These measures align with sustainability goals and lower operational costs.

Figure 1. Location mapping for sustainable platform and green hydrogen production at site

The present chapter is structured as follows: starting with problem statement, a review of related studies is presented, followed by a mathematical formulation; integrated modelling of green hydrogen and platforms is then covered and finally, a potential assessment of energy harvesting is evaluated. Subsequently, the study's findings and discussion will be continued after the conclusion.

Problem Statement

As solar energy is abundant and sustainable in the World, it is one of the most favourable choices among all renewable energy sources. Several strategies are available with promising outcomes to overcome the variable character of this green source, even though solar energy plays a crucial role in supplying clean energy despite its inherent intermittent nature. The core motive behind the study was to produce green hydrogen with PHS backup to lower greenhouse gas emissions, cut back on fuel purchases and to promote increased energy independence. Following the primary goals of this study:

- Pre-feasibility techno-economic analysis, sensitivity analysis and optimal hybrid system design for hybrid system configurations taking into account green energy sources, hydrogen production, seasonal load demand, diesel prices, project lifespan, interest rate, PV derating and uncertainty in parameter estimations.
- Viability check of the best independent green railway platform possible for the site under consideration, concentrating on the production of hydrogen based on technical (renewable fraction and unmet load percentage) and economic (least NPC and LCOE) factors.

Related Works Review

The present issue of climate change demands a swift switch from energy systems reliant on fossil fuels to ones that rely on renewable sources. Green hydrogen is priced at INR 492/kg in India. It will cost INR 1.1 lakh to run a hydrogen-powered motor that travels from Bengaluru to Chennai using 228 kg of renewable hydrogen (ET EnergyWorld, 2022). Fuel cells need 27% more money to operate than diesel engines because they must convert hydrogen into electricity. Fuel cell and storage fees are extra expenses. With its low energy and CO2 intensities, rail transportation is a viable strategy for increasing energy diversification and reducing emissions. Yet, much like other infrastructure, train requires expensive expenditures (International Energy Agency, 2023).

High levels of usage are necessary for rail projects to be financially and ecologically feasible. The success of rail building projects depends on maintaining this equilibrium. With a target of 50% renewable energy by 2031, Brazil, the largest energy market in Latin America, has made renewables a priority in its energy expansion plan 2021–2031. 85% of power consumption is now met by renewables, with 88% projected by 2030 (ETN News, 2023). At 63%, hydropower is the most common source of energy. There are 15 GW of installed wind power and another 4.6 GW planned for 601 wind farms by 2023. With a freight efficiency of 12 times and a passenger efficiency of 3 times greater than that of highways, rail-based transportation is incredibly environmentally beneficial (Metro Rail News, 2023). With IGBT-based technology, regenerative braking, and efficient operation that maximizes energy savings, MELPL's first 12000 HP WAG12 electric locomotive in India is a significant achievement. Even while the production and application of green hydrogen from renewable energy sources like solar and wind power has advanced significantly, these stand-alone systems have certain drawbacks of their own. Following are the recent research aligned to it:

This research not only signifies a shift towards energy sustainability but also embodies a commitment to reducing environmental impact. By adopting solar and wind renewable sources, Hatia railway station will become a greener and more energy-efficient facility, setting an example for cleaner and more eco-friendly energy practices in the region. Through a detailed exploration of its potential, challenges, and collaborative pathways, this research seeks to contribute to the understanding of how harnessing renewable energy sources can create a resilient, vibrant, and sustainable future for Hatia and beyond. Technical complexity demands careful thought, such as controlling energy storage and maximizing the integration of many sources. Economic viability must be assessed in light of the long-term advantages, taking into account both the setup and maintenance costs. In addition, it is imperative to develop governmental frameworks and laws that facilitate the use of hybrid systems and assure a smooth transition towards greener energy.

(i). Green Electrification of Railways System

The Indian Railways plans to speed the transition to hydrogen trains by issuing tenders for 35 hydrogen-powered trains. The goal is to run these green trains on heritage lines, with an emphasis on retrofitting fuel cells made from hydrogen into existing DEMU trains (Economic Times, 2023). Siemens Mobility has secured contracts for nine hydrogen fuel cell trains and 61 battery trains from Germany and Denmark. The Mireo Plus B and Mireo Plus H trains were demonstrated. emphasizing the complementary nature of batteries for short lengths and hydrogen for longer. non-electric parts (Rail Tech, 2023). Japan plans to achieve carbon neutrality by 2050, with the transport sector accounting for 17.4% of CO2 emissions. Trains produce less carbon than vehicles and planes, but reaching Carbon neutrality in railroads necessitates a shift to 100% clean energy (Japan Times, 2023). The study of (Gangwar et al., 2014) highlighted the significance of IR's policy option in traction selection to accomplish environmentally friendly transportation, while accounting for broader societal goals and resource sharing, given India's developmental stage and resource limitations. A continued study used a life cycle assessment to evaluate the Mumbai Suburban Railway's environmental effect. By giving future environmental evaluations of public transportation in India a useful baseline, the study helps planners and decision makers prioritize sustainable mobility solutions (Shinde et al., 2018).

In order to overcome capacity issues in current RPSs brought on by rising passenger demand, (Yang et al., 2019) article investigated the idea of a HRPS. The suggested hybridization has the ability to reduce electricity prices by 9%, as demonstrated by the case study of the Sarry railway power substation. In (Li et al., 2020), a unique multi-criteria decision-making framework for high-speed railway DPVPS site selection is introduced. Its efficacy is validated by a case study in Hebei Province, which gives railroads a useful tool for choosing DPVPS sites. Air pollutants (NO_2, PM2.5) were measured inside and outside two UK railway stations with varying diesel train usage. A random-forest model predicted the impact of specific diesel-powered trains on air quality (Font et al., 2020). Railways are essential for transportation, but their global expansion raised questions about energy use and carbon emissions. Solutions are available thanks to developments in energy harvesting, which make use of acoustic, thermal, magnetic, wind and solar energy sources (Zuo et al., 2022). Another study suggested utilizing the RBE from ERSs to charge EVs. The strategy combines RPFCs with stationary HESSs. The efficacy of the method in actual system scenarios is confirmed by simulations (Kaleybar et al., 2023). The measurement of non-exhaust emissions from railway wheel-rail contact is the main objective of this work. Experiments conducted on a wheel-rail test bench indicate that PM10 emissions vary from 0.33 to 5.57 g/km, with notable contributions from PM2.5 and PM1 (Fruhwirt et al., 2023). Continued study explored energy flexibility strategies for a net-zero metro railway system, employing forced shutdown, charging, peak shaving, and more. Without changing capital costs, these measures reduce operational costs and increase net present value (Kumar & Cao, 2023).

African railway development is examined in (Bouraima et al., 2023) comprehensive study of the literature, which identifies gaps and obstacles. The results show a limited uptake of electric trains, a lack of technological integration, and information gaps in terms of passenger demand, cost-effectiveness, and environmental impact. The study assessed energy consumption and emissions using simulation and optimization, utilizing a multi-criteria analysis to identify the best alternative for accomplishing a sustainable mode shift (Gurri et al.. 2023). In (Kamel et al., 2023), suggested a smart integrated power source for railway traction that combines Li-ion batteries with fuel hydrogen cells. In accordance with the operational modes and battery condition, the control system controls both power sources. The performance of two three-phase machines is improved. In order to transition to a low-carbon future, China is actively modernizing its high-speed railway traction power supply infrastructure. The goals of operation control, form evolution, and motivations are all examined in this work (Li et al., 2023). Reviewing decarbonization tactics for IPA, (Pivetta, et al., 2024) research highlighted the potential of hydrogen as a fuel, energy transport and feedstock. There are a number of obstacles to overcome, such as the high cost of hydrogen-powered cars and grid restrictions for onshore power supplies.

(ii). Planning and Operation of Energy Efficient Railway Transportation Corridor

Hydrogen trains, like the one in Berlin, provide a low-emission alternative to non-electric rail lines. Hydrogen's adaptability is critical, with methods of manufacture ranging from ecologically harmful 'grey' hydrogen to cleaner 'green' hydrogen. Cost is a challenge, particularly for green hydrogen, and major government investment is required to develop facilities (Sky. 2022). The IDTechEx research emphasizes the need to transition to untethered electric trains in response to rising diesel costs and environmental concerns. Advances in Li-ion battery manufacturing outfitted with multi-megawatt-hour batteries (IDTechEx, 2023). California's latest In-Use Locomotive Rule requires that all locomotives manufactured after 2035 that serve the state be zero-emission. The regulation considers overhead wire electrification and hydrogen to be zero-emission technology (Streets blog, 2023). The Indian Railways is ramping up its electrification efforts, to achieve a Net Zero Carbon Emitter by 2030. As of June 30, 2023, 59,096 route km of the Broad-Gauge network has been electrified, with 14 states/UTs having reached 100% electrification (Financial Express, 2024).

Further, the failed attempt to privatize the UK's rail system has resulted in higher public subsidies, the storage of idle electric trains, and a lackluster punctuality record (Casson, 2004). A study of (Xiong et al., 2012) examined improvement way for the electric system's energy efficiency in contemporary high-speed rails using a section of the Hefei-Wuhan railway as a case study. The problem of powering the EDSS built isolated and sparsely inhabited plateau railways is the subject of this research (Gangwar et al., 2014). In order to improve continuous operation, it presents the RESPS, a reasonably priced solution with an inventive topology, an industrial design, and a piecewise linear PI control method. This study assessed the transmission losses in the widely used 2 × 25 kV railway electrification system (Riego-Martinez et al., 2020). The use of power electronic devices for improved efficiency and controllability, the integration of renewable energies and on-board or wayside energy storage are some of the new technologies that are emphasized (Arboleya et al., 2020). The significance of protective relay quality for secure railway electrified operations is examined in this study. Relay dynamic performance is precisely and effectively tested with a dynamic simulation test platform that uses a RTDS (Shen et al., 2021).

In (Baron, 2021), the study highlighted Europe's sluggish and uneven progress toward unrestricted, free terminal access and points out that certain terminal management companies are pursuing entrepreneurial ventures and diversifying into retail and real estate. In (Zhu et al., 2023), authors evaluated scenarios, constraint circumstances and optimization algorithms, offering the railway industry a resource to improve development that is ecologically friendly, efficient, and of high quality. In order to improve the manual railway designs now in use in metropolitan areas, (Gao et al., 2023) study presented the FORA framework. When compared to the manual route, the enhanced dynamic programming technique significantly reduces the amount of demolition material, carbon dioxide emissions and overall construction expenses. An improved open-loop based stability analysis technique for vehicle-grid systems with multiple vehicles accessed is presented in (Meng et al., 2023). Further, (Rahman et al., 2023) presented a mobile robotic sensing platform-based health evaluation technique for railway track maintenance. With programs like RTS Phase-II, India is aggressively encouraging renewable energy, especially solar plants. Despite obstacles, RTS proliferation can be enhanced by working with stakeholders, appropriate business models and supportive regulations, allowing India to lower emissions and contribute to global climate goals (Kumar et al., 2023). Further, the model, which focused on offshore wind generation, eliminates 4.19 MtCO2-eq of emissions and provides 60% of total consumption from renewable sources by reducing the predicted energy demand by 50% (Gil-Garcia et al., 2024).

(iii). Control and Management of Railway Electrical Infrastructure

A synthetic methodology for the capacity and usage analysis of intricately interconnected train networks is presented in (Francesco et al., 2016). It seeks to offer a thorough study of suburban rail networks as well as a sound methodology for wide network feasibility or strategic research. In continued study, power flow is managed by the MGSC to maximize efficiency and minimize costs. Real data and simulation presented effective the suggested control architecture (Gabbar et al.. 2017). The suggested strategy's efficacy in optimizing energy distribution and boosting overall energy savings in multi-ESS systems is demonstrated by simulation and experimental findings (Zhu et al., 2020). Through storage configuration and scheduling tactics, the results show enhanced PV permeability, decreased carbon dioxide emissions, and increased system flexibility using the Lanzhou-Xinjiang Railway (Ning et al., 2021). A train station equipped with solar and wind power as well as an energy storage system is part of the design. SCs are charged at stations using a pantograph and are utilized for energy recovery during regenerative braking (Cabrane & Lee, 2022). Further, study analyzed topologies for high-power AC-DC converters and on-board power electronic traction transformers, providing a thorough assessment of arrangements for MVDC railway electrification systems (Sharifi et al., 2022).

In order to precisely estimate net provided power in electric railway system substations under prolonged feeding conditions, a method is proposed in (Cho et al., 2022) study. In (Kamel et al., 2022), a sSOP that connects the local low voltage distribution grid to train electrification is introduced. The sSOP improves braking efficiency and makes it easier to employ renewable energy sources by including an energy storage system and a rail and grid energy management system. This study of (Foroutan et al., 2023) used NDEA to assess the effectiveness of 19 Iranian railway stations. In contrast to conventional models, the output-oriented SBM model with VRS detects efficient and inefficient stations and provides insights for performance improvement. Renewable energy sources and electric buses are integrated in an inventive energy management concept for the Metrobus charging station in Istanbul. The technology shown potential profitability by reducing carbon emissions significantly through grid and solar energy optimization (Saray et al., 2024). In the event of overhead line failures, (Piraino et al., 2023) looked into fuel cell and battery-powered powertrains for a conventional electric train operating on the Reggio Calabria - Catanzaro Lido track.

MATHEMATICAL FORMULATION

A circuit-based simulation model of a PV cell such that variations in temperature and irradiance can also be used to approximate the electrical behaviour of the photovoltaic cell. The solar energy generation is represented by (1) (Loy-Benitez et al., 2021),

$$P_{SP} = A_{SP} + \eta S_P + f + IR + \eta p \tag{1}$$

where, $PS_{P:}$ power generated by a PV unit, $AS_{P:}$ individual PV unit area, $\eta SP_{:}$ power conditioning efficiency, f: derating factor, IR: solar radiation and ηp: $_m$odule reference efficiency.

Powered repositories are a one kind of hydroelectric energy storage is hydro power. It is a setup where two water reservoirs at varying heights work together to provide power when water flows through a turbine and descends from one to the other. Power is also needed for the system to pump water back into the top reservoir. Because it has the ability to store energy and release it when needed, it functions similarly to a energy storage battery. The pumped hydro storage energy is represented by (2) (Loy-Benitez et al., 2021),

$$W_S = \frac{m_S \times v \times (\mathrm{P}_{out} - \mathrm{P}_{in})}{\eta_S} \tag{2}$$

where, W_S: pump's required mechanical power, m_S: mass flow rate of the fluid, $\eta S_{:}$ isentropic efficiency of the pump and v: specific volume of the fluid.

The energy produced from the PV modules exceeds the demand from the railways. The electrolyzer having two electrodes can effectively create H_2 from RES. The equation for the electrolyser is represented by (3) (Loy-Benitez et al., 2021),

$$f_H = \eta_{FE} \frac{i_C A_C}{zF} \tag{3}$$

where, f_H: H_2 production rate, z: number of moles of electrons transferred in the reaction, F: Faraday constant (9.6485×10^4 C.mol^{-1}), i_C: current density, A_C: effective cell area and $\eta F_{E:}$ Faraday efficiency.

A converter that steps up or steps down DC voltage from one side of the converter to the other when powered by an associated controller and gate-signal generator is represented by the bidirectional DC-DC converter block. The bidirectional flow of energy is represented by the equation (4) - (5) (del Puerto-Flores et al., 2011),

$$-Lz = \mu z1 + \mathrm{r}z - vA_C \tag{4}$$

$$Cz_1 = \mu z - iD_C \tag{5}$$

where, x: average inductor current, z_1: average capacitor voltage, L: inductance, C: capacitance, v_{AC}: voltage of the ac-line source, r: parasitic inductor resistance and i_{DC}: current of the dc-port.

The average net present cost of electricity generation for a PV over the course of its lifetime is measured by the LCOE. It is utilized for reliable comparisons between various electricity producing techniques and for the planning of investments. The LCOE is represented by the equation (6) (Loy-Benitez et al., 2021),

$$LCOE = \frac{TAC}{E_T} \tag{6}$$

where, E_T: the electricity generated per year, TAC: contains the economic constants related to capital costs of investment, lifetime, replacement cost, maintenance and operation costs of each component.

The study of capital and operational costs in the production of hydrogen, varies depending on the kind of hydrogen. The costs associated with transporting and storing hydrogen are not included in LCOH. It can mathematically expressed by (12) (Global Energy Perspectives, 2023),

$$LCOH = \frac{\sum_{i=1}^{n} \frac{TC_n}{(1+i)^n}}{\sum_{i=1}^{n} \frac{H_n}{(1+i)^n}} \tag{7}$$

where, TC_n: total dispenses of the project in year n and H_n: generated H_2 for the same duration.

According to the net present cost rule, investments should only be made in projects that show a positive net present cost. Furthermore, it is not advisable to proceed with any project or investment that has a negative net present value. The equation (8) shows the divisible method for getting a solution to this (Grubbström, 2022),

$$NPC_x = \sum_{a=1}^{x} \frac{\Delta T}{T} e^{-ip\Delta T} = \frac{\Delta T}{T} \cdot \frac{1-e^{-pT}}{e^{p\Delta T}-1} \tag{8}$$

where, T: Time period and NPC_x: net present cost.

The percentage of annual profit before interest and tax, or annual profit before interest and tax during the operation period, to the project's total investment is known as return of interest. The return of interest is determined by the equation (9) (Li et al., 2021),

$$ROI = \frac{P_R}{N} \tag{9}$$

where, *ROI*: rate of interest, *N*: the total investment of project and P_R: the annual profit before interest and tax of project.

The predicted yearly rate of increase for an investment is known as IRR. The same formula used to compute net present value is used to calculate IRR; however, NPV is set to zero and the IRR is calculated using the equation (10) (Prol & Steininger, 2020),

$$-\delta d \cdot \frac{K_{\delta d} \cdot (1 - K_{\delta d^N})}{1 - K_{\delta d}} \tag{10}$$

where, $K_{\delta d:}$ the discount factor of the constant part of the backup charge.

Continuous CO_2 emissions from diesel generators throughout their useful lives are their main environmental impact. The two biggest sources of emissions worldwide are the generation of heat and electricity. This is represented by equation (11) (Ghimire et al., 2024),

$$E_{CO_2} = \frac{Molecular\ \ weight\ of\ CO_2}{Molecular\ weight\ of\ carbon} \tag{11}$$

where, E_{CO_2} : carbon dioxide emissions per year.

The percentage of power produced by a specific resource is referred to as its penetration. It can be used to describe either the amount consumed or the percentage of total electricity generated. The percentage of power produced by a specific resource is referred to as its penetration. It can be used to describe either the amount consumed or the percentage of total electricity generated. The renewable penetration is estimated by (12) (Grubbström, 2022),

$$R_P = (R_{EG}/T_{ED}) \times 100 \tag{12}$$

where, R_P: renewable penetration, R_{EG}: renewable energy generation and T_{ED}: total energy demand.

Study Approach

Current System

The electric needs of Rameswarm nagar colony,basargar,hatia, 8855+QGP, Hatia Railway Colony Type II, Maharatoli, Ranchi, Jharkhand 834003, India are met with 2.3 kW of generator capacity. Our operating costs for energy are currently \$20,898 per year.

Proposed System

We propose adding 1,500 kW of PV and 254 kWh of battery capacity. This would reduce your operating costs to \$13,384/yr. Your investment has a payback of N/A years and an IRR of N/A%.

Integrated Modeling of Green Hydrogen and Railway Platform

The green energy consisting GEN/PV/battery hybrid systems can be presented by a block diagram, which shows the components and their interconnections. This diagram provides a visual overview of how the generator, electrolyzer, solar panels and PH storage work together to generate and store green energy in a self-sufficient manner. The electric load been installed is for the community purpose and the battery is a pumped hydro battery and the converter is a system converter. The system modeling is illustrated in Figure 2 and flowchart for RE based platform electrification and green hydrogen production is presented in Figure 3.

Potential Assessment of Energy Harvesting

The Hatia's geographical location situates it in the midst of an area replete with substantial solar irradiance as given in Table 2. The maps provided by NASA of a visual representation of the area's geography and topography and help orient viewers to the region's physical features and landmarks. The location it shows is of Hatia railway station platform. The location has a co-ordinate of 23° 18′ 34″ N, 85° 18′ 31″ E. The consistent and abundant sunlight that graces the region throughout the year lays a robust groundwork for the harnessing of solar energy through photovoltaic systems. This unique solar potential renders Hatia a coveted hub for efficient and productive solar power generation. Against the backdrop of Jharkhand's energy landscape, Hatia's proactive embrace of renewable energy sources emerges as a transformative contributor. By orchestrating green energy initiatives, Hatia has the potential to significantly diminish the state's dependency on conventional fossil fuels. This proactive shift not only aligns with global trends but also ushers in reduced greenhouse gas emissions and mitigated environmental footprints.

The characteristics of solar GHI resource analysis (presented in Figure 4), which shows the amount of solar radiation received by a surface horizontal to the ground, is important for assessing the solar energy potential of a site and designing the appropriate integrated system.

Figure 2. Schematic of the integrated green railway platform and hydrogen production

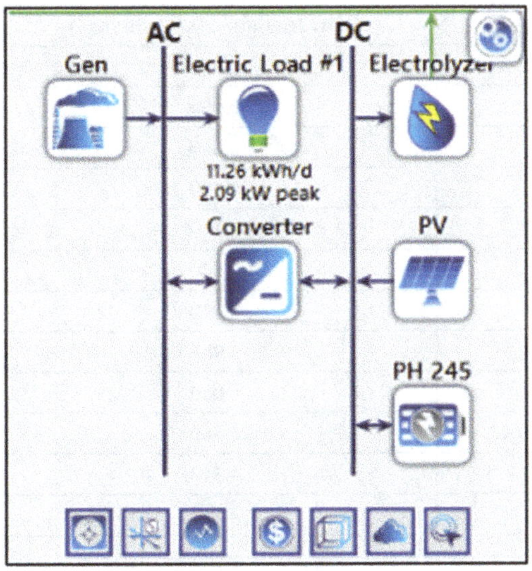

Figure 3. Flowchart for RE based platform electrification and green hydrogen production

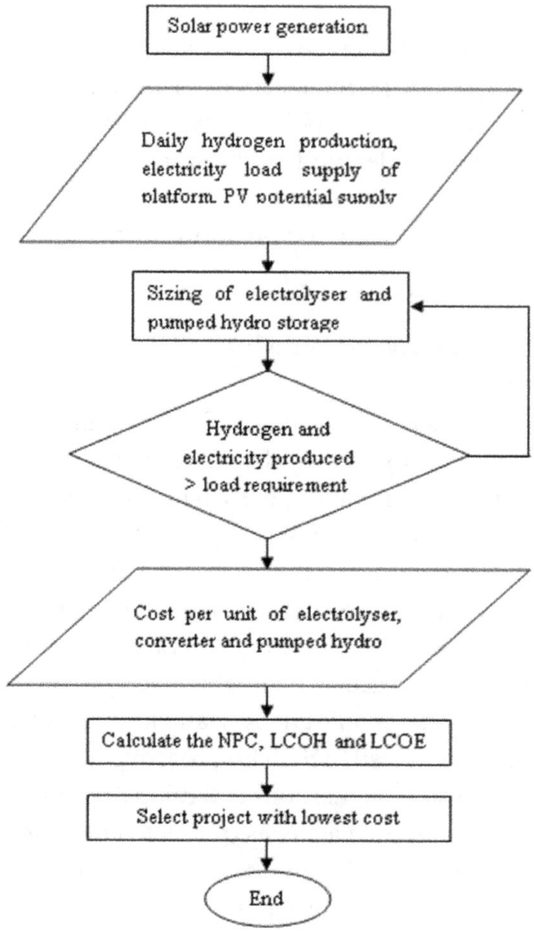

Table 2. Meteorological resources for sustainable platform and green hydrogen production

Month	Clearness Index	Daily Radiation (kWh/m²/day)	Daily Temperature (°C)
Jan	0.616	4.300	15.930
Feb	0.6275	5.050	19.710
Mar	0.600	5.650	25.260
Apr	0.590	6.190	30.320
May	0.557	6.140	32.280
Jun	0.434.	4.840	29.830
Jul	0.373	4.120	26.510
Aug	0.376	3.980	25.850
Sep	0.420	4.060	25.060
Oct	0.544	4.560	22.640
Nov	0.605	4.340	18.950
Dec	0.624	4.130	15.870

Figure 4. Meteorological characteristics for green energy powered railway platform

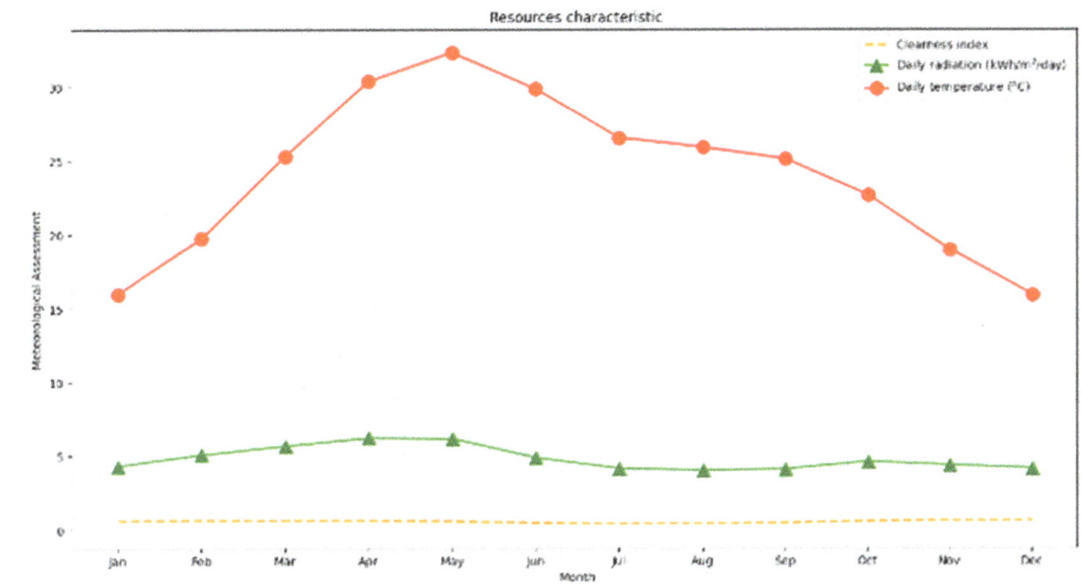

For the clearness index, the peak month is February with index of 0.627 and lowest month is September with 0.418 and the daily radiation data is measured in kWh/m²/day having peak month for the month of April with a value of 6.190 and the lowest month for the month of August having a value of 3.980. The performance and efficiency of renewable energy systems that is sensitive to temperature changes, such as solar panels and batteries. The site has the maximum temperature in the month of May with a value of 32.280 and the lowest temperature downfall in the month of December having daily temperature value of 15.870 respectively given in Table 3.

RESULTS AND DISCUSSION

Different countries have put in place a several technology and agreements to preserve environmental sustainability. Following section briefs power consumption and generation:

(i). Energy Consumption: Electrical energy consumption summary play a crucial role in understanding and managing energy consumption pattern. This information is invaluable for implementing energy-saving measures and optimizing energy usage. In Table 3-4, data presents the proposed system electrical summary including the production, consumption by the load, excess electricity achieved, renewable fraction rates. The electricity produced by the generic flat PV is 2,266,982 kWh/year and that by the autosize genset is 0.575kWh/year. The total production 2,266,983 kWh/year and the consumption rate is 408,140 kWh/year. The excess electricity produces by the system is 1,858,173 kWh/year. Figure 5 presents the total electrical production throughout the year.

(ii). Optimization of Green Energy Generation: Renewable penetration refers to the percentage of total energy demand that is met by renewable sources such as solar, electolyzer, generator. The renewable penetration metric is important for assessing the progress and effectiveness of renewable energy integration. A higher renewable penetration percentage signifies a greater reliance on renewable energy

Table 3. Summary of energy production for the proposed railway platform model

Production	kWh/Year (%)
Generic flat plate PV	2,266,982 (100%)
Autosize Genset	0.575 (0.0000254%)
Total	2,266,983(100%)

Table 4. Consumption overview of energy for the proposed railway platform model

Consumption	kWh/Year (%)	Consumption	kWh/Year (%)
AC primary load	4,109 (1.01%)	Unmet electric load	0 (0%)
DC primary load	0 (0%)	Capacity storage	0 (0%)
Deferrable load	0 (0%)	Renewable fraction	100 (%)
Excess electricity	1,858,173 (82.0%)	-	-
Total	408,140 (100%)	Maximum renewable penetration	2,902 (%)

sources and a reduced dependence on fossil fuels or non-renewable energy sources. Table 5 and Figure 5 present the renewable penetration matrices.

(a). Solar Power Generation: In the present study, the gereric flat plate PV is used for improve efficiency, the systems contains anti-reflective coatings and tracking devices that optimize solar absorption. In terms of repair, generic flat plate PV systems are noted for their resilience and extended lifespan, requiring minimal upkeep. The rated capacity is 1500kW. The capital cost of PV panels refers to the initial investment required to purchase and install the panels, which in this case is $2250. This cost covers the equipment and installation expenses associated with setting up the PV system.The replacement cost of the PV panels is also $2250, indicating the cost of replacing the panels after a certain period. The O&M cost of the PV panels is $10 per kW/year and the derating factor is 80%.

PV panel performance refers to the percentage of total energy produces by the PV panel. The PV output is pivotal in evaluating stored energy effectiveness. The PV output for assessing the progress and

Table 5. Renewable energy penetration and generation for the proposed railway platform model

Parameters	Value
Nominal renewable capacity divided by total nominal capacity (%)	99.8
Usable renewable capacity divideed by total capacity (%)	99.8
Total renewable production divided by load (%)	555
Total renewable production divided by generation (%)	100
One minus total nonrenewable production divided by load (%)	100
Renewable output divided by load (%)	2,902
Renewable output divided by total generation (%)	100
One minus nin renewable output divided by total load (%)	100

Figure 5. Characteristics for renewable energy penetration for the railway platform model

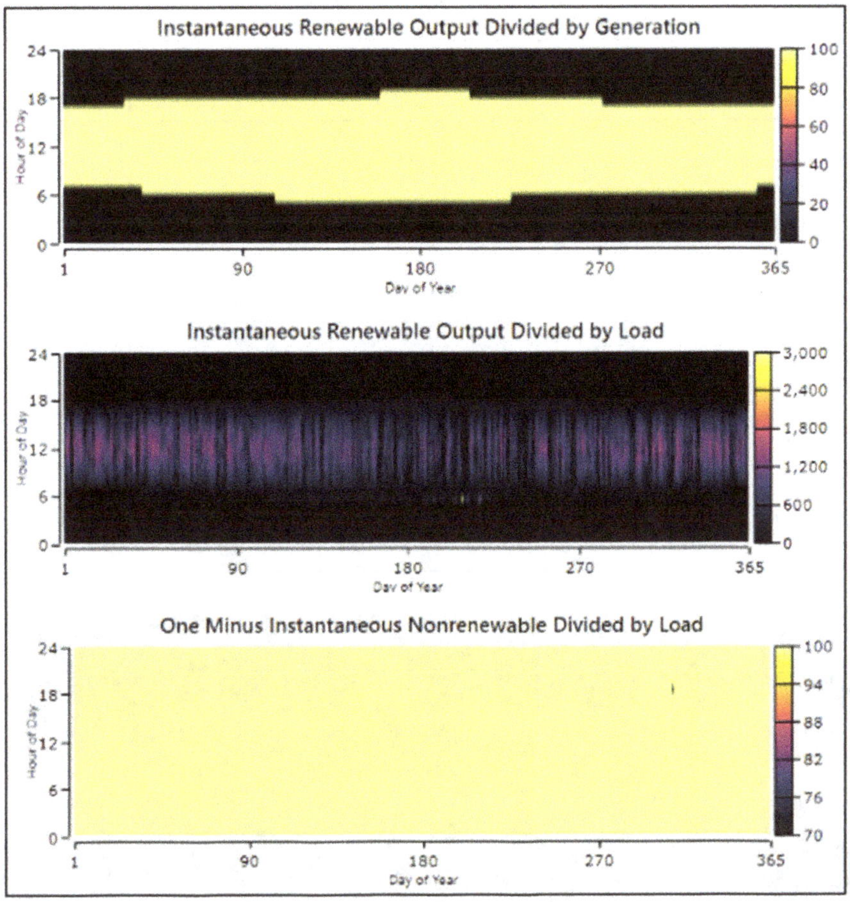

effectiveness of stored energy details. Table 6 presents, the PV details of total production, maximum output, capacity factor, respectively; while in Figure 6 presents the PV performance matrix.

(c). Power Converter Performance: The generic system converter is essential for optimizing functionality and providing compatibility among multiple subsystems. Its capacity to convert and adapt to individual requirements emphasizes its importance as a vital building element in the design and operation of multidimensional systems that span several disciplines. The converter capital costs \$300 to purchase and install, with a \$300 replacement cost. Its annual O&M cost is \$0.0 per kW, which covers all continuing expenditures during its lifespan. The converter has a lifetime of 15 years, after which it is assumed that it will need to be replaced due to wear and tear or obsolescence. The converter has an efficiency of 95%, meaning that it is able to convert 95% of the input energy into the desired output, with only 5% being lost as waste heat. In the proposed study, the generic converter capacity 2.23kW. Table 7 presents converter capacity matrices and hours of operation details respectively. Figure 7 presents the inverter and rectifier characteristics.

(d). Pumped Hydro Storage: The generic pumped hydro storage system is a widely adopted approach for large-scale energy storage, its scalability in cost, extended cycle life and quick reaction to changes in energy demand. The proposed system includes a 245 kWh pumped hydro storage unit capable of holding

Table 6. Description of PV module for sustainable platform and green hydrogen production

Operational parameters	Results
Rated power capacity (kW)	1,500
Mean power output (kW)	259
Mean power output (kWh/day)	6,211
Capacity factor (%)	17.3
Total energy production (kWh/year)	2,266,982
Minimum power output (kW)	0
Maximum power output (kW)	1,492
PV penetration (%)	55,170
Hours of operation (hours/year)	4,378
Levelized cost of energy ($/kWh)	0.0000900

Figure 6. Power output characteristics of solar PV module for sustainable platform

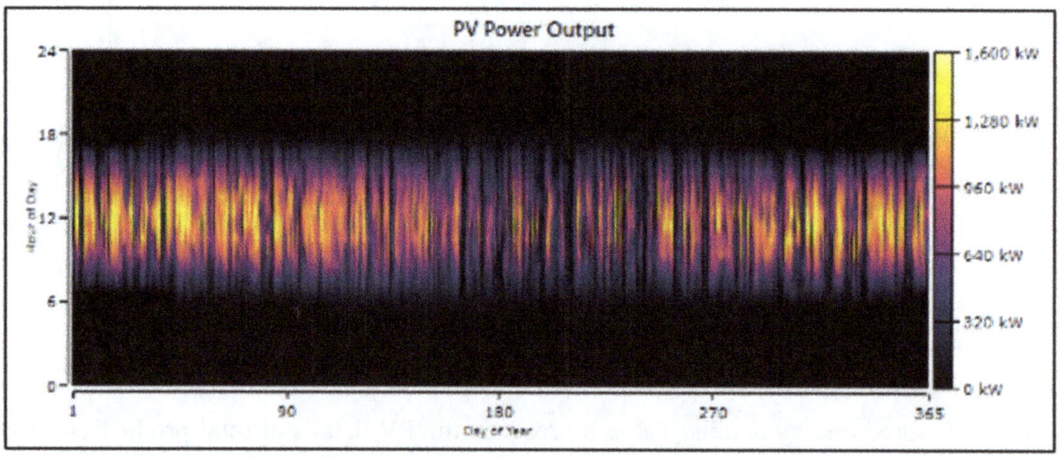

Table 7. Power conversion mechanism for sustainable platform and green hydrogen production

Quantity	Inverter	Rectifier
Rated power capacity (kW)	2.23	2.23
Mean power output (kW)	0.469	0
Minimum power putput (kW)	0.0207	0
Maximum power output (kW)	2.03	0
Capacity factor (%)	21.0	0
Hours of operation (hour/year)	8,760	0
Energy out (kWh/year)	4,109	0
Energy in (kWh/year)	4,325	0
Power losses (kWh/year)	216	0

Figure 7. Power converter operating characteristics for sustainable platform

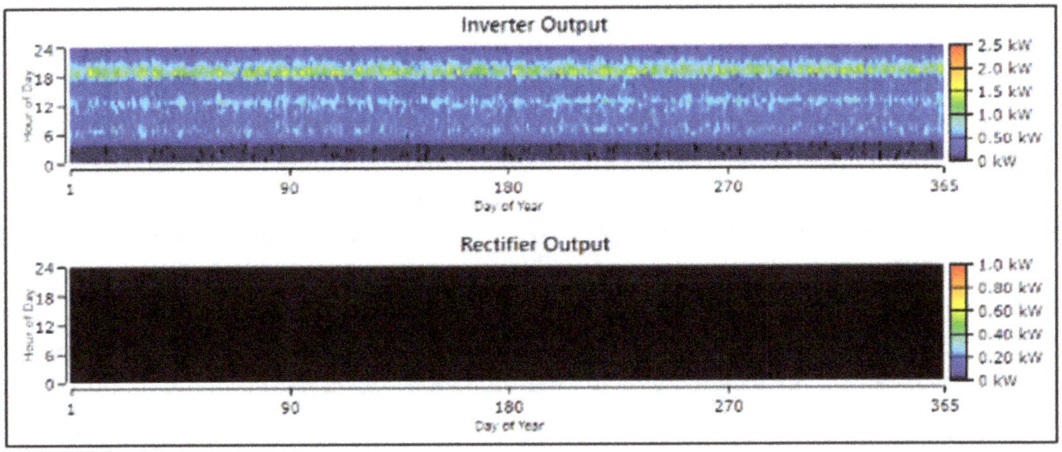

1000 m³ of water. Figure 8 shows the storage details, with a nominal voltage of 240V. The system may charge and discharge at a maximum current of 91.6 A. The storage has a capital cost, replacement cost, O&M charges of 22,000$, 500$ 2000$ respectively and with 7 years predicted lifespan. This design suggests the pumped hydro storage system's specification and economical characteristics, emphasizing its suitability for large-scale energy management.

Figure 8. Operating characteristics of pumped hydro energy storage for sustainable platform

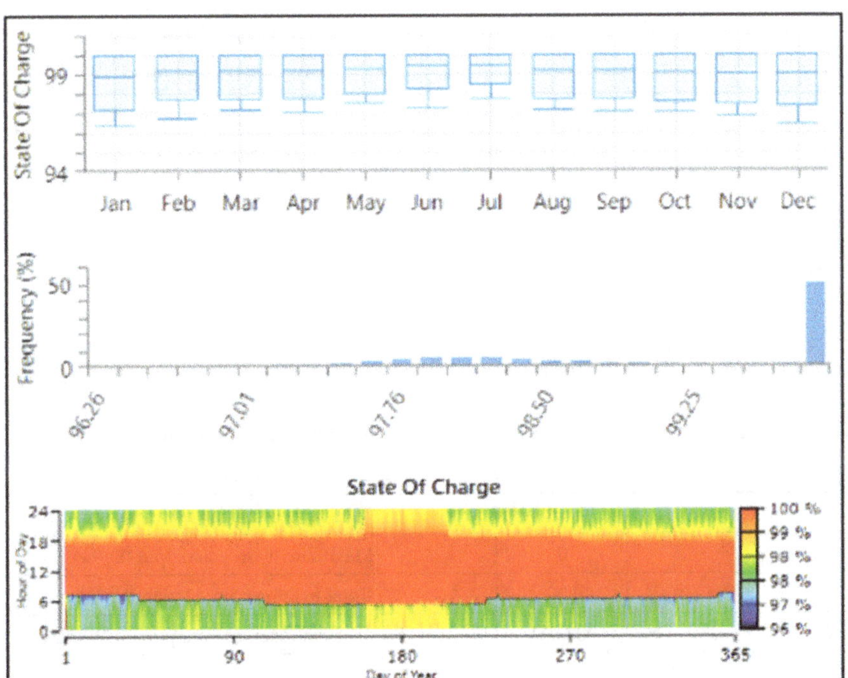

Pumping water from the bottom to the top of the reservoir using excess power produced during periods of low demand, utilizing electricity produced from green sources (mainly intermittent ones), hydroelectric power plant lakes or to use water electrolysis to produce gaseous H_2. Storage refers to the percentage of total energy demand that is stored and the renewable sources such as solar, electolyzer, generator. Storage summary for assessing the progress and effectiveness of stored energy details. Table 8 and Figure 8 illustrate PHS details and state of charge characteristics, respectively.

(e). Hydrogen Electrolyzer: The generic electrolyzer, noted for its versatility and effectiveness in enabling electrolysis operations, has an impressive 85% efficiency in turning electrical energy into chemical energy. This unique characteristic enhances to its attractiveness, making it a cost-effective and ecologically beneficial alternative. Whether used for renewable energy storage or industrial operations, the electrolyzer's modular design and high efficiency contribute greatly to its versatility and sustainability, further advancing clean energy technology and environmental conservation. In the proposed study, the capital and replacement cost of the system is 4,226$, 4,226$, 0.65$ respectively. The output details of the hydrogen electrolyzer are shown in Table 9 and portrayed in Figure 9, respectively. The lifespan of the system is 15 years and the efficiency of the system is 85%.

This graph displays the input power to the electrolyzer over a specific time period. This metric represents the highest input power observed during a specific period. The capacity factor is the ratio of actual output over the rated capacity of the electrolyzer. It indicates how effectively the electrolyzer is being utilized. A high-capacity factor indicates efficient operation, while a low factor suggests underutilization. In Table 9 present the rated power capacity and mean power output, respectively.

Table 8. Energy storage with pumped hydro integrated system for sustainable platform

Operational Parameters	Results	Operational Parameters	Results
Battery (quantity)	1.00	Lifetime throughput (kWh)	15,221
String size	1.00	Expected life (year)	7
Strings in parallel (strings)	1.00	Average energy cost ($/kWh)	0
Bus voltage (V)	240	Energy in (kWh/year)	2,410
Autonomy (hour)	542	Energy out (kWh/year)	1,957
Storage wear cost ($/kWh)	0	Storage depletion (kWh/year)	5.51
Nominal capacity (kWh)	254	Losses (kWh/year)	458
Usable nominal capacity (kWh)	254	Annual throughput (kWh/year)	2,174

Table 9. Parameters of hydrogen electrolyzer for sustainable platform and green hydrogen

Operational Parameters	Results	Operational Parameters	Results
Rated power capacity (kW)	100	Mean hydrogen output (kg/hr)	0.994
Mean power input (kW)	46.1	Minimum hydrogen output (kg/hr)	0
Minimum input power (kW)	0	Maximum hydrogen output (kg/hr)	2.15
Maximum input power (kW)	100	Total hydrogen production (kg/hr)	8,707
Total input energy (kWh/year)	404,031	Specific hydrogen consumption (kWh/kg)	46.4
Capacity factor (%)	46.1	-	

Figure 9. Characteristics of electrolyzer for sustainable platform and green hydrogen

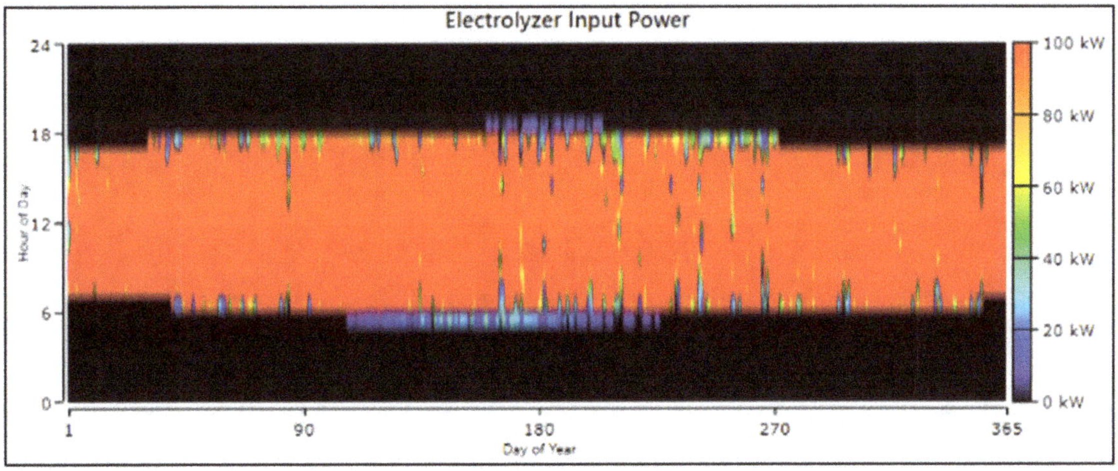

(e). DC Operating Performance: The DC operating capacity daily profile depicts a visual representation of the fluctuation in operating capacity for a specific device or system across the span of a day. Typically applied within the domain of renewable energy systems or related devices, this profile specifically showcases the DC capacity of a component, such as a solar panel, battery, or electrolyzer, as it evolves throughout various periods within the day. The trajectory or line depicted on the graph visually conveys the dynamic alteration in the DC operating capacity of the targeted device or system over the course of the day. Figure 10 presents the DC operating capacity daily profile characteristic, where the data shows the DC capacity throughout the year.

(g). AC Operating Performance: The AC required operating capacity daily profile provides a visual representation of how much the operational capacity of a device or system is required for changes over the duration of a day. This profile holds particular significance in the realm of renewable energy systems and associated equipment. It effectively showcases the AC capacity of a specific component, such as a solar inverter, wind turbine, or generator, as it varies across different time segments within the day. Figure 11 presents the AC operating capacity daily profile characteristic where the data shows the AC capacity throughout the year.

(ii). Energy-economic Optimization: The charts comprise of inflation rates, which are important factors to consider when evaluating the financial feasibility and profitability of energy projects and investments. In Tables 10-11, the rate chart is represented where the discount, inflation, annual capacity shortage, project lifetime rate analysis, respectively.

The summary of costs associated with different renewable energy sources provides an overview of the expenses involved in generating, storing, and distributing energy from sources such as generator, solar panel. Analyzing the cost summaries allows individuals and organizations to make informed decisions about the feasibility and viability of different renewable energy sources. By comparing the expenses associated with various technologies, they can determine which option aligns best with their budget, energy requirements, and long-term goals. Table 12 and Figure 12 shows the capital cost and the replacement cost and O&M. The total cost of the system is $621,940.31 The IRR is 78%, ROI is 75% and the simple payback year of the project is 1.3 years.

Figure 10. DC operating daily profile for sustainable platform and green hydrogen production

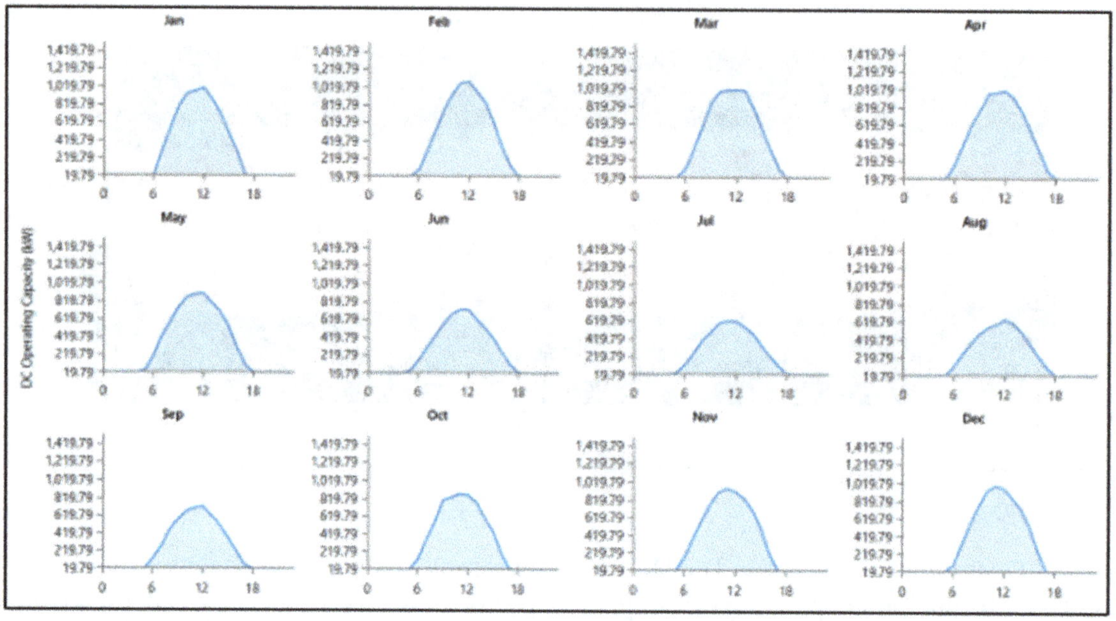

Figure 11. AC operating daily profile for sustainable platform and green hydrogen production

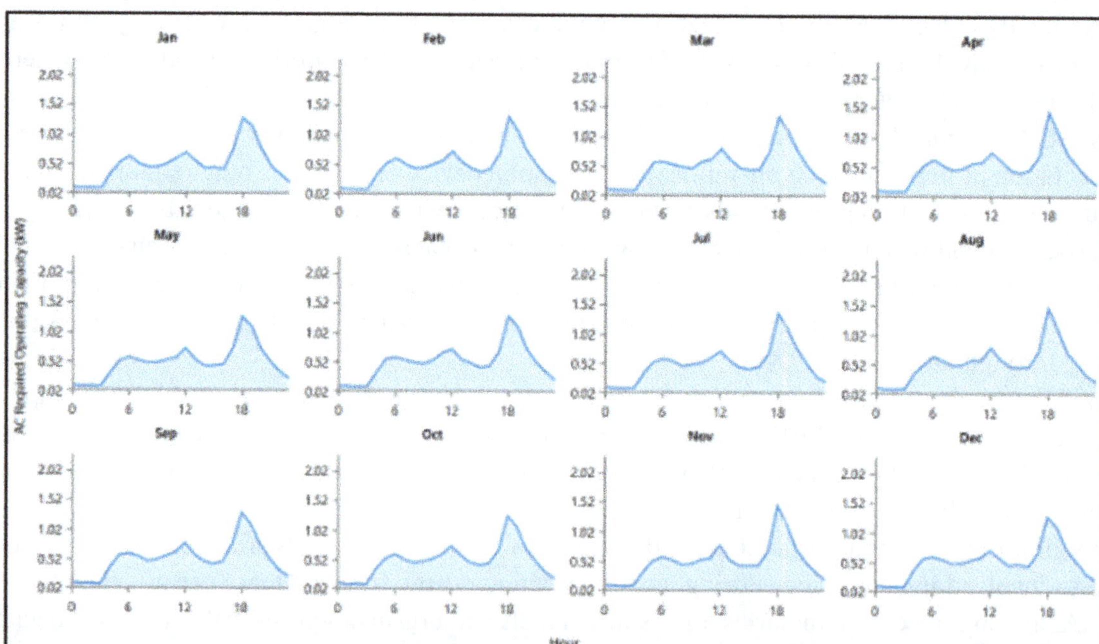

The highest capital and replacement cost is for the system is electrolyzer which capital cost is around $422,600 and replacement cost is $179,298.12.

Table 10. Rate chart details for sustainable platform and green hydrogen production

Financial Aspect	Value
Discount rate (%)	8.00
Inflation rate (%)	2.00
Annual capacity storage (%)	0.00
Project lifetime (years)	25.00

Table 11. Cost summary details for sustainable platform and green hydrogen production

Component	Capital ($)	Replacement ($)	O&M ($)	Fuel ($)	Salvage ($)	Total ($)
Autosize Genset	$1,150.00	$0.00	$0.892	$27.33	$275.03	$903.19
Generic 245kWh Pumped Hydro	$22,000.00	$710.29	$25,855.03	$0.00	$51.33	$48,513.99
Generic Electrolyzer	$422,600.00	$179,298.12	$840.29	$0.00	$33,745.72	$568,992.69
Generic flat plate PV	$2,500.00	$0.00	$129.28	$0.00	$0.00	$2,629.28
System Converter	$670.30	$284.39	$0.00	$0.00	$53.53	$901.17
Total	$448,920.44	$180,292.80	$26,825.49	$27.33	$34,125.61	$621,940.31

Figure 12. Economical optimization for sustainable platform and green hydrogen production

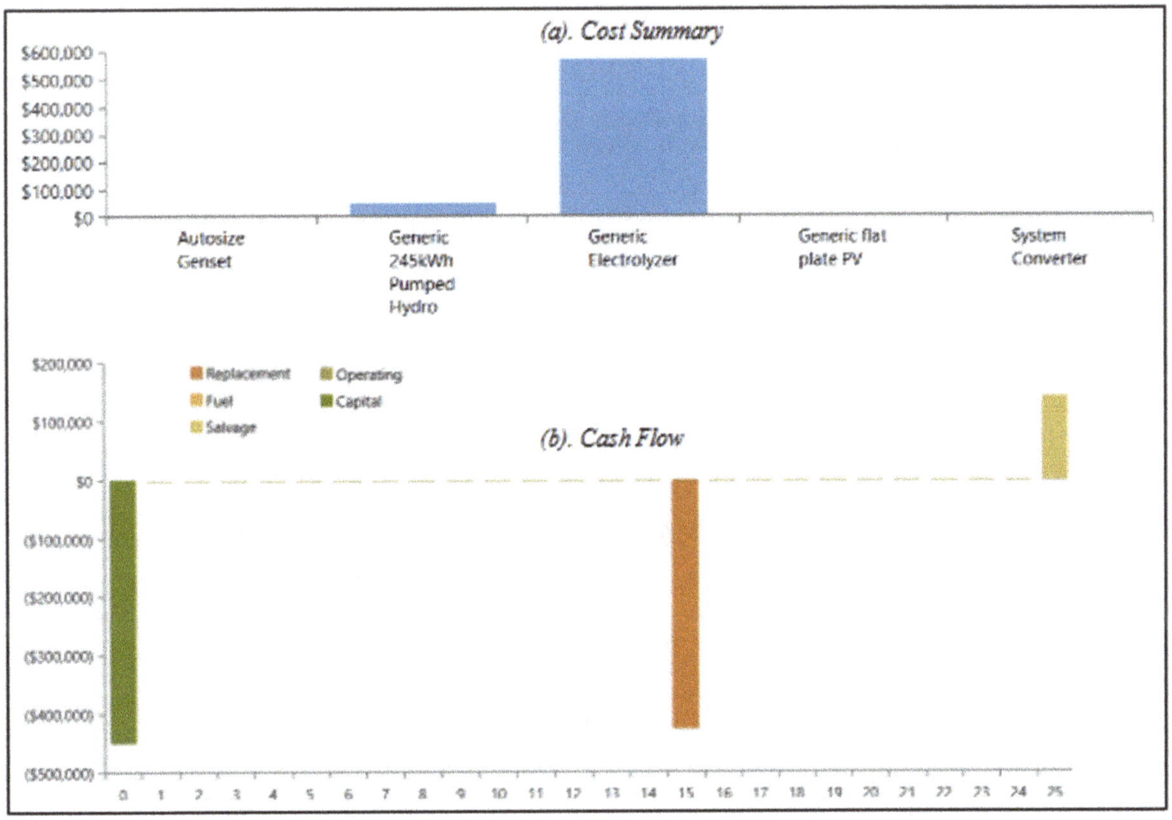

The cash flow statement is a financial statement that provides valuable insights into the sources and uses of cash during a specific period. By analyzing the cash flow statement, the gain in a clear understanding of the sources of cash inflows. One important aspect of analyzing the cash flow statement is assessing the net cash flow. This is the difference between the total cash inflows and the total cash outflows during a given period. A positive net cash flow indicates that the organization generated more cash than it used, while a negative net cash flow suggests that more cash was used than generated.The proposed study system has high capital cost & give simple payback in 1.36 years. Figure 13 cash flow characteristic represents the flow of system costing including capital, replacement, salvage, operating and fuel. The optimization results provide valuable insights into the performance and efficiency of a renewable energy system, enabling system designers and operators to identify opportunities for improvement and maximize the energy output of the system. By analyzing the optimization results, operators can unfold the opportunities to improve in various aspects of the system. This may include identifying components or processes that are not operating at their full potential, detecting inefficiencies or bottlenecks within the system, or recognizing areas where adjustments can be made to increase energy production.

(iii). Environmental Impact Assessment: Emissions in a system are the gases, particles, or pollutants released into the environment as a result of numerous activities. In the proposed study, the emission of CO_2 is around 0.6 kg/year and the other emission substances are represented in Table 13, respectively.

Table 12. Optimization result for sustainable platform and green hydrogen production

Integrated System					PV (KW)	Gen (KW)	PH245	Electrolyzer (KW)	Converter (KW)	Dispatch	NPC ($)	COE ($)	Operating cost ($/year)	Initial Capital ($)	Ren Frac (%)	Total fuel (L/year)
				-	1,500	2.30	1	NR*	2.24	CC	52,950	0.997	2,060	26,322	100	0.264
		-		-	1,500	2.30	NR*	NR*	1.50	CC	143,371	2.70	10,773	4,100	27.6	1,269
-				-	NR*	2.30	1	NR*	0.583	LF	246,484	4.64	17,262	1,150	0.855	1,789
-		-	-	-	NR*	2.30	NR*	NR*	NR*	CC	271,311	5.11	20,898	447,656	0	2,457
	-				1,500	NR*	1	100	1.85	CC	620,833	11.69	13,400	448,920	100	0
					1,500	2.30	1	100	2.23	CC	621,940	11.71	13,384	426,700	100	0.264
		-			1,500	2.30	NR*	100	1.50	CC	712,364	13.41	22,097	446,700	27.6	1,269
-					NR*	2.30	1	100	2.05	CC	773,709	14.57	25,321	446,366	0	1,457
-		-			NR*	2.30	NR*	100	0.100	CC	840,344	15.82	32,223	423,780	0	2,457

NR*: NR means not reported.

Table 13. Emissions reduction observations for the proposed sustainable platform

Quantity	Value (kg/Year)
Carbon Dioxide	0.692
Carbon Monoxide	0.00436
Unburned Hydrocarbons	0.00190
Particular Matter	0.0000260
Sulfur Dioxide	0.00169
Nitrogen Oxides	0.00410

CONCLUSION

A thorough technical design and economic evaluation framework for a large-scale, electrically powered green hydrogen generation facility is presented in this paper. Many technologies can be used to create hydrogen from a variety of sources. Achieving net zero carbon emissions in the future calls for extensive use of renewable energy sources, particularly solar energy. Additionally, the effect of platform placement and the location of renewable energy harvesting on the cost of electricity demand has been investigated. The PV panel, battery, inverter, and electrolyzer energy system shows encouraging sustainability. It produces 2,266,982 kWh of energy annually with very little loss, making it an environmentally beneficial and efficient power source. The system's dependability is enhanced by the battery, which can run for 542 hours with losses of just 458 kWh annually, and the inverter, which can run for up to 2.03 kW with losses of 216 kWh annually. The electrolyzer supports renewable energy applications by producing 8,707 kg annually. Even with a $58,640.42 overall cost, the low release of toxic gases—specifically, 0.692 kg/year of carbon dioxide - demonstrates the advantages for the environment. This integrated strategy strikes a good mix between environmental stewardship, cost-effectiveness and energy efficiency, making it an excellent option for sustainable.

REFERENCES

Arboleya, P., Mayet, C., Mohamed, B., Aguado, J. A., & de La Torre, S. (2020). A review of railway feeding infrastructures: Mathematical models for planning and operation. *eTransportation*, *5*, 100063. doi:10.1016/j.etran.2020.100063

Baron, N., (2021). *Railway terminal regulation*. Academic Press.

Bouraima, M. B., Alimo, P. K., Agyeman, S., Stuno, P. D., Lartey-Young, G., Ehebrecht, D., & Qiu, Y. (2023). Africa's railway renaissance and sustainability: Current knowledge, challenges, and prospects. *Journal of Transport Geography*, *106*, 103487. doi:10.1016/j.jtrangeo.2022.103487

Cabrane, Z., & Lee, S. H. (2022). Control and Management of Railway System Connected to Microgrid Stations. *IEEE Access : Practical Innovations, Open Solutions*, *10*, 40445–40455. doi:10.1109/ACCESS.2022.3165832

Casson, M. (2004). The future of the UK railway system: Michael Brooke's vision. *International Business Review*, *13*(2), 181–214. doi:10.1016/j.ibusrev.2003.10.002

Cho, G. J., Lee, C., & Kim, J. (2022). Equivalent load estimation method for determining railway power requirements under extended feeding conditions using a simplex algorithm. *International Journal of Electrical Power & Energy Systems*, *137*, 107770. doi:10.1016/j.ijepes.2021.107770

del Puerto-Flores, D., & Scherpen, J. M. (2011). Power-Based Controller for a Bidirectional AC-DC Power Converter. *IFAC Proceedings Volumes, 44*(1), 9751-9756.

Economic Times. (2023). *Indian Railways set to float tender for 35 hydrogen trains*. https://economictimes.indiatimes.com/industry/transportation/railways/indian- railways-could-float-tender-for-35-hydrogen-trains/articleshow/98226195.cms

Energy World, E. T. (2022). *Is India ready for the switch to Green Hydrogen in passenger trains?* https://energy.economictimes.indiatimes.com/news/renewable/is-india-ready-for-the-switch-to-green-hydrogen-in-passenger-trains/96535240

ET EnergyWorld. (2023). *India's first Hydrogen train: GreenH Electrolysis, Medha Servo to develop production facility in Haryana.* https://energy.economictimes.indiatimes.com/news/renewable/indias-first-hydrogen-train-greenh-electrolysis-medha-servo-to-develop-production-facility-in-haryana/105897700

Financial Express. (2024). *Budget 2024 1 Indian Railways' green revolution: As India moves towards 100% railway electrification, let's take a look at progress so far.* https://www.financialexpress.com/budget/budget-2024-1-indian-railways-green-revolution-as- india-moves-towards-100-railway-electrification-lets-take-a-look-at-progress-so-far- 3372124/

Font, A., Tremper, A. H., Lin, C., Priestman, M., Marsh, D., Woods, M., Heal, M. R., & Green, D. C. (2020). Air quality in enclosed railway stations: Quantifying the impact of diesel trains through deployment of multi-site measurement and random forest modelling. *Environmental Pollution, 262*, 114284. doi:10.1016/j.envpol.2020.114284 PMID:32443191

Foroutan, S. L. F., & Bamdad, S. (2023). Efficiency measurement of railway passenger stations through network data envelopment analysis. *Research in Transportation Business & Management, 466*, 100767. doi:10.1016/j.rtbm.2021.100767

Francesco, R., Gabriele, M., & Stefano, R. (2016). Complex railway systems: Capacity and utilisation of interconnected networks. *European Transport Research Review, 8*(4), 1–21. doi:10.1007/s12544-016-0216-6

Fruhwirt, D., Sturm, P., Nöst, T., Leonhardt, P., Bode, G., Michael, S., & Rodler, J. (2023). PM emissions from railways Results of tests on a wheel-rail test bench. *Transportation Research Part D, Transport and Environment, 122*, 103858. doi:10.1016/j.trd.2023.103858

Gabbar, H. A., Othman, A. M., & Singh, K. (2017). Control architecture of resilient interconnected microgrids (RIMGs) for rauway infrastructures. *Energy Procedia, 141*, 594–603. doi:10.1016/j.egypro.2017.11.079

Gangwar, M., & Sharma, S. M. (2014). Evaluating choice of traction option for a sustainable Indian Railways. *Transportation Research Part D, Transport and Environment, 33*, 135–145. doi:10.1016/j.trd.2014.08.025

Gao, Y., Zhang, T., Wang, Q., Yang, S., Zhuo, J., Tang, J., Zhu, Y., Wang, P., & He, Q. (2023). Low-construction-emissions Optimization for Railway Alignment Fine- grained Designs within Densely Populated Urban Areas. *Construction & Building Materials, 366*, 130179. doi:10.1016/j.conbuildmat.2022.130179

Ghimire, R., Niroula, S., Pandey, B., Subedi, A., & Thapa, B. S. (2024). Techno-economic assessment of fuel cell-based power backup system as an alternative to diesel generators in Nepal: A case study for hospital applications. *International Journal of Hydrogen Energy, 56*, 289–301. doi:10.1016/j.ijhydene.2023.12.174

Global Energy Perspectives. (2023). *Outlook for Green and Blue Hydrogen Market.* https://www.gep.com/blog/strategy/Green-and-blue-hydrogen-current-levelized-cost-of-production-and-outlook

Grubbström, R. W. (2022). On the net present value distribution affected by a random walk process. *International Journal of Production Economics, 250*, 108686. doi:10.1016/j.ijpe.2022.108686

Gurri, S., Santacaterina, E., Guarrera, M., & Dalla Chiara, B. (2023). Driving modal shift on low-traffic railway lines through technological innovation: A case study in Piedmont (Italy) including hydrogen fuel-cells as an alternative. *Transportation Research Procedia, 69*, 99–106. doi:10.1016/j.trpro.2023.02.150

IDTechEx. (2023). *Battery Electric & Hydrogen Fuel Cell Trains 2023-2043.* https://www.idtechex.com/en/research-report/battery-electric-and-hydrogen-fuel-cell-trains- 2023-2043/930

India Briefing. (2024). *How India Plans to Achieve Cost-Effective Green Hydrogen Production.* https://www.india-briefing.com/news/india-strives-for-cost-effective-green-hydrogen-production-30876.html/

International Energy Agency. (2023) *Rail* https://www.iea.org/energy-system/transport/rail

Japan Times. (2023). *Just how green are Japan's trains?* https://www.japantimes.co.jp/news/2023/07/02/national/japan-trains-cars-carbon-emissions/

Kaleybar, H. J., Brenna, M., Foiadelli, F., & Zaninelli, D. (2023). Electric Railway System Integrated with EV Charging Station Realizing Train-to-Vehicle Technology. *IEEE Transactions on Vehicular Technology.*

Kamel, T., Amor, P., Polater, N., Zhang, Y., Tian, Z., Hillmansen, S., & Tricoli, P. (2023). Development of a smart hybrid drive system with advanced logistics for railway applications. *International Journal of Hydrogen Energy.*

Kamel, T., Tian, Z., Zangiabadi, M., Wade, R., Pickert, V., & Tricoli, P. (2022). Smart soft open point to synergically improve the energy efficiencies of the interconnected electrical railways with the Cow voltage grids. *International Journal of Electrical Power & Energy Systems, 142*, 1082880. doi:10.1016/j.ijepes.2022.108288

Kumar, G. M. S., & Cao, S. (2023). Leveraging energy flexibilities for enhancing the cost- effectiveness and grid-responsiveness of net-zero-energy metro railway and station systems. *Applied Energy, 333*, 120632. doi:10.1016/j.apenergy.2022.120632

Kumar, V., Jethani, J. K., & Bohra, L. (2023). Combating elimate change through renewable sources of electricity-A review of rooftop solar projects in India. *Sustainable Energy Technologies and Assessments, 60*, 103526. doi:10.1016/j.seta.2023.103526

Li, C., Xu, C., & Li, X. (2020). A multi-criteria decision-making framework for site selection of distributed PV power stations along high-speed railway. *Journal of Cleaner Production, 277*, 124068. doi:10.1016/j.jclepro.2020.124086

Li, X., Zhu, C. & Liu, Y. (2023). Traction power supply system of China high-speed railway under Jow-carbon target: Form evolution and operation control. *Electric Power Systems Research*, 223.

Li, Z., Wang, Q., Fang, M., & Luo, Z. (2021). Thermodynamic and economic analysis of a new 600 MWe coal-fired power plant integrated with CaO-based carbon capture system. *International Journal of Greenhouse Gas Control*, *109*, 103386. doi:10.1016/j.ijggc.2021.103386

Loy-Benitez, J., Safder, U., Nguyen, H. T., Li, Q., Woo, T., & Yoo, C. (2021). Techno-economic assessment and smart management of an integrated fuel cell-based energy system with absorption chiller for power, hydrogen, heating, and cooling in an electrified railway network. *Energy*, *233*, 121099. doi:10.1016/j.energy.2021.121099

Meng, X., Zhang, Q., Hu, G., Zhang, G., Zhang, Y., Liu, F., & Liu, Z. (2023). Multi-vehicle accessed railway vehicle-grid system stability analysis and optimization based on OLTC. *Control Engineering Practice*, *141*, 105693. doi:10.1016/j.conengprac.2023.105693

Metro Rail News. (2023). *Sustainable & Green Mobility: Green Initiatives of Metro & Railways in India* https://www.metrorailnews.in/sustainable-green-mobility-green-initiatives-of-metro-railways-in-india/

News, E. T. N. (2023) *No looking back: Energy transition in Brazil and Latin America* https://etn.news/buzz/brazil-latin-america-energy-transition-2023

Ning, F., Ji, L., Ma, J., Jia, L., & Yu, Z. (2021). Research and analysis of a flexible integrated developmentmodel of railway system and photovoltaic in China. *Renewable Energy*, *175*, 853–867. doi:10.1016/j.renene.2021.04.119

Piraino, F., Genovese, M., Pagnotta, L., Caposciutti, M., Dei, L.F.N. & Fragiacomo, P., (2023). Integrated hydrogen and battery energy systems as emergency backup in electric trains. *Energy Conversion ana Management: X*, *18*, 100382.

Pivetta, D., Dall'Armi, C., Sandrin, P., Bogar, M., & Taccani, R. (2024). The role of hydrogen as enabler of industrial port area decarbonization. *Renewable & Sustainable Energy Reviews*, *189*, 113912. doi:10.1016/j.rser.2023.113912

Prol, J. L., & Steininger, K. W. (2020). Photovoltaic self-consumption is now profitable in Spain: Effects of the new regulation on prosumers' internal rate of return. *Energy Policy*, *146*, 111793. doi:10.1016/j.enpol.2020.111793

Rahman, M., Liu, H., Masri, M., Durazo-Cardenas, 1. & Starr, A. (2023). A railway track reconstruction method using robotic vision on a mobile manipulator: A proposed strategy. Computers in Industry, 148, 103900.

Rail Tech. (2023). *Will battery or hydrogen trains be the future? The vision of Siemens Mobility.* https://www.railtech.com/rolling-stock/2023/05/04/will-battery-or-hydrogen-trains- be-the-future-the-vision-of-siemens-mobility/?gdpr-accept

Raj, K., Lakhina, P., & Stranger, C. (2022). *Harnessing green hydrogen opportunities for deep decarbonisation in India.* NITI Aayog, RMI. Available: https://www. niti. gov. in/sites/default/files/2022-06/Harnessing_Green_Hydrogen_V21_DIGITAL_29062022. pdf

Riego Martinez, J., Perez-Alonso, M., & Duque-Perez, O. (2020). Influence of the rai electrification system topology on the energy consumption of train trajectories. *IET Renewable Power Generation*, *14*(18), 3589–3598. doi:10.1049/iet-rpg.2020.0119

Saray, M., Saray, M., Kazan, C., & Guner, S. (2024). Optimization of renewable energy usage in public transportation: Mathematical model for energy management of plug-in PV-based electric metrobuses. *Journal of Energy Storage*, *78*, 109946. doi:10.1016/j.est.2023.109946

Sharifi, S., Ferencz, I. F., Kamel, T., Petreus, D., & Tricoli, P. (2022). Medium-voltage DC electric railway systems: A review on feeding arrangements and power converter topologies. *IET Electrical Systems in Transportation*, *12*(4), 223–237. doi:10.1049/els2.12054

Shen, R., Han, Z., Liu, S., & Gao, S. (2021), October. Design and Application of Dynamic Test Platform for Railway System Protective Relays. *International Conference on Electrical and Information Technologies for Rail Transportation*, 66-73.

Shinde, A. M., Dikshit, A. K., Singh, R. K., & Campana, P. E. (2018). Life cycle analysis based comprehensive environmental performance evaluation of Mumbai Suburban Railway, India. *Journal of Cleaner Production*, *188*, 989–1003. doi:10.1016/j.jclepro.2018.04.022

Sky. (2022). *The future of energy may lie with hydrogen, but the journey to get there won't be easy.* https://news.sky.com/story/the-future-of-energy-may-lie-with-hydrogen-but-the-journey-to-get-there-wont-be-easy-12732319

Streets blog. (2023). *California Needs Leadership on Electric Rail.* https://cal.streetsblog.org/2023/07/06/california-needs-leadership-on-electric-mil

Yang, N., Pouget, J., Letrouvé, T., Jecu, C., & Joseph-Auguste, L. (2019). Techno-economic design methodology of hybrid energy systems connected to electrical grid: An application of hybrid railway power substation. *Mathematics and Computers in Simulation*, *158*, 107–119. doi:10.1016/j.matcom.2018.06.012

Yao, P., & Zhu, Z. (2020). Industrial design and application of a railway electric special power supply. *CPSS Transactions on Power Electronics and Applications*, *5*(4), 317–328. doi:10.24295/CPSSTPEA.2020.00026

Zhu, C., Lu, J., & Li, X. (2023). Review of Studies on Energy-Efficient Train Operation in High- Speed Railways. *IEEJ Transactions on Electrical and Electronic Engineering*, *18*(3), 451–462. doi:10.1002/tee.23741

Zhu, F., Yang, Z., Lin, F., & Xin, Y. (2020). Decentralized cooperative control of multiple energy storage systems in urban railway based on multiagent deep reinforcement learning. *IEEE Transactions on Power Electronics*, *35*(9), 9368–9379. doi:10.1109/TPEL.2020.2971637

Zuo, J., Dong, L., Yang, F., Guo, Z., Wang, T., & Zuo, L. (2022). Energy harvesting solutions for railway transportation: A comprehensive review. *Renewable Energy*.

LIST OF ABBREVIATIONS

BHEL: Bharat Heavy Electric Limited
CHC: Coordinated hierarchical control
DPVPS: Distributed PV power station
EDSS: Electric distribution supply system
EETO: Energy-efficient train operation
EMR: Energetic macroscopic representation
ERSs: Electric railway systems
ESSs: Energy storage systems
EVs: Electric vehicles
FESP: Flywheel energy storage platform
FOR A: Fine-grained optimization for railway alignment
GHI: Global horizontal irradiance
HESSs: Hybrid energy storage systems
HIL: High-Fidelity
HOG: Head on generation
HRPS: Hybrid railway power substation
IMPC: Internal model predictive control
IPA: Industrial port area
IR: Indian Railways
IRR: Internal rate of return
LCOE: Levelized cost of electricity
LCOH: Levelized cost of production/hydrogen
MGSC: Microgrid supervisory controller
MVDC: Medium voltage DC
NDEA: Network data envelopment analysis
O&M: Operation and maintenance
PV: Photovoltaic
RBE: Regenerative braking energy
RESPS: Railway electric special power supply
RIMG: Resilient interconnected microgrid
RPFC: Railway power flow controllers
RPS: Railway power substations
RTDS: Real-time digital simulator
SBM: Slack based measure
SC: Supercapacitor
sSOP: Smart soft open point
TSIIS: Traction-storage-information integrated station
VRS: Variable return to scale

Chapter 10
Unveiling the Potential and Hurdles of Blockchain in Green Hydrogen Management:
A Literature Review

Lynda Ait Bachir
https://orcid.org/0000-0002-8141-1551
University Center Aflou El Cherif Bouchoucha, Algeria

ABSTRACT

Discovering how blockchain technology may solve the core problems of the green hydrogen supply chain is the primary goal of this research. The research reviews the existing literature and concludes that blockchain's intrinsic properties, including decentralization, immutability, privacy, and transparency, have the potential to spur revolutionary improvements in the green hydrogen supply chain. Blockchain technology promotes confidence among stakeholders and makes international commerce easier by improving green hydrogen's certification and traceability, streamlining its transportation, storage, consumption, and marketing strategies, and guaranteeing regulatory compliance. As a result, green hydrogen ecosystems become more efficient and sustainable.

INTRODUCTION

In the continued pursuit of environmentally friendly power sources, green hydrogen is emerging as a frontrunner. (Nicolin & Nicolin, 2023) As a clean and sustainable renewable energy source, it has the potential to significantly impact the fight against climate change and carbon emissions. (Habash et al., 2023) Serving as an eco-friendly alternative to traditional fossil fuels, green hydrogen emits only water vapour instead of harmful by-products. (UG Student, Department of Chemical Engineering, Priyadarshini College of Engineering, Nagpur (M.H), India. et al., 2023) The versatility of this technology opens the door to a greener, more sustainable future across various sectors such as transportation, power generation, and industry. (Elshafei & Mansour, 2023) Green hydrogen presents a compelling opportunity to

DOI: 10.4018/979-8-3693-1014-4.ch010

meet the escalating energy demands of our world without compromising environmental sustainability for future generations. (Olabi et al., 2023)

The green hydrogen sector is rapidly emerging as a crucial player in the transition to a sustainable energy future. However, the sector faces challenges related to cost, efficiency, and environmental impact. Ensuring sustainability in the green hydrogen industry requires a commitment to transparency and traceability throughout the entire production and supply chain.(Azadnia et al., 2023)

Transparency and traceability are paramount in establishing the credibility of green hydrogen as an environmentally friendly solution.(Nicolin & Nicolin, 2023) Stakeholders, including consumers, investors, and policymakers, require a clear understanding of the supply chain processes for green hydrogen, covering production, energy sources, distribution, consumption, transportation and associated emissions. (Ufrj - Universidade Federal Do Rio De Janeiro et al., 2023) This transparency ensures accountability and fosters trust, crucial elements for the sector's long-term viability.(Hoffmann et al., 2023)

The supply chains for green hydrogen are complex,(Azadnia et al., 2023) Due to these challenges, the scaling and integration of green hydrogen into the global energy mix are not as straightforward as they could be.

Technological applications for the Fourth Industrial Revolution provide solutions that enhance transparency and enable tracking of the carbon footprint in green hydrogen supply chains. These solutions encompass the use of blockchain and IoT technologies, big data analytics, artificial intelligence, cloud computing, and more. (M. Akhtar, 2022)

Enter blockchain technology, which has gained widespread recognition for its groundbreaking impact in the financial sector, largely attributed to its decentralized, transparent, and secure nature. (Chuadhry et al., 2023) This raises a crucial question: How can blockchain technology help streamline the intricate supply chains of green hydrogen? This chapter explores the analysis and review of literature on blockchain technology and its potential to enhance the efficiency and sustainability of the green hydrogen supply chain.

1. GREEN HYDROGEN: ELEMENT OF THE FUTURE - UNDERSTANDING ITS NATURE AND POTENTIAL

1.1 The Fundamental Nature of Hydrogen

Hydrogen, abbreviated as "H," is the universe's most abundant and simplest element. Among all molecules, hydrogen, which exists in nature as a pair of hydrogen atoms (H_2), is the lightest. In its most basic form, it is a tasteless, odourless, and colourless gas. Most hydrogen on Earth is found in hydrocarbon or water molecules. Hydrogen and oxygen are the most common forms of water or H_2O. (*Green Hydrogen*, 2020)

Hydrogen has the highest energy density of today's most-used fuels, including diesel, natural gas, and gasoline.(Armaroli & Balzani, 2011) Hydrogen energy is usually measured in kilogrammes (kg) rather than volume (like natural gas) due to its very low density at room temperature. One kilogramme of hydrogen, for instance, has about the same amount of energy as two and a half kilogrammes of petrol, or one gallon.

1.2 Hydrogen's Multifaceted Applications and Advances in Technology

Understanding the fundamental importance of hydrogen across many scientific fields is essential before exploring its many uses and technical advancements. Hydrogen's vital significance in geoscience is shown by its natural existence on Earth, which is both a neglected component and a crucial part of Earth's operations. (Zgonnik, 2020) Hydrogen is essential for more than just energy transport because of its central role in many scientific processes, most notably hydrogen bonding. Many areas of scientific investigation and knowledge rely on it. (G.-J. Zhao & Han, 2012) The idea of a hydrogen economy uses hydrogen's multidisciplinary character by bringing together energy systems, materials science, and engineering experts. (Goltsov et al., 2006) It bodes well for hydrogen's future role as a foundational component of sustainable development plans.

The versatility of hydrogen and its possible uses in many industries have made it a hot topic. From energy storage to medical therapies, the discovery of hydrogen's diverse uses has sparked extensive research and development. The evolution of molecular hydrogen has led to its practical application in therapies and diagnostic tests, such as the breath hydrogen test. (Dixon et al., 2013) Additionally, the discovery of high-capacity hydrogen storage materials has paved the way for advancements in automotive applications and energy storage. (J. Yang et al., 2010) (Balat, 2008) Theoretical discoveries of high-capacity hydrogen storage metal tetrahydrides have also expanded the possibilities for hydrogen utilisation. (W. Zhao et al., 2022)

Improved basal plane active two-dimensional catalysts for hydrogen evolution have been discovered to enhance our grasp of hydrogen's potential as an energy carrier. (Zhang et al., 2019) These discoveries have implications for renewable energy production and storage, as hydrogen can play a crucial role in addressing energy sustainability challenges. (Zhu et al., 2022) (White et al., 2011) (Edwards et al., 2007) Hydrogen can facilitate ecologically friendly chemical reactions, as shown by the rapid synthesis and identification of covalent organic framework photocatalysts for the formation of hydrogen peroxide. (Pan, 2019)

1.3 Toward Clean Hydrogen: Pathways to Sustainability

The pathways for hydrogen production encompass diverse technologies and economic environmental considerations.(Arcos & Santos, 2023) As the global energy landscape continues to evolve, the choice of hydrogen production pathways will significantly influence the overall sustainability and environmental impact of energy systems:

The production of hydrogen can be classified into different colours based on the method used. These colours include Aqua, White, grey, turquoise, blue, and green, each representing a specific production pathway and associated technologies. (Hermesmann & Müller, 2022)

Grey hydrogen is created by "steam reforming" from fossil fuels such as natural gas or methane. This approach's lack of carbon capture and storage causes it to produce more carbon dioxide than its green, turquoise, or blue hydrogen alternatives. The most harmful kind of hydrogen, called black or brown hydrogen, is made from bituminous or lignite coal, but grey hydrogen is less carbon-intensive. Why? Because the carbon dioxide and carbon monoxide that is produced when black or brown hydrogen is produced cannot be recovered. (World Economic Forum, 2021)

In contrast, blue hydrogen—an ecologically friendly alternative—offers a method that still uses natural gas's methane—a process known as a greenhouse gas—but the industry promotes it as a solution.

When first released into the atmosphere, it warms the planet at a rate more than a hundred times faster than carbon dioxide. (Blaine, 2021)

One non-carbon emitting technique is the extraction of Aqua Hydrogen from oil sands (natural bitumen) and conventional oil fields. All energy conversion occurs underground, which significantly improves the process's efficiency. This technique could be eco-friendly for producing hydrogen from commercially unrecoverable oil deposits, resulting in water purification. (Doug, 2019)

"White hydrogen" is the term for hydrogen that occurs in nature. It occurs naturally as a gas in many forms, including volcanic gases, geysers, hydrothermal systems, and deep inside the oceanic crust or continental crust. It appears in a broad variety of geological settings and rock types. (Energy Observer, 2023)

Green hydrogen, which goes by a few different names, including "clean hydrogen," "renewable hydrogen," and "low-carbon hydrogen." is the most eco-friendly choice because it helps achieve decarburization goals and reduces emissions of greenhouse gases. It is obtained from renewable sources through water electrolysis. (Arcos & Santos, 2023) Turquoise hydrogen, generated via electrolysis with carbon capture utilization storage, presents a promising pathway with potential environmental benefits, although its large-scale implementation and associated challenges require further exploration. (Ajanovic et al., 2022)

Steam reforming may produce Grey hydrogen efficiently and cheaply, but this process leaves a devastating carbon imprint. Although it uses renewable energy sources to produce green hydrogen, electrolysis wastes over 70% of the energy that goes into it. A carbon-free, plentiful, and infrastructure-light source, naturally occurring white hydrogen seems to be a potential one. Evaluating global white hydrogen supplies is challenging since there has been so little study. The solution to a carbon-free civilization and a better future for Earth and humanity could lie in a hydrogen economy built on white and green energy. (Energy Observer, 2023)

The financial feasibility of diverse hydrogen hues is fundamental. The competitiveness and cost-effectiveness of grey, blue, and green hydrogen greatly influence their integration and acceptance in many sectors, such as manufacturing, transportation, and power production. (Germscheidt et al., 2021) (Arcos & Santos, 2023) (J. Huang et al., 2023)

Based on the Hydrogen Science Coalition, the definition of four critical conditions of green hydrogen within the context of the "Hydrogen Square" (HydS) model can be outlined as follows: (Hydrogen Science Coalition, 2022)

- **Definition of green Hydrogen**: This involves specifying what constitutes 'clean' hydrogen, focusing on the methods and conditions under which hydrogen is produced, and ensuring it aligns with low carbon and environmental standards.
- **Green Hydrogen Inputs:** This refers to the raw materials and energy sources used in hydrogen production. For hydrogen to be considered clean, these inputs must be environmentally friendly, potentially involving renewable energy sources or low-emission methods.
- **Low Production Emissions**: The process of producing hydrogen must have minimal emissions. This includes direct emissions from the production process and indirect emissions associated with the entire production lifecycle, from sourcing inputs to the final output.
- **Independent Process Audit:** This condition requires a thorough and independent review of the hydrogen production process. This audit ensures that the production adheres to the established

standards of cleanliness and efficiency, and it likely involves assessing the process's environmental impact, emissions, and sustainability.

The Hydrogen Square model emphasises the interconnectedness of the four stages of hydrogen energy systems: production, storage, safety, and utilisation. This model underlines that the cleanliness and efficiency of hydrogen must be considered at each stage of its lifecycle. Additionally, the model accounts for sub-system processes like purification and compression, stressing that their costs (energy consumption and recovery rate) are crucial in selecting production pathways and the overall Life Cycle Assessment (LCA) of hydrogen. (Dawood et al., 2020)

Many methods and resources that may be used to manufacture hydrogen are renewable. The current method of producing hydrogen is not sustainable since it produces 830 Mt/a of CO_2 because it is almost dependent on the reforming and gasification of fossil hydrocarbons, such as coal (23% of the total) and natural gas (76%) (García Martín, 2022). Methods for producing hydrogen, such as water electrolysis(Bertuccioli et al., 2014), and methane pyrolysis (Weger et al., 2017) (Schneider et al., 2020) (Muradov, 2001) are gaining popularity. Renewable sources provide the majority of the energy and raw materials, and these technologies may be considered practical for decreasing environmental effects. (Mehmeti et al., 2018) Biomass, hydropower, tidal, geothermal, and wave energy are all examples of RES that may be harnessed. The potential of these sources on a global scale is enormous, surpassing even the total energy consumption of the globe (Bouckaert et al., 2021). Due to geographical, societal, and technological limitations, only a portion of this potential can be realised. (Wind, 2017) Of that, many technologies are vying for the same renewable energy and raw material supply.

1.4 Innovating the green Hydrogen Energy Sector: From Historical Roots to Future Challenges

The development of the hydrogen energy sector as a response to environmental concerns and the depletion of fossil fuels in the 20th century Initially, the world's economy heavily relied on oil and natural gas, which led to significant technological advancements but also posed a threat of irreversible environmental damage. (Goltsov et al., 2006) The conception of an ecologically clean hydrogen energy system emerged in the mid-1970s, led by a conscientious segment of the global scientific community. This was in response to the environmental crisis, the diminishing of fossil fuel resources, and the energy crisis at the time. (N, 1974) (Veziroglu, 1975a) (Veziroglu, 1975c) (Legasov, 1979.) (Bockris & Klerer, 1976) (Bockris & Klerer, 1978)

Hydrogen energy, known for not producing harmful effluents or contributing to the greenhouse effect, became a focus of the International Association for Hydrogen Energy (IAHE), established at the University of Miami. (Veziroglu, 1975b) The IAHE played a significant role in promoting hydrogen energy through publications and conferences. By the mid-1980s, a comprehensive hydrogen energy sector conception covered hydrogen production, delivery, utilisation, material reliability and system safety issues. (Goltsov et al., 2006) Hydrogen energy research expanded globally, with over 50 countries conducting studies and some initiating national programs. (Goltsov et al., 2006) The term "hydrogen economy" gained traction, describing a system where hydrogen is a crucial energy source. The impressive developments over 48 years include commercialisation efforts in hydrogen technologies and forming a global hydrogen movement, with research extending to prospects and regional technical aspects.

Hydrogen's capacity to be created from renewable sources, such as solar and wind power makes it a promising fuel for the 21st century. It would help with the need for sustainable energy solutions. (Laurent & Bazarkina, 2019) The buzz surrounding the green hydrogen sector gained momentum following the United Nations Climate Change Conference of Parties (COP 26 and COP 28), where the participating nations agreed to achieve net zero carbon emissions by the year 2070. (GRAAF, 2022) (Jonas Moberg, 2023)

The global green hydrogen market was valued at USD 676 million in 2022 and is projected to reach USD 7314 million by 2027, growing at 61.0% cagr during the forecast period. The market's growth is attributed to the lowering cost of producing renewable energy from all sources, development of electrolysis technologies and high demand.(Marketsandmarket, 2023) However, challenges in the hydrogen industry, including production, storage, and distribution, must be addressed to fully realize its benefits as an energy source. These Green Hydrogen Industry Challenges include technology readiness and cost, infrastructure development, market off-takers, financing, supply chain bottlenecks, and policy frameworks. (Meda et al., 2023) (Osman et al., 2022) (Rasul et al., 2022) (Laurent & Bazarkina, 2019)

- **Technology Readiness and Cost:** The evaluation of technology readiness and cost becomes critical to the industry's performance when it develops traction and is widely used:
    1. Advancements in electrolyze technologies are needed to improve efficiency, and durability, and reduce capital costs.(Risbud et al., 2023)
    2. Improving fuel cell efficiency can enhance the overall performance and cost-effectiveness of green hydrogen systems.
    3. The cost of green hydrogen production, primarily through electrolysis, needs to be reduced to compete with conventional fuels. This requires advancements in technology, economies of scale, and the utilization of low-cost renewable energy sources.
    4. Technology readiness is a key determinant of the successful deployment of green hydrogen solutions. The Green Hydrogen Industry faces challenges in advancing technologies from lower TRLs to higher, ensuring that innovations move seamlessly from the laboratory to practical, scalable applications.(Ismoyo et al., 2023)
    5. Policy support and market competitiveness: Conventional hydrogen generation techniques, which can have pre-existing infrastructures and less upfront expenditures, pose a threat to green hydrogen. Strong measures promoting green hydrogen's market competitiveness are required to encourage its adoption. (Farrell, 2023)
- **Infrastructure Development:** The creation of a strong infrastructure is necessary before green hydrogen can be produced and used:
    1. To scale up green hydrogen, large-scale hydrogen production facilities, such as electrolyzes run by renewable energy, must be established. (Toure et al., 2023)
    2. Establishing efficient and appropriate infrastructure for hydrogen storage, transportation and distribution is essential for the widespread use of hydrogen as an alternative to fossil fuels. (M. Yang et al., 2023)
    3. There are difficulties in retrofitting current infrastructure to support the generation and distribution of green hydrogen. It takes meticulous planning and money to upgrade storage systems, pipelines, and facilities.(Bollmann et al., 2023)

- **Green Hydrogen Market Off-takers**: The importance of market off-takers involvement is growing as the business picks up steam. For the market to flourish sustainably, these off-takers face obstacles that must be overcome.
    1. The market's growth depends on the creation of a significant demand for green hydrogen. Market demand can be increased by involving important players in industries including transportation, manufacturing, and electricity production and by offering incentives for green hydrogen adoption. (Green Hydrogen Opportunities for Emerging and Developing Economies, 2022)
    2. Market participants must be persuaded of the advantages and sustainability of green hydrogen. Building trust and encouraging client acceptance can be achieved by showcasing the green hydrogen technologies' dependability, effectiveness, and environmental benefits. (Müller & Eichhammer, 2023)
- **Financing and Investment**(Kumar et al., 2023):
    1. Green Hydrogen scaling necessitates large infrastructure and technology implementation expenses. Capital-intensive obstacles can be addressed by luring in both public and private investment, offering financial incentives, and creating creative financing schemes.
    2. Clear cost reduction pathways supported by focused research and development efforts can promote investment.
- **Supply Chain Bottlenecks:** Supply chain bottlenecks are a major Green Hydrogen Industry Challenge as it gains traction as a major participant in the global shift to sustainable energy. A complicated supply chain must run smoothly for green hydrogen generation and distribution to be successful. The present discourse delves into the difficulties presented by supply chain obstructions in the Green Hydrogen Sector and looks at possible remedies to guarantee the industry's continued expansion.
    1. Green Hydrogen production requires the availability and sourcing of certain raw materials, such as catalysts for electrolysis and renewable energy sources, which can present supply chain difficulties. Creating sustainable and varied supply networks is crucial.
    2. To prevent bottlenecks and enable scaling up, it is imperative to provide a sufficient supply of essential equipment, components, and infrastructure for green hydrogen systems, including fuel cells and electrolyzes.(Bollmann et al., 2023)
- **Policy Framework:** The necessity of clear and encouraging policy frameworks is fundamental to the industry's success.
    1. Governments are essential in establishing the conditions necessary for green hydrogen scaling. The establishment of supportive policies, regulations, and market frameworks that include long-term targets, financial incentives, and carbon pricing mechanisms can encourage investments and promote the adoption of green hydrogen.
    2. Global market expansion and supply chain resilience can be ensured by harmonizing laws and regulations at the international level, which can promote cross-border trade and cooperation in green hydrogen technologies.

When it comes to finding new solutions for a sustainable and eco-friendly energy future, clean hydrogen is right at the front. Hydrogen is a promising energy source, and the Hydrogen Science Coalition has identified four requirements for clean hydrogen: a precise definition, eco-friendly inputs, minimal emissions during production, and impartial audits. At the heart of these standards is the monitoring of the carbon footprint, a crucial metric for evaluating and guaranteeing the ecological soundness of hydrogen generation.

Applications of the Fourth Industrial Revolution that help measure the carbon footprint in green hydrogen supply chains use cutting-edge tech that makes things more transparent and makes people more responsible for their impact on the environment. Some examples of these uses are:

- The Internet of Things (IoT) enables connected devices and sensors to efficiently monitor various stages of the supply chain in real-time and provide data on production, storage, and distribution processes, hence enabling the effective tracking of carbon footprints. (Pechlivanis, 2023)
- Big data analytics, applied to the vast amounts of data generated by the hydrogen supply chain, enable effective carbon footprint management by revealing patterns, improving operations, and evaluating environmental impacts. (Matter, 2023)
- Predictive modelling using artificial intelligence (AI) can maximise the production and distribution of green hydrogen. This paves the way for preventative measures to boost efficiency and cut down on carbon emissions. (Katterbauer et al., 2023)
- Cloud computing offers many advantages to the green hydrogen supply chain, including secure data storage and interchange, increased collaboration, and easier access to data related to carbon footprint assessments.(Yenugula et al., 2023)
- Every transaction in the green hydrogen supply chain can be tracked using blockchain technology, which creates secure and transparent ledgers. This makes the record reliable and unchangeable. (Azadnia et al., 2023)

Blockchain technology stands out among these solutions as a potential instrument that may efficiently validate the sustainability credentials of clean hydrogen generation while also giving transparency and security. As a result; we can investigate how blockchain technology might transform the certification and monitoring procedures for clean hydrogen.

2 . EXPLORING BLOCKCHAIN TECHNOLOGY IN GREEN HYDROGEN SUPPLY CHAIN MANAGEMENT

Green hydrogen produced through renewable energy sources has gained significant attention as a clean and sustainable energy carrier with the potential to decarbonise various sectors. However, the complexities and challenges within the green hydrogen supply chain pose significant barriers to its widespread adoption and integration into existing energy systems.

The advent of blockchain technology has been a game-changer in several sectors, including the banking, healthcare, and energy industries, among many others. This part aims to provide an in-depth exploration of blockchain technology by delving into its definition, types, layers and stages of work, and the structure of the blockchain cluster. The analysis will draw upon a systematic review of current research on blockchain technology to offer a comprehensive understanding of its architecture and consensus mechanisms. This part will also examine the challenges and opportunities associated with implementing blockchain in the energy sector based on existing literature.

2.1 Unravelling the Potential of Blockchain Technology

Although "blockchain" is often associated with digital currency like Bitcoin, its uses go much beyond that. (Zheng et al., 2017) It is thus, crucial to comprehend the fundamental principles underlying this innovative technology.

Blockchain is a technique of preserving and processing data that makes it impossible for the system to be changed, hacked, or manipulated by illegitimate sources. All transactions on the blockchain network are recorded in an encrypted distributed ledger called a blockchain. (Barde et al., 2023) (N. Gupta & Agrawal, 2023) The "nodes" in a network are the physical components that enable data transmission between devices connected via the internet. They trade information, blocks, and financial transactions. (Nasir & Song, 2019) Each node contributes to the network's upkeep, verifies blocks and transactions, and aids with security measures. (Hanif & Song, 2019), (Paksoy & Deveci, 2023) Because no one entity is in charge under this decentralised system, all network transactions and communications are more secure. (Cayirci & Rong, 2018) Moreover, there is diversity among nodes. Light nodes depend on full nodes for transaction information, whereas full nodes keep a full copy of the chain and participate in verification procedures. (Chang, 2018)

There are three leading blockchains: public blockchains, consortium blockchains, and private blockchains. (Komalavalli et al., 2020) (H. Huang et al., 2020) (Andreev et al., 2018) (Anatolievna et al., 2020) Public blockchains are open to anyone and allow for decentralised transactions and interactions without a trusted intermediary. (Aggarwal & Kumar, 2021) Consortium blockchains are also open but limited to a specific group of participants, providing greater transaction confidentiality and faster processing. (Xu et al., 2019) On the other hand, Private blockchains have closed access and are used for private sharing and data exchange among a specific group of individuals. Different types of blockchains serve different purposes, and the needs and objectives of an application or organisation dictate the sort of blockchain that is most appropriate.

Blockchain technology consists of three main phases: Blockchain 1.0 is the first iteration of digital money; it emphasises cryptocurrency payments and decentralisation. In blockchain 2.0, sometimes called the digital finance stage, smart contracts provide new economic, financial, and market applications. (H. Huang et al., 2020)

Blockchain 3.0, the next generation of the digital society, offers decentralised solutions to a wide range of businesses, not only banking. Additionally (Li et al., 2021):

- A blockchain-based multi-stage signing method enhances transaction security and traceability using multi-stage signing and authentication. (Xu et al., 2019)
- Another method involves constructing a blockchain with main and slave chains to store and verify application data in a project application process. (Bazaee et al., 2020)
- Furthermore, the maturity levels of blockchain technology are classified into emerging, identified, defined, operational, and mature stages, which can provide solutions to organisational problems.
- Web of Things (IoT), machine learning, and artificial intelligence (AI) are some of the new technologies that will be a part of blockchain 3.0. The goal of this integration is to build better, more efficient systems.
- One primary goal is fixing the scalability problems that plagued earlier blockchain iterations. It requires the introduction of novel structures and algorithms, such as sidechains or sharding, to process a greater volume of transactions efficiently.

- Improving the cryptographic methods to enhance security and privacy is crucial in blockchain 3.0. This might include more advanced encryption methods and privacy-preserving techniques like zero-knowledge proofs.
- Interoperability between Different Blockchains enables seamless interaction between blockchain platforms, allowing data transfer and value across diverse blockchain networks. Proof of Stake (PoS) and Delegated Proof of Stake (DPoS) are two more energy-efficient consensus processes that might help alleviate some environmental issues surrounding blockchain technology and cryptocurrencies like Bitcoin.
- User-Friendly Interfaces and Experiences: Enhancing the user interface and experience to encourage wider adoption among non-technical users. It includes simplifying interactions with blockchain systems and making them more accessible to the public.
- Working within the bounds of evolving regulatory landscapes to ensure compliance with legal standards across different countries and regions.
- Developing community-driven governance models to manage and evolve the blockchain system democratically.

Blockchain 3.0 represents a maturing and diversification of blockchain technology, aiming to solve real-world problems across multiple industries while addressing the limitations of earlier blockchain generations.

Blockchain technology comprises five essential layers integral to its functionality and transformative potential: Infrastructure or hardware, data, network, consensus, and application.

1. **Infrastructure Layer (Hardware Layer)**: This foundational layer consists of the physical hardware that supports the blockchain system, including computers, servers, protocols, and other devices that store and process data. (Cao et al., 2021) In this layer, VMware, Docker, and various data transfer media contribute to the sending and receiving of data within the network.
2. **Data Layer:** Data Layer: contains protocols and cryptographic and digital signature algorithms that encrypt data and transactions and send them to the first layer or decrypt and send them to the third layer. At this level, the data blocks are structured and stored in a chronological chain of blocks, each securely linked to the previous block via cryptographic hashes. This layer's inner workings are subsequently described in depth: (Zheng et al., 2017)

One of the most fundamental building blocks of the blockchain, the block records a series of transactions. Using a cryptographic address (HASH) and a serial number, every block in the chain is connected to the one before. Two parts make up a block: the header and the body. (Azzi et al., 2019)

The list of transactions linked to the block and their sizes are included in the block's body. The head of the block contains data specific to:

- Conditions for issuing the block (a set of rules that must be followed to verify validity);
- The hash of the preceding block's header is used to confirm the block's validity and its transactions, and it also forms a link in the chain by including the previous block's hash into the current block's header. Removing or altering a block of transactions after it has been added to the blockchain is impossible. If one block were to be modified, it would render all blocks useless after it.
- Timestamp;

- Merkle Tree Root (a mathematical data structure used for digital security and efficient data validation);
- Nonce (a usually 32-bit random number used in mining operations, i.e., solving a complex problem using a computer), the purpose of using nonce is to find a hash (hash) for the new block that corresponds with certain conditions. These conditions depend on the protocol used in the blockchain, and the most famous condition is that the new block hash must start with a certain number of zeroes. To achieve this, miners start with a nonce value of zero (or any other value) and calculate the block hash using this nonce. If the hash does not start with the required number of zeroes, they increase the nonce value by one and calculate the hash again. This process is repeated until nonce result in the required hash. This process consumes significant time and computing resources and is part of Proof of Work, aiming to ensure network security and add blocks randomly and correctly. When a suitable nonce is found, the block is accepted and added to the blockchain, and the miner is usually rewarded in digital currency.
- The target is the value the potential block hash, calculated using the selected nonce, must meet to be accepted. The target is typically a relatively small number. Once a block miner finds a nonce that produces a smaller hash than the target, the block is accepted and added to the blockchain.

The transaction: represents any operation that occurs within the blockchain. It could be a digital currency transfer transaction, smart contract activation, or any other process.

When specific criteria are satisfied, such as the terms of payment or secrecy, the smart contract may be programmed to automatically carry out the contract's terms and actions on every node in the network. (Tejashwini et al., 2023) All nodes in the blockchain network need to reach a consensus on the inputs, outputs, and states impacted by smart contracts before a transaction can be executed. With smart contracts, transactions may be executed automatically, cutting out intermediaries. (Azzi et al., 2019).

The chain is a sequence of interlinked blocks. Blocks are connected to ensure the continuity of the transaction sequence.

The digital signature may confirm users' identities and the legitimacy of transactions. Encryption keys are used to produce digital signatures.

A to ensure that all transactions are legitimate; the network uses the consensus technique to come to a unanimous decision. Methods like Proof of Stake (POS) and Proof of Work (POW) are only two of several available for reaching a consensus. The private key and public key are used in digital signatures. The private key allows the user to sign transactions, while the public key allows anyone to verify the signature.

1. **Network Layer:** Network Layer: This layer manages data transmission across the blockchain network. It involves the protocols and nodes that ensure data is consistently distributed across the network, allowing the ledger to be maintained without a central authority. (Angrish et al., 2018) The stability of the whole network is ensured by the propagation of transactions/blocks in the network, which is defined in this layer.
2. **The Layer of Consensus:** The protocols that enable blockchain nodes to reach a consensus over the ledger's current state are specified. Many consensus techniques are used to keep the distributed ledger in sync and genuine, including Proof of Work, Proof of Stake, and Delegated Proof of Stake. (S. Gupta et al., 2020)

3. **Application Layer:** At its highest level, the blockchain facilitates user interaction via its many applications. Here, it will find the user interfaces, decentralised applications (DApps), and smart contracts that let users interact with the blockchain for various tasks, such as managing supply chains and conducting financial transactions. (Delimatsis, 2018)

Each layer plays a distinct role in enabling secure, visible, transparent, and decentralised digital transactions within the blockchain ecosystem. (Koppers, 2020) Moreover, functioning cohesively provides the robustness and flexibility that make blockchain a revolutionary technology.

The working model of blockchain involves one party initiating a transaction by creating a block and storing transaction details such as timestamps and links to the previous blocks. (T. Yang, 2020) which is then validated by computers in the network and added to the chain. Each participant in the network has control over the blockchain, and the blocks cannot be altered without the consensus of the entire network. (Bobbert & Ozkanli, 2020) The blocks in the blockchain are interconnected and protected using cryptographic functions. (G. Xu et al., 2022)

2.2 The Green Hydrogen Value Chain: Technological، Economic، and Social Perspectives

A standard green hydrogen supply chain consists of many steps. Wind, solar, and hydropower are some of the renewable energy sources that produce power (Kim et al., 2019). The next step is to employ electrolysers to create hydrogen using the power that is generated by the renewable energy source. Multiple forms of transportation may then be used to transport and distribute the generated hydrogen to the end users.

The green hydrogen supply chain has evolved in response to the increasing global focus on reducing carbon emissions and transitioning towards a low-carbon economy Hydrogen gas and oxygen are produced during the electrolysis of water when renewable energy sources like solar or wind power are used to produce green hydrogen. Conventional ways of producing hydrogen often involve the use of fossil fuels, which in turn contribute to the emissions of greenhouse gases. This approach provides an environmentally friendly alternative. A sea change is occurring in the energy scene as we move towards green hydrogen generation; this change might spur innovation and sustainability in many other sectors.

When examining the green hydrogen supply chain, it is crucial to consider the technological advancements needed to enhance efficiency and effectiveness in production, to meet the growing demand for renewable hydrogen. Areas rich in renewable energy resources will have to import renewable hydrogen to regions with low production capability, therefore this is especially crucial for them. (IRENA, 2023)

2.2.1 Standardising Green Hydrogen Guarantees of Origin and Establishing a Certification System

The expansion of the global hydrogen trade depends on the capacity to categorize hydrogen fuels according to their respective carbon footprints and sustainability standards. To guarantee the veracity and openness of green hydrogen generation, it is crucial to emphasise the significance of regulatory frameworks and international standards. The legislative environment, in which the green hydrogen supply chain operates, together with other national and international rules, often influences decisions about these issues. (Velazquez Abad & Dodds, 2020) So The International Renewable Energy Agency suggests:(IRENA, 2023)

- Existing certification methods have obvious limitations when it comes to facilitating international commerce in hydrogen. Addressing issues like design gaps, standards, ecolabelling, and greenhouse gas footprint computation is crucial for facilitating fair commerce and comparison across borders.
- Harmonisation and Standardisation Are Necessary: To make hydrogen commerce easier, we need standards in places like eco-labelling, the computation of greenhouse gas footprints, and the requirements for compatible regulatory frameworks. If the hydrogen market is to operate well, these factors must be standardised.
- Establishing a trustworthy quality infrastructure and a transparent chain of custody are critical steps in building confidence in hydrogen certifications. This facilitates the trade of hydrogen and, maybe, its derivatives, as well as other eco-friendly commodities like ammonia and steel.

2.2.2 The Transportation of Green Hydrogen Presents Unique Challenges Due To ITS

Gorji, (2023); Omid et al. (2024):

- Storage and Compression: Because hydrogen's energy density is low at atmospheric pressure, it has to be either compressed or liquefied before transportation.
- Ensuring Safety and Controlling Leakage: Hydrogen presents unique safety concerns due to its extreme flammability and potential dangers when not handled properly.
- Compression and liquefaction are energy-intensive processes, which drive up transportation costs.
- Problems with Existing Infrastructure: In many places, the necessary infrastructure for hydrogen transportation is still in the works.

To satisfy the present energy demand on a worldwide scale, facilities for the storage and movement of large quantities of environmentally friendly hydrogen are essential.

How to Transport Hydrogen Most Efficiently:

- For short to medium distances, one of the most effective means of transporting hydrogen is via pipes (AIT BACHIR, 2021). It can save transit costs in the long run, but it involves a lot of investment in infrastructure. Compressed gas via pipeline (CGH2-PL) had the lowest global warming potential (GWP) (1.57 kgCO2-eq/kgH2) and was therefore determined to be the most environmentally friendly alternative among the aforementioned paths in a thorough cradle-to-gate life cycle evaluation. (M. S. Akhtar et al., 2021)
- Compressed or liquefied hydrogen may be transported using specialised vehicles for shorter distances or in areas without pipeline access.
- Transport by Sea: Liquefied hydrogen may be transported by sea for international commerce. Apt for long-distance travel, it requires cutting-edge cooling technology and rigorous safety protocols.
- To make transportation and storage easier, hydrogen may be transformed into chemical substances such as ammonia or methanol. Although this approach requires extra conversion steps, it does away with storage and transportation problems. Liquid ammonia is a potential future hydrogen and energy transporter because of the encouraging findings of its direct use in DAFCV for GWP (1.62 kgCO2-eq/kgH2). (M. S. Akhtar et al., 2021)

The integration of green hydrogen into existing gas infrastructure and the development of dedicated hydrogen pipelines are crucial aspects that require careful planning and investment to ensure the seamless transportation of green hydrogen from production sites to end-users.(M. S. Akhtar et al., 2021)

Storage indeed plays a critical role in the green hydrogen supply chain. It involves several key aspects(He et al., 2021):

- Given hydrogen's low density and strong flammability, safe and effective storage of the gas is of the utmost importance.
- Compression, liquefaction, and chemical storage (such as metal hydrides or liquid organic hydrogen carriers) are common storage strategies.
- Energy Losses: The overall efficiency of hydrogen as a transporter of energy is affected by the energy losses that are connected with each storage option.
- Large-scale deployment of green hydrogen, in particular, necessitates the development or adaptation of storage infrastructure.
- When calculating the economics of producing and distributing green hydrogen, the cost of storage is an important consideration.

The intermittent nature of renewable energy sources necessitates the storage of excess green hydrogen for future use, requiring innovative storage technologies and infrastructure (Moulebe et al., 2022). The lack of sufficient storage capacity is a significant risk factor in the green hydrogen supply chain, emphasising the need for strategic investments in storage infrastructure to mitigate supply-demand uncertainties and ensure the stability of green hydrogen supply.(Azadnia et al., 2023)

Hydrogen storage, however, presents several difficulties, some of which will be detailed below. (*Green Hydrogen*, 2021)

- Because hydrogen has a low specific gravity, storing it compressed requires a lot of energy.
- Requirements for temperature and pressure while storing solid hydrogen.
- The design, the legality, the social implications, and the considerable expense.
- Safety considerations arise from the possibility of chemical reactions and the short storage life of components such as fibre, metals, polymers, etc.
- Hydrogen stored in bulk at geographical features runs the risk of contamination, necessitating additional purification steps before final usage.

Indeed, a complicated part of the supply chain is ensuring that green hydrogen reaches end-users and different industrial sectors. Several things contribute to its complexity(Müller & Eichhammer, 2023):

- Varying End-User Demands: Hydrogen volume, purity, and delivery format demands vary between consumers.
- Hydrogen distribution presents unique hurdles when it comes to integrating with the current energy infrastructure in a way that is both safe and efficient.
- The development of a market for hydrogen needs both technical improvements and legislative backing, especially in industries that have not previously used hydrogen.
- To ensure broad adoption, it is critical to make green hydrogen affordable and available to a variety.
- Developing after-sales service for hydrogen technology.(*Green Hydrogen*, 2021)

- The establishment of a robust distribution network, including hydrogen refuelling stations and industrial supply chains, requires coordinated efforts among stakeholders to overcome logistical and regulatory barriers.
- The challenges associated with establishing hydrogen fuel stations and central distribution sites, highlight the need for comprehensive planning and investment in infrastructure to support the widespread adoption of green hydrogen as a clean energy carrier. (Monteiro & Brito, 2023)

Green hydrogen supply chains are already complicated, and the social and economic factors only make things worse. Some examples are:

- The public's view and support of green hydrogen as a viable long-term energy solution is critical. Green hydrogen has many potential applications, and it is important to inform and involve people about these possibilities.(Hoffmann et al., 2023)
- Ensuring that green hydrogen can compete economically with other energy sources is a difficulty when it comes to economic viability. Costs of manufacturing, market price, and possible subsidies are all factors to be considered.(Müller & Eichhammer, 2023) (*Green Hydrogen*, 2021)
- Employment and Skill Development: Jobs in conventional energy industries may be lost as we move towards a green hydrogen economy, but new possibilities will be created in green energy. Strategies for skill development and workforce transformation are crucial. (Hamukoshi et al., 2022)
- It is critical to establish legislative and regulatory frameworks that promote investment in environmentally friendly hydrogen technology and infrastructure. (Bade et al., 2024)
- Green hydrogen supply chain risk factors, such as high investment costs and the creation of policies and regulations, may be better understood and addressed by identifying and analysing them. (Azadnia et al., 2023)
- Policymakers and decision-makers may benefit from a more holistic view of the hydrogen supply chain's sustainability if they rank the many economic, technical, environmental, and social elements that contribute to it. (Manzardo et al., 2023)

Additional challenges require careful consideration and the importance of understanding the social and economic implications of green hydrogen value chains, particularly in terms of job creation, community engagement, and equitable distribution of benefits. The different stages of the value chain have distinct impacts on various stakeholders, necessitating a holistic approach to address social and economic challenges and ensure the inclusive and sustainable development of the green hydrogen supply chain. (Vallejos-Romero et al., 2022)

2.3 Blockchain-Enhanced Green Hydrogen Supply Chain From Production to Utilisation

In the quest to establish a sustainable energy future, the integration of blockchain technology in green hydrogen supply chain management marks a transformative step. This innovative approach revolutionizes how we produce, distribute, and consume hydrogen, addressing key challenges for transitioning and solving the problems with current and future green hydrogen systems to an irreversible, efficient, trustworthy, transparent, and entirely secure digital platform. " All transactions are guaranteed to be transparent, private, immutable, settled, decentralized, and scalable" (Monrat et al., 2019) (Xinyi et al., 2018) (Zheng et al., 2018) (Savian et al., 2023)

The use of blockchain technology in the green hydrogen industry has been the subject of very few survey or review publications. After reviewing the literature, we have concluded that the findings do not warrant moving forward with a full-scale energy blockchain platform implementation. So, to help researchers in this area, our study provides a thorough systematic literature analysis that maps out all the energy blockchain applications, guidelines, suggestions, and orientations.

2.2.1 Standardisation and Certification of Green Hydrogen Using Blockchain Technology

The following are some methods in which blockchain technology might help with the difficulties of creating a certification system and standardising green hydrogen Guarantees of Origin:(Wannack, 2022) (Almutairi et al., 2022) (Wannack, 2023)

- **Unchangeable recordings of Original:** Blockchain technology offers a transparent and immutable ledger system. The blockchain can record the carbon footprint and other sustainability indicators for each batch of green hydrogen that is created. The verifiability and origin tracking of this information guarantees that green hydrogen's environmental claims are genuine.
- **Streamlining Certification Processes:** Streamlining Certification Processes: Blockchain technology has the potential to standardize and streamline the certification process for green hydrogen, by automating the certification procedure according to predetermined sustainability criteria is possible with the use of smart contracts. Moreover, eliminating room for prejudice and human mistakes, this technology ensures that certifications are more consistent and dependable.
- **Strengthening Global Trade of Hydrogen and Derivatives:** Because blockchain standardizes data formats and verification techniques across borders in real-time, it allows a more smooth worldwide clean hydrogen market and its derivatives . It streamlines international commerce by allowing certification data to be shared between regulatory organizations and nations, which builds trust in hydrogen certificates and guarantees conformity with international standards and customs clearance
- **Harmonization of Standards and Ecolabeling:** A uniform approach to green hydrogen certification may be achieved by keeping standardized ecolabeling information and helping to harmonize multiple national and international requirements.

2.2.2. Blockchain IN Transportation and Distribution of Green Hydrogen

Using blockchain technology, we can address the following issues with green hydrogen transportation in a new and exciting way (Barja-Martinez et al., 2021) (Khezami et al., 2022) (Ferraro et al., 2023):

- Hydrogen transit, including storage and compression conditions, may be monitored in real-time using blockchain technology. All the way through the transit procedure, this guarantees safety and integrity.
- It is feasible to react promptly to safety hazards or leakages in the hydrogen transportation process by capturing sensor data on the blockchain.

- Through the effective management and automation of logistical operations, blockchain technology may optimize transportation costs. By automating the execution of agreements, smart contracts may save operating expenses by eliminating the need for human interaction.
- The most efficient routes for hydrogen transportation may be determined with the use of blockchain technology, AI, and the Internet of Things, guaranteeing that the routes are both cost-effective and ecologically beneficial.
- Alternative hydrogen carriers, such as ammonia or methanol, may be traced and verified using blockchain technology. Through their conveyance, this guarantees that these carriers remain pure and undamaged.
- The integration of green hydrogen into the current gas infrastructure may be facilitated by blockchain technology. The new pipelines and infrastructure's operating data, maintenance schedules, and compliance records may be securely and transparently managed on this platform.
- By offering a centralized platform for regulatory compliance, customs clearance, and certification across numerous countries, blockchain technology may streamline the international commerce of green hydrogen.

2.2.3 Innovations and Efficiency in Storage Blockchain-Enhanced Green Hydrogen

Innovative options for controlling the storage of green hydrogen are offered by blockchain technology, which addresses various challenges:

- Saving data on gas density, pressure, and temperature for efficient and safe storage. Operators may be alerted to any safety hazards or storage conditions that are not ideal in real-time via this monitoring.
- Blockchain systems can monitor power use throughout the liquefaction, compression, and storage procedures by connecting to Internet of Things (IoT) devices and smart sensors. Areas of energy waste and potential for efficiency gains may be better pinpointed with the use of this data.
- Oversight of infrastructure development projects and their funding. Improvements in budgeting and cost analysis were made possible by the safekeeping of information pertaining to infrastructure investments, maintenance plans, and operating expenses.
- Monitor the production, storage, and final usage of hydrogen during its entire lifespan. Documenting the purification operations and checking the hydrogen purity are all part of this. Blockchain technology aids in reducing contamination concerns and preserving the purity of stored hydrogen by keeping a transparent record of its path.
- Data pertaining to safety standards and regulatory requirements may be stored and managed via blockchain systems. Hydrogen storage facilities may be better guaranteed to adhere to all applicable safety and environmental requirements with the use of this consolidated database of regulatory information.
- Innovative storage options, such as metal hydrides or liquid organic hydrogen carriers, are encouraged to be developed by blockchain technology. New storage solutions may be developed more quickly with the use of blockchain technology, which allows for data exchange and cooperation among academics and industry participants.

2.2.4 Managing the Green Hydrogen Consumption Chain

The integration of blockchain technology with a green hydrogen supply chain allows for the creation of an open, secure, and convenient information-sharing platform and offers innovative solutions for consumption and marketing (Ratanasopitkul, 2018) (Wang et al., 2022) (Wannack, 2022) (Kirli et al., 2022) (Wannack, 2023):

- Blockchain can provide a decentralized marketplace for green hydrogen, making it easier for industries to access and adopt hydrogen technologies.
- Green hydrogen distribution stations can be unified into a shared network utilizing blockchain technology.
- Adaptable supply chain by tracking and analyzing real-time data on hydrogen demand from various end-users. This can help in adjusting production and distribution to meet specific volume, purity, and delivery format requirements.
- Maximizing Station Efficiency: Making stations available to the public during idle times increases their usage efficiency.
- Track the performance and maintenance needs of hydrogen technologies, enabling better after-sales service. This can be achieved through a blockchain-based registry that records product history, maintenance records, and customer feedback.
- A blockchain network can coordinate efforts among stakeholders to overcome logistical and regulatory barriers for data exchange in managing a shared robust hydrogen Network. This includes managing hydrogen refuelling stations and industrial supply chains more efficiently and be used that hydrogen distribution aligns with the grid's capacity and safety standards.
- Classification of hydrogen stations into public, special, and self-use categories. Public stations are open to all, while special and self-use stations can be made available to the public during specific times.
- Ensuring secure and transparent transactions while setting fair and reliable hydrogen pricing.
- Blockchain can facilitate safe and efficient integration with existing energy infrastructures by paying in comprehensive planning and investment for hydrogen fuel stations and central distribution sites.
- By ensuring transparency and accountability in project development and execution, blockchain can support the widespread adoption of green hydrogen as a clean energy carrier.
- Enhancing Consumer Experience: Offering a user-friendly platform to find the nearest and most cost-effective hydrogen stations and Ensuring the quality of distributed hydrogen and the reliability of the service through blockchain monitoring.
- Ensuring secure and transparent transactions while setting fair and reliable hydrogen pricing.
- Optimizing the Use of Hydrogen Stations. Ex: approach two-way dynamic pricing mechanism for hydrogen filling stations enhanced by blockchain technology.
- Providing reliable data to station owners and consumers for making informed decisions about hydrogen usage and distribution.
- Effective monitoring and regulation of the network by governmental bodies to ensure compliance with standards.
- Encouraging more participants to add their stations to the network through financial or regulatory incentives.

- Operational Transparency of operations and transactions, enhancing trust and reliability among all stakeholders.
- Utilization of smart contracts for automated transactions, billing, and record-keeping, further streamlining the hydrogen distribution process.
- Building Appropriate Interfaces with Preexisting User Systems This includes studying the hydrogen process's interface needs, outlining those needs for data, and building designs for nomination, portfolio management, trading, and billing systems' interfaces.

2.2.5 Blockchain's Role in Green Hydrogen Ecosystems

Blockchain technology can effectively address various challenges in green hydrogen supply chains (Bazaee et al., 2020) (Wannack, 2023) (Wannack, 2022):

- The public may be better informed about the long-term viability of green hydrogen with the aid of blockchain technology.
- Stakeholders may monitor the whole hydrogen production and distribution lifecycle with the use of supply chain risk management, which helps to spot inefficiencies and hazards.
- A Comprehensive Perspective for Policymakers and Decision-makers: for use in analysing and making decisions. May be used to evaluate green hydrogen's societal, economic, and ecological effects.
- The comprehensive monitoring and documentation of various stakeholders' contributions guarantee that all players in the supply chain, including local communities, are acknowledged and reimbursed properly, which promotes community engagement and equitable benefits.
- Generates stable employment prospects by launching blockchain-focused businesses. It helps build the workforce in the expanding GREEN hydrogen industry by luring qualified individuals and offering training and certification programs.
- Facilitates the development of novel business models, which in turn encourages the launch of new businesses.
- Helps boost value creation by improving economic viability and technological capabilities.
- Encourages R&D while easing the flow of information between academic institutions and businesses. Within the context of the sector coupling framework, this cooperative setting encourages scientific advancement and is in harmony with strategic innovation management.

CONCLUSION

In light of blockchain's potential to enhance efficiency, traceability, and transparency in the renewable energy sector, its application in managing the green hydrogen supply chain has garnered significant interest. The distributed and immutable ledger system provided by blockchain technology has the potential to revolutionize the management of green hydrogen supply chains. This is due to its ability to:

1- Improves operations related to transportation, trade, and transactions throughout the supply chain for the hydrogen market.
2- Supporting social, environmental, and financial sustainability in a green hydrogen market.

3- Make green hydrogen supply chains far more standardized, certified, and trustworthy. Helps the green hydrogen industry expand by making sustainability claims more credible, standardizing internationally, and easing efficient worldwide commerce.
4- Automating transactions, invoicing, and record-keeping using smart contracts significantly streamlines the hydrogen distribution process.
5- Blockchain network maximizes resource consumption, reduces waste, and increases efficiency, which brings us to point number six: improved use of resources.
6- A blockchain-based system has the potential to promote the use of green hydrogen, an eco-friendly and sustainable energy source, which may help with environmental sustainability.
7- Development and upkeep of a hydrogen distribution network by providing a sustainable financial model for owners of hydrogen stations.
8- A green hydrogen market that is more responsive and dynamic, able to adjust to changes in supply and demand.

Ultimately, blockchain technology aids in the adoption of a green hydrogen economy, which may assist with decarbonization, innovation, and sustainability. However, it also requires stakeholders to collaborate, invest strategically, and prepare in advance to manage the intricacies of the supply chain. Green hydrogen holds great promise as a sustainable and environmentally friendly energy carrier, but several obstacles must be overcome before its full potential can be realized. Policymakers, industry players, and researchers could all benefit from a deeper understanding of these implications and consequences. Academic literature and studies shed light on the obstacles and potential solutions within the green hydrogen supply chain, providing a balanced view of the challenges and rewards of transitioning towards a hydrogen economy with minimal carbon emissions.

REFERENCES

Abreu, V. H. S. D., Coralli, A., Proença, L. F. C., Santos, A. S. (2023). An Overview of Life Cycle Assessment of Renewable Hydrogen. *ENSUS2023 - XI Encontro de Sustentabilidade Em Projeto*, 224–240. doi:10.29183/2596-237x.ensus2023.v11.n2.p224-240

Aggarwal, S., & Kumar, N. (2021). Core components of blockchain. In *Advances in Computers* (Vol. 121, pp. 193–209). Elsevier, doi:10.1016/bs.adcom.2020.08.010

Ait Bachir, L. (2021). *Algerian Green Hydrogen Production Opportunities and challenges in light of a sustainable energy system*. Academic Press.

Ajanovic, A., Sayer, M., & Haas, R. (2022). The economics and the environmental benignity of different colors of hydrogen. *International Journal of Hydrogen Energy*, *47*(57), 24136–24154. doi:10.1016/j.ijhydene.2022.02.094

Akhtar, M. (2022). Industry 4.0 Technologies Impact on Supply Chain Sustainability. In T. Bányai, Á. Bányai, & I. Kaczmar (Eds.), *Supply Chain—Recent Advances and New Perspectives in the Industry 4.0 Era*. IntechOpen. doi:10.5772/intechopen.102978

Akhtar, M. S., Dickson, R., & Liu, J. J. (2021). Life Cycle Assessment of Inland Green Hydrogen Supply Chain Networks with Current Challenges and Future Prospects. *ACS Sustainable Chemistry & Engineering*, *9*(50), 17152–17163. doi:10.1021/acssuschemeng.1c06769

Almutairi, K., Hosseini Dehshiri, S. J., Hosseini Dehshiri, S. S., Hoa, A. X., Arockia Dhanraj, J., Mostafaeipour, A., Issakhov, A., & Techato, K. (2022). Blockchain Technology Application Challenges in Renewable Energy Supply Chain Management. *Environmental Science and Pollution Research International*, *30*(28), 72041–72058. doi:10.1007/s11356-021-18311-7 PMID:34989989

Anatolievna, K. E., & Vladimirovna, B. V. (2020). *The main features of blockchain technologies classification*. doi:10.5281/ZENODO.4291075

Andreev, R. A., Andreeva, P. A., Krotov, L. N., & Krotova, E. L. (2018). Review of Blockchain Technology: Types of Blockchain and Their Application. *Intellekt. Sist. Proizv.*, *16*(1), 11. doi:10.22213/2410-9304-2018-1-11-14

Angrish, A., Craver, B., Hasan, M., & Starly, B. (2018). A Case Study for Blockchain in Manufacturing: "FabRec": A Prototype for Peer-to-Peer Network of Manufacturing Nodes. *Procedia Manufacturing*, *26*, 1180–1192. doi:10.1016/j.promfg.2018.07.154

Arcos, J. M. M., & Santos, D. M. F. (2023). The Hydrogen Color Spectrum: Techno-Economic Analysis of the Available Technologies for Hydrogen Production. *Gases*, *3*(1), 1. Advance online publication. doi:10.3390/gases3010002

Armaroli, N., & Balzani, V. (2011). The Hydrogen Issue. *ChemSusChem*, *4*(1), 21–36. doi:10.1002/cssc.201000182 PMID:21226208

Azadnia, A. H., McDaid, C., Andwari, A. M., & Hosseini, S. E. (2023). Green hydrogen supply chain risk analysis: A european hard-to-abate sectors perspective. *Renewable & Sustainable Energy Reviews*, *182*, 113371. doi:10.1016/j.rser.2023.113371

Azzi, R., Chamoun, R. K., & Sokhn, M. (2019). The power of a blockchain-based supply chain. *Computers & Industrial Engineering*, *135*, 582–592. doi:10.1016/j.cie.2019.06.042

Bade, S. O., Tomomewo, O. S., Meenakshisundaram, A., Ferron, P., & Oni, B. A. (2024). Economic, social, and regulatory challenges of green hydrogen production and utilization in the US: A review. *International Journal of Hydrogen Energy*, *49*, 314–335. doi:10.1016/j.ijhydene.2023.08.157

Balat, M. (2008). Potential importance of hydrogen as a future solution to environmental and transportation problems. *International Journal of Hydrogen Energy*, *33*(15), 4013–4029. doi:10.1016/j.ijhydene.2008.05.047

Barde, S., Barde, A., Barde, S., & Barde, A. (2023). *Blockchain Technology: Worldwide Supplier of Digital Services and Technology – Tactics, Methods, Limitations, and Performance*. Https://Services.Igi-Global.Com/Resolvedoi/Resolve.Aspx?Doi=10.4018/978-1-6684-7808-0.Ch012

Barja-Martinez, S., Aragüés-Peñalba, M., Munné-Collado, Í., Lloret-Gallego, P., Bullich-Massagué, E., & Villafafila-Robles, R. (2021). Artificial intelligence techniques for enabling Big Data services in distribution networks: A review. *Renewable & Sustainable Energy Reviews, 150*, 111459. doi:10.1016/j.rser.2021.111459

Bazaee, G., Hassani, M., & Shahmansouri, A. (2020). Identifying blockchain technology maturity's levels In the oil and gas industry with a grounded theory approach. *Petroleum Business Review*. doi:10.22050/pbr.2020.256181.1138

Bertuccioli, L., Chan, A., Hart, D., Lehner, F., Madden, B., & Standen, E. (2014). Development of water electrolysis in the European union. Final report. *Fuel Cells and Hydrogen Joint Undertaking*.

Blaine, F. (2021, August 12). *Touted as clean, 'blue' hydrogen may be worse than gas or coal | Cornell Chronicle*. https://news.cornell.edu/stories/2021/08/touted-clean-blue-hydrogen-may-be-worse-gas-or-coal

Bobbert, Y., & Ozkanli, N. (2020). LockChain Technology as One Source of Truth for Cyber, Information Security and Privacy. In K. Arai, S. Kapoor, & R. Bhatia (Eds.), *Intelligent Computing* (Vol. 1230, pp. 336–347). Springer International Publishing. doi:10.1007/978-3-030-52243-8_24

Bockris, J. O., & Klerer, J. (1976). Energy: The Solar Hydrogen Alternative. *Journal of the Electrochemical Society, 123*(8), 284C. doi:10.1149/1.2133054

Bockris, J. O., & Klerer, J. (1978). *Solar-Hydrogen Energy Systems An Authoritative Review of Water-Splitting Systems by Solar Beam and Solar Heat: Hydrogen Production, Storage and Utilisation*. Pergamon Press.

Bollmann, J., Pitchaimuthu, S., & Kühnel, M. F. (2023). Challenges of Industrial-Scale Testing Infrastructure for Green Hydrogen Technologies. *Energies, 16*(8), 3604. doi:10.3390/en16083604

Bouckaert, S., Pales, A. F., McGlade, C., Remme, U., Wanner, B., Varro, L., D'Ambrosio, D., & Spencer, T. (2021). *Net zero by 2050: A roadmap for the global energy sector*. Academic Press.

Cao, B., Wang, Z., Zhang, L., Feng, D., Peng, M., & Zhang, L. (2021). *Blockchain Systems, Technologies and Applications: A Methodology Perspective* (arXiv:2105.03572). arXiv. http://arxiv.org/abs/2105.03572

Cayirci, E., & Rong, C. (2018). Blocknetwork for Big Information Fusion and NoSQL Knowledgebase Management. *2018 IEEE Seventh International Conference on Communications and Electronics (ICCE)*, 379–384. 10.1109/CCE.2018.8465729

Chang, H.-S. (2018). *Blockchain network and method of operation thereof.* Academic Press.

Chuadhry, M. A., Bhatti, M. G., & Shah, R. A. (2023). Impact of Blockchain Technology in Modern Banking Sector to Exterminate the Financial Scams: Impact of Blockchain Technology in Modern Banking Sector to Exterminate the Financial Scams. *Sukkur IBA Journal of Computing and Mathematical Sciences, 6*(2), 27–38. doi:10.30537/sjcms.v6i2.1170

Dawood, F., Anda, M., & Shafiullah, G. M. (2020). Hydrogen production for energy: An overview. *International Journal of Hydrogen Energy, 45*(7), 3847–3869. doi:10.1016/j.ijhydene.2019.12.059

Delimatsis, P. (2018). *When Disruptive Meets Streamline: International Standardization in Blockchain* (*SSRN* Scholarly Paper 3184824). doi:10.2139/ssrn.3208237

Dixon, B. J., Tang, J., & Zhang, J. H. (2013). The evolution of molecular hydrogen: A noteworthy potential therapy with clinical significance. *Medical Gas Research*, *3*(1), 10. doi:10.1186/2045-9912-3-10 PMID:23680032

Doug, F. (2019, September 16). *How is cutting carbon like winning a world war? Applying unconventional thinking to an increasingly urgent problem.* Explore UCalgary. https://explore.ucalgary.ca/climate-change-carbon-capture-reducing-CO2

Elshafei, A. M., & Mansour, R. (2023). Green Hydrogen as a Potential Solution for Reducing Carbon Emissions: A Review. *Journal of Energy Research and Reviews*, *13*(2), 1–10. doi:10.9734/jenrr/2023/v13i2257

Energy Observer. (2023). *Quel potentiel pour l'hydrogène naturel?* Energy Observer. https://www.energy-observer.org/fr/ressources/hydrogene-naturel

Farrell, N. (2023). Policy design for green hydrogen. *Renewable & Sustainable Energy Reviews*, *178*, 113216. doi:10.1016/j.rser.2023.113216

Ferraro, M., Gallo, P., Ippolito, M. G., Massaro, F., Sanseverino, E. R., Ruffino, S., Santamaria, M., & Sciumè, G. (2023). A test bench for the production of green hydrogen and its traceability and certification using blockchain technology. *Proceedings - 2023 IEEE International Conference on Environment and Electrical Engineering and 2023 IEEE Industrial and Commercial Power Systems Europe, EEEIC / I and CPS Europe 2023*. Scopus. 10.1109/EEEIC/ICPSEurope57605.2023.10194789

García Martín, J. (2022). The future of hydrogen: Seizing today's opportunities. *Economía Industrial*, *424*, 183–184.

Germscheidt, R. L., Moreira, D. E. B., Yoshimura, R. G., Gasbarro, N. P., Datti, E., dos Santos, P. L., & Bonacin, J. A. (2021). Hydrogen Environmental Benefits Depend on the Way of Production: An Overview of the Main Processes Production and Challenges by 2050. *Advanced Energy and Sustainability Research*, *2*(10), 2100093. doi:10.1002/aesr.202100093

Goltsov, V. A., Veziroglu, T. N., & Goltsova, L. F. (2006). Hydrogen civilization of the future—A new conception of the IAHE. *International Journal of Hydrogen Energy*, *31*(2), 153–159. doi:10.1016/j.ijhydene.2005.04.045

Gorji, S. A. (2023). Challenges and opportunities in green hydrogen supply chain through metaheuristic optimization. *Journal of Computational Design and Engineering*, *10*(3), 1143–1157. doi:10.1093/jcde/qwad043

GRAAF. (2022, December). *Hydrogen's decade.* IMF. https://www.imf.org/en/Publications/fandd/issues/2022/12/hydrogen-decade-van-de-graaf

Green Hydrogen: A Guide to Policy Making. (2020). https://typeset.io/papers/green-hydrogen-a-guide-to-policy-making-yd42vr1uos

Green Hydrogen: Challenges for Commercialization - IEEE Smart Grid. (2021). https://smartgrid.ieee.org/bulletins/february-2021/green-hydrogen-challenges-for-commercialization

Green hydrogen opportunities for emerging and developing economies: Identifying success factors for market development and building enabling conditions (OECD Environment Working Papers 205; OECD Environment Working Papers, Vol. 205). (2022). doi:10.1787/19970900

Gupta, N., & Agrawal, R. (2023). Applications of Blockchain in Cyber Security Industry. In R. Agrawal & N. Gupta (Eds.), *Blockchain Applications in Cybersecurity Solutions* (pp. 48–61). Bentham Science Publishers. doi:10.2174/9789815080599123010006

Gupta, S., Sinha, S., & Bhushan, B. (2020). Emergence of Blockchain Technology: Fundamentals, Working and its Various Implementations. SSRN *Electronic Journal*. doi:10.2139/ssrn.3569577

Habash, D., Abdelsalam, E., Alnawafah, H., Almomani, F., Salameh, T., & Al Makky, A. (2023). Green Hydrogen: A Novel Hybrid Solar Chimney Power Plant Integrated with Electrolysis Station. *2023 Advances in Science and Engineering Technology International Conferences (ASET)*, 1–6. 10.1109/ASET56582.2023.10180725

Hamukoshi, S. S., Mama, N., Shimanda, P. P., & Shafudah, N. H. (2022). An overview of the socio-economic impacts of the green hydrogen value chain in Southern Africa. *Journal of Energy in Southern Africa*, *33*(3), 3. Advance online publication. doi:10.17159/2413-3051/2022/v33i3a12543

Hanif, M., & Song, H. (2019). Blocks' Network: Redesign Architecture Based on Blockchain Technology. *2019 IEEE International Conference on Industrial Internet (ICII)*, 34–39. 10.1109/ICII.2019.00017

He, G., Mallapragada, D. S., Bose, A., Heuberger, C. F., & Gençer, E. (2021). Hydrogen Supply Chain Planning With Flexible Transmission and Storage Scheduling. *IEEE Transactions on Sustainable Energy*, *12*(3), 1730–1740. doi:10.1109/TSTE.2021.3064015

Hermesmann, M., & Müller, T. E. (2022). Green, Turquoise, Blue, or Grey? Environmentally friendly Hydrogen Production in Transforming Energy Systems. *Progress in Energy and Combustion Science*, *90*, 100996. doi:10.1016/j.pecs.2022.100996

Hoffmann, C., Ziemann, N., Penske, F., & Müsgens, F. (2023). The value of secure electricity supply for increasing acceptance of green hydrogen First experimental evidence from the virtual reality lab. *2023 19th International Conference on the European Energy Market (EEM)*, 1–9. 10.1109/EEM58374.2023.10161844

Huang, H., Tian, J., Min, G., & Miao, W. (2020). Introduction to blockchains. In *Blockchains for Network Security: Principles, technologies and applications* (pp. 1–21). IET Digital Library. doi:10.1049/PBPC029E_ch1

Huang, J., Balcombe, P., & Feng, Z. (2023). Technical and economic analysis of different colours of producing hydrogen in China. *Fuel*, *337*, 127227. doi:10.1016/j.fuel.2022.127227

Hydrogen Science Coalition. (2022, December 21). Clean Hydrogen Definition—Hydrogen Science Coalition %. *Hydrogen Science Coalition*. https://h2sciencecoalition.com/briefings/clean-hydrogen-definition/

IRENA. (2023). *Creating a global hydrogen market: Certification to enable trade*. IRENA.

Ismoyo, K., Muhajirin, Iskendar, Ajidarmo, K., Triyanto, & Muis, A. (2023). Analysis of technology readiness level of dual fuel harbor tug boat research achievement. *IOP Conference Series. Earth and Environmental Science, 1166*(1), 012004. doi:10.1088/1755-1315/1166/1/012004

Jonas Moberg. (2023, June 22). *Finance for a just renewable energy and green hydrogen economy roundtable—OECD.* https://www.oecd.org/environment/cc/cefim/green-hydrogen/financeforajustrenewableenergyandgreenhydrogeneconomyroundtable.htm

Katterbauer, K., Shehri, A. A., Qasim, A., & Yousif, A. (2023). Sustainable hydrogen production utilizing artificial intelligence – a McKee reservoir study. *84th EAGE Annual Conference & Exhibition*, 1–5. 10.3997/2214-4609.202310062

Khezami, N., Gharbi, N., Neji, B., & Braiek, N. B. (2022). Blockchain Technology Implementation in the Energy Sector: Comprehensive Literature Review and Mapping. *Sustainability (Basel), 14*(23), 15826. doi:10.3390/su142315826

Kim, J. H., Hansora, D., Sharma, P., Jang, J.-W., & Lee, J. S. (2019). Toward practical solar hydrogen production – an artificial photosynthetic leaf-to-farm challenge. *Chemical Society Reviews, 48*(7), 1908–1971. doi:10.1039/C8CS00699G PMID:30855624

Kirli, D., Couraud, B., Robu, V., Salgado-Bravo, M., Norbu, S., Andoni, M., Antonopoulos, I., Negrete-Pincetic, M., Flynn, D., & Kiprakis, A. (2022). Smart contracts in energy systems: A systematic review of fundamental approaches and implementations. *Renewable & Sustainable Energy Reviews, 158*, 112013. doi:10.1016/j.rser.2021.112013

Koppers, F. (2020). *Lock for e.g. Chain, bracket or cable, forms closed circuit sensor device.* Academic Press.

Kumar, S., Arzaghi, E., Baalisampang, T., Garaniya, V., & Abbassi, R. (2023). Insights into decision-making for offshore green hydrogen infrastructure developments. *Process Safety and Environmental Protection, 174*, 805–817. doi:10.1016/j.psep.2023.04.042

Laurent, T., & Bazarkina, E. (2019). *Natural hydrogen the fuel of the 21st century.* Academic Press.

Legasov, V. (1979). *Nuclear hydrogen energy and technology. Part 2.* Atomno-vodorodnaya energetika i tekhnologiya.

Li, W., Qiu, W., Yin, K., & Hu, M. (2021). *Blockchain-based multi-stage signing method, device, computer apparatus, and storage medium.* Academic Press.

Manzardo, A., Ren, J., Toniolo, S., & Scipioni, A. (2023). Critical factors and cause-effect analysis for enhancing the sustainability of hydrogen supply chain. In A. Scipioni, A. Manzardo, & J. Ren (Eds.), *Hydrogen Economy* (2nd ed., pp. 67–111). Academic Press. doi:10.1016/B978-0-323-99514-6.00010-8

Marketsandmarket. (2023). *Green Hydrogen Market, Industry Size Forecast Report.* MarketsandMarkets. https://www.marketsandmarkets.com/Market-Reports/green-hydrogen-market-92444177.html

Matter, U. (2023). *Big Data Analytics: A Guide to Data Science Practitioners Making the Transition to Big Data* (1st ed.). Chapman and Hall/CRC. doi:10.1201/9781003378822

Mehmeti, A., Angelis-Dimakis, A., Arampatzis, G., McPhail, S. J., & Ulgiati, S. (2018). Life Cycle Assessment and Water Footprint of Hydrogen Production Methods: From Conventional to Emerging Technologies. *Environments (Basel, Switzerland)*, *5*(2), 2. Advance online publication. doi:10.3390/environments5020024

Monrat, A. A., Schelen, O., & Andersson, K. (2019). A Survey of Blockchain From the Perspectives of Applications, Challenges, and Opportunities. *IEEE Access : Practical Innovations, Open Solutions*, *7*, 117134–117151. doi:10.1109/ACCESS.2019.2936094

Monteiro, E., & Brito, P. S. D. (2023). Hydrogen supply chain: Current status and prospects. *Energy Storage*, *5*(7), e466. doi:10.1002/est2.466

Moulebe, L. P., Touati, A., Akpoviroro, E. O., & Rabbah, N. (2022). Study on performance of a green hydrogen production system integrated with the thermally activated cooling. *Acta Innovations*, *47*(47), 5–19. doi:10.32933/ActaInnovations.47.1

Müller, V. P., & Eichhammer, W. (2023). Economic complexity of green hydrogen production technologies—A trade data-based analysis of country-specific industrial preconditions. *Renewable & Sustainable Energy Reviews*, *182*, 113304. doi:10.1016/j.rser.2023.113304

Muradov, N. (2001). Hydrogen via methane decomposition: An application for decarbonization of fossil fuels. In International Journal of Hydrogen Energy (Vol. 26, Issue 11, pp. 1165–1175). doi:10.1016/S0360-3199(01)00073-8

N, V. T. (1974). Dynamics of a Universal Hydrogen Fuel System. *Hydrogen Energy*, *0*, 1309–1326.

Naik, O., Jaiswal, S., Bhuyar, K., & Kodape, S. (2023). Green Hydrogen Production as a Sustainable Initiative for Alternative Energy Source: A Review. *International Journal of Basic Sciences and Applied Computing*, *9*(10), 1–5. doi:10.35940/ijbsac.I0503.0691023

Nicolin, B. A., & Nicolin, I. (2023). Green hydrogen as an environmentally-friendly power source. *INCAS BULLETIN*, *15*(2), 141–147. doi:10.13111/2066-8201.2023.15.2.13

Olabi, A. G., Abdelkareem, M. A., Mahmoud, M. S., Elsaid, K., Obaideen, K., Rezk, H., Wilberforce, T., Eisa, T., Chae, K.-J., & Sayed, E. T. (2023). Green hydrogen: Pathways, roadmap, and role in achieving sustainable development goals. *Process Safety and Environmental Protection*, *177*, 664–687. doi:10.1016/j.psep.2023.06.069

Omid, M. A., Şahin, M. E., & Cora, Ö. N. (2024). Challenges and Future Perspectives on Production, Storage Technologies, and Transportation of Hydrogen: A Review. *Energy Technology (Weinheim)*. Advance online publication. doi:10.1002/ente.202300997

Paksoy, T., & Deveci, M. (Eds.). (2023). *Smart and sustainable operations and supply chain management in industry 4.0* (1st ed.). CRC Press. doi:10.1201/9781003180302

Pan, Y. (2019). Theoretical discovery of high capacity hydrogen storage metal tetrahydrides. *International Journal of Hydrogen Energy*, *44*(33), 18153–18158. doi:10.1016/j.ijhydene.2019.05.145

Pechlivanis, C. (2023). Implementing IoT Technology in Practice: Monitoring the Supply Chain for Sustainable Operation. *WSEAS Transactions on Systems*, *22*, 349–359. doi:10.37394/23202.2023.22.38

Ratanasopitkul, P. (2018). Blockchain – Revolutionize Green Energy Management. *2018 International Conference and Utility Exhibition on Green Energy for Sustainable Development (ICUE)*, 1–6. 10.23919/ICUE-GESD.2018.8635666

Risbud, M., Kaur, G., Dhawale, D. S., Zhu, H., Haque, N., & Giddey, S. (2023). Electrolyzer technologies for hydrogen economy. In *Hydrogen Economy* (pp. 459–485). Elsevier. doi:10.1016/B978-0-323-99514-6.00003-0

Savian, F., Tomasin Da Silva, L., Sellitto, M., Sauer Oliveira, H. L., Neri, A., & Butturi, M. A. (2023). Green Hydrogen, Industrial Symbiosis, And Blockchain: Enhancing Sustainability And Resilience In Supply Chains. ECMS 2023 Proceedings Edited by Enrico Vicario, Romeo Bandinelli, Virginia Fani, Michele Mastroianni, 350–356. doi:10.7148/2023-0350

Schneider, S., Bajohr, S., Graf, F., & Kolb, T. (2020). State of the Art of Hydrogen Production via Pyrolysis of Natural Gas. In ChemBioEng Reviews (Vol. 7, Issue 5, pp. 150–158). doi:10.1002/cben.202000014

Tejashwini, P. S., Bharath Gowda, B., Shukla, A., Chiranth, J. S., & Harish, V. (2023). Trusted Crowdfunding Platform using Smart Contracts. *2023 World Conference on Communication & Computing (WCONF)*, 1–7. 10.1109/WCONF58270.2023.10234998

Toure, I., Payman, A., Camara, M. B., & Dakyo, B. (2023). A Review on Electrolyser and Hydrogen production from wind Energy. *2023 11th International Conference on Smart Grid (icSmartGrid)*, 1–6. 10.1109/icSmartGrid58556.2023.10171040

Vallejos-Romero, A., Cordoves-Sánchez, M., Cisternas, C., Sáez-Ardura, F., Rodríguez, I., Aledo, A., Boso, Á., Prades, J., & Álvarez, B. (2022). Green Hydrogen and Social Sciences: Issues, Problems, and Future Challenges. *Sustainability (Basel)*, *15*(1), 303. doi:10.3390/su15010303

Velazquez Abad, A., & Dodds, P. E. (2020). Green hydrogen characterisation initiatives: Definitions, standards, guarantees of origin, and challenges. *Energy Policy*, *138*, 111300. doi:10.1016/j.enpol.2020.111300

Veziroglu, T. (1975a). A universal hydrogen energy system and world parameters. *Proceedings of a Symposium on Introduction to Hydrogen Energy.*

Veziroglu, T. (1975b). Hydrogen Energy. *Hydrogen Energy*. Advance online publication. doi:10.1007/978-1-4684-2607-6

Veziroglu, T. (1975c). Solar–hydrogen energy system and hydrogen production. Academic Press.

Wang, L., Jiao, S., Xie, Y., Xia, S., Zhang, D., Zhang, Y., & Li, M. (2022). Two-way dynamic pricing mechanism of hydrogen filling stations in electric-hydrogen coupling system enhanced by blockchain. *Energy*, *239*, 122194. doi:10.1016/j.energy.2021.122194

Wannack, V. (2022). Blockchain Based Hydrogen Market (BBH2)—A Paradigm-Shifting, Innovative Solution for a Climate-Friendly and Sustainable Structural Change. *Open Conference Proceedings, 2*, 59–63. 10.52825/ocp.v2i.147

Wannack, V. (2023). Blockchain Technology as a Catalyst for a Sustainable and Efficient Green Hydrogen Supply Chain. In K. Arai (Ed.), *Proceedings of the Future Technologies Conference (FTC) 2023, Volume 3* (pp. 367–372). Springer Nature Switzerland. 10.1007/978-3-031-47457-6_23

Weger, L., Abánades, A., & Butler, T. (2017). Methane cracking as a bridge technology to the hydrogen economy. *International Journal of Hydrogen Energy, 42*(1), 720–731. Advance online publication. doi:10.1016/j.ijhydene.2016.11.029

White, T. A., Whitaker, B. N., & Brewer, K. J. (2011). Discovering the balance of steric and electronic factors needed to provide a new structural motif for photocatalytic hydrogen production from water. *Journal of the American Chemical Society, 133*(39), 15332–15334. doi:10.1021/ja206782k PMID:21875106

Wind, G. (2017). *Sun onto the Grid. A Manual for Policy Makers*. Insight Series.

World Economic Forum. (2021, July 27). *Grey, blue, green – why are there so many colours of hydrogen?* World Economic Forum. https://www.weforum.org/agenda/2021/07/clean-energy-green-hydrogen/

Xinyi, Y., Yi, Z., & He, Y. (2018). Technical Characteristics and Model of Blockchain. *2018 10th International Conference on Communication Software and Networks (ICCSN)*, 562–566. 10.1109/ICCSN.2018.8488289

Xu, X., Weber, I., & Staples, M. (2019). Varieties of Blockchains. In X. Xu, I. Weber, & M. Staples, Architecture for Blockchain Applications (pp. 45–59). Springer International Publishing. doi:10.1007/978-3-030-03035-3_3

Yang, J., Sudik, A., Wolverton, C., & Siegel, D. J. (2010). High capacity hydrogen storage materials: Attributes for automotive applications and techniques for materials discovery. *Chemical Society Reviews, 39*(2), 656–675. doi:10.1039/B802882F PMID:20111786

Yang, M., Hunger, R., Berrettoni, S., Sprecher, B., & Wang, B. (2023). A review of hydrogen storage and transport technologies. *Clean Energy*, *7*(1), 190–216. doi:10.1093/ce/zkad021

Yenugula, M., Sahoo, S. K., & Goswami, S. S. (2023). Cloud computing in supply chain management: Exploring the relationship. *Management Science Letters*, *13*(3), 193–210. doi:10.5267/j.msl.2023.4.003

Zgonnik, V. (2020). The occurrence and geoscience of natural hydrogen: A comprehensive review. *Earth-Science Reviews*, *203*, 103140. doi:10.1016/j.earscirev.2020.103140

Zhang, J., Wu, J., Zou, X., Hackenberg, K., Zhou, W., Chen, W., Yuan, J., Keyshar, K., Gupta, G., Mohite, A., Ajayan, P. M., & Lou, J. (2019). Discovering superior basal plane active two-dimensional catalysts for hydrogen evolution. *Materials Today*, *25*, 28–34. doi:10.1016/j.mattod.2019.02.014

Zhao, G.-J., & Han, K.-L. (2012). Hydrogen Bonding in the Electronic Excited State. *Accounts of Chemical Research*, *45*(3), 404–413. doi:10.1021/ar200135h PMID:22070387

Zhao, W., Yan, P., Li, B., Bahri, M., Liu, L., Zhou, X., Clowes, R., Browning, N. D., Wu, Y., Ward, J. W., & Cooper, A. I. (2022). Accelerated Synthesis and Discovery of Covalent Organic Framework Photocatalysts for Hydrogen Peroxide Production. *Journal of the American Chemical Society*, *144*(22), 9902–9909. doi:10.1021/jacs.2c02666 PMID:35635501

Zheng, Z., Xie, S., Dai, H., Chen, X., & Wang, H. (2017). An Overview of Blockchain Technology: Architecture, Consensus, and Future Trends. *2017 IEEE International Congress on Big Data (BigData Congress)*, 557–564. 10.1109/BigDataCongress.2017.85

Zheng, Z., Xie, S., Dai, H. N., Chen, X., & Wang, H. (2018). Blockchain challenges and opportunities: A survey. *International Journal of Web and Grid Services*, *14*(4), 352. doi:10.1504/IJWGS.2018.095647

Zhu, Q., Johal, J., Widdowson, D. E., Pang, Z., Li, B., Kane, C. M., Kurlin, V., Day, G. M., Little, M. A., & Cooper, A. I. (2022). Analogy Powered by Prediction and Structural Invariants: Computationally Led Discovery of a Mesoporous Hydrogen-Bonded Organic Cage Crystal. *Journal of the American Chemical Society*, *144*(22), 9893–9901. doi:10.1021/jacs.2c02653 PMID:35634799

Chapter 11
Utilizing Waste Water for Sustainable Energy Generation via IoT-Integrated Technologies

Sabyasachi Pramanik
https://orcid.org/0000-0002-9431-8751
Haldia Institute of Technology, India

ABSTRACT

Innovative technologies including geothermal desalination, anaerobic digestion, microbial fuel cells, and internet of things integration may be able to aid with the combined challenges of wastewater treatment and energy sustainability. Anaerobic digestion converts organic waste into biogas, a renewable energy source that may be used to produce heat and electricity. Microbial fuel cells (MFCs) use the metabolic processes of microbes to convert organic materials into electricity, allowing for the simultaneous treatment of wastewater and the generation of renewable energy. Geothermal desalination reduces the need for fossil fuels and uses heat from geothermal sources to turn brackish or salty water into freshwater, which is less harmful to the environment. IoT connection allows geothermal desalination systems and bioenergy from wastewater processes to collect, monitor, control, and optimise data in real-time.

INTRODUCTION

Treatment of wastewater is essential for both human safety and environmental health. The facilities purify the water and maintain its purity, but they need energy, which raises the cost of operation and contributes to carbon emissions. Anaerobic digestion, microbial fuel cells, and geothermal desalination are a few examples of sustainable energy sources that researchers are investigating. The incorporation of IoT into various processes for improved efficiency and environmental advantages is also covered in this chapter. Utilizing microorganisms to break down organic material in wastewater without the need of oxygen, anaerobic digestion creates biogas as a byproduct. Through this process, organic waste is transformed into biogas, a sustainable energy source for the creation of both power and heat. Anaerobic

DOI: 10.4018/979-8-3693-1014-4.ch011

digestion is used in wastewater treatment facilities to manage organic waste and recover useful energy, which lessens the need for non-renewable resources (Llácer-Iglesias et al., 2021).

Microbial Fuel Cells (MFCs) are an innovative technology that has promise for producing sustainable energy from wastewater. They create electricity by harnessing the metabolic processes of microorganisms, enabling continuous power production. MFCs have the ability to provide simultaneous pollution removal and energy recovery in wastewater treatment systems. For use in actual applications, research strives to enhance MFC performance and scalability. Geothermal desalination is a cutting-edge technique that transforms brackish or saltwater into freshwater by combining geothermal energy with desalination procedures. This strategy addresses freshwater shortages while simultaneously producing renewable energy, resulting in high energy efficiency and decreased dependency on fossil fuels (Chandrasekhar et al., 2020).

By providing real-time monitoring, data gathering, and analysis of parameters in anaerobic digestion, MFCs, and geothermal desalination systems, IoT technology increases the efficiency of bioenergy production. Operators can optimise energy output and water treatment efficiency, detect problems, and monitor process variables remotely (Boopathi & Myilsamy, 2021; Haribalaji et al., 2021; Sampath et al., 2022). IoT-based control and automation systems provide better operational management, lower energy use, and predictive maintenance, which reduces costs and helps the environment. In order to generate sustainable energy from wastewater, this chapter examines anaerobic digestion, microbial fuel cells, and geothermal desalination methods. The advantages and difficulties of incorporating IoT for effective resource utilization and resource optimization are covered. The future of wastewater treatment and resource recovery may be more sustainable and energy-efficient if these improvements are understood (Karn et al., 2021).

Due to rising energy needs and environmental concerns, it is essential to develop and apply novel methods for the sustainable energy production from wastewater. We can lessen our influence on the environment and support a low-carbon economy by using energy potential in wastewater treatment procedures. With less dependence on outside energy sources, lower operating costs, and reduced carbon emissions, this strategy provides a solution to balance energy use in wastewater treatment facilities. Diversifying the energy mix via sustainable wastewater energy production lessens reliance on fossil fuels and its negative effects on the environment. By decentralizing production and lowering sensitivity to outside interruptions, it also provides energy security and resilience. The use of IoT in the production of bioenergy from wastewater has opportunities for improving system efficiency and resource use. According to Guo et al. (2019), this data-driven strategy promotes proactive decision-making, early identification of operational abnormalities, and focused optimization techniques, all of which lead to an increase in energy efficiency and cost-effectiveness.

Technology breakthroughs, cost-effectiveness, scalability, and regulatory frameworks are all necessary for sustainable energy production from wastewater. For acceptance and long-term sustainability, stakeholder participation, governmental backing, and public awareness are essential. The convergence of IoT, geothermal desalination, microbial fuel cells, anaerobic digestion, and microbial fuel cells offers a viable route to wastewater management that is both ecologically responsible and energy efficient. These technologies may minimize environmental effects, operating costs, and support a sustainable energy future by converting wastewater treatment facilities from energy consumers to energy producers (Aftab et al., 2020).

For the preservation of the environment, public health, and the well-being of communities, wastewater treatment and energy sustainability are essential. While energy sustainability reduces non-renewable energy usage and supports renewable and clean energy alternatives, wastewater treatment eliminates hazardous contaminants. Addressing these issues and establishing a sustainable future need an understanding of these factors. For a number of reasons, including environmental protection, public health, resource recovery, circular economy, energy efficiency, and lowering carbon footprints, wastewater treatment is crucial. (Anitha et al., 2023; Boopathi, 2023c, 2023b; Koshariya et al., 2023a; Palaniappan et al., 2023; Sathish et al., 2023) Untreated wastewater may affect aquatic ecosystems, contaminate water bodies, reduce oxygen levels, and have a detrimental influence on biodiversity. Safe discharge is guaranteed, water resources are safeguarded, ecosystems are preserved, and ecological equilibrium is maintained. Poor wastewater treatment may also endanger the public's health since it can harbor microbes and pathogens that can lead to illnesses that are transmitted via the water. Wastewater treatment supports a sustainable, circular civilization by removing important resources including energy, nutrients, and water (Meneses-Jácome et al., 2016).

Utilizing technology like energy recovery systems, cutting-edge aeration methods, and renewable energy integration to optimize energy sustainability may lower energy needs, operating costs, decrease greenhouse gas emissions, and fight climate change. The Sustainable Development Goals of the United Nations, which place a high priority on access to clean water and sanitation as well as the promotion of cheap, clean energy, depend on energy sustainability and wastewater treatment (Radeef & Ismail, 2021).

Wastewater has enormous bioenergy potential for generating sustainable energy and is a result of home, industrial, and agricultural activity. It has organic material from food scraps, agricultural runoff, and human waste that may be turned into renewable energy sources. A commonly used process called anaerobic digestion decomposes organic material without the need of oxygen, reducing complex chemicals to simpler ones and producing biogas. As long as there is organic matter available, microbial fuel cells (MFCs) can continuously produce power from wastewater. According to Liu et al. (2013), hydrothermal carbonization (HTC) transforms organic matter into hydrochar, a sustainable energy source with fewer carbon emissions.

Algae may be grown in nutrient-rich wastewater, which can be utilized as a feedstock for biofuels. Sludge-to-Energy systems reduce volume, save disposal costs, and recover energy from organic material by converting wastewater sludge into heat, electricity, or biofuels. Bioenergy may be produced from wastewater, decreasing reliance on fossil fuels, greenhouse gas emissions, and fostering a circular economy. By converting organic waste into renewable energy sources through technologies like anaerobic digestion, microbial fuel cells, hydrothermal carbonization, algae cultivation, and sludge-to-energy, environmental problems can be solved, energy sustainability can be achieved, and a resource-efficient, circular society can be promoted (Chandrasekhar et al., 2020).

In this chapter, cutting-edge technologies for wastewater treatment and energy production are discussed, including anaerobic digestion, microbial fuel cells, geothermal desalination, and IoT integration. Organic matter is converted into biogas by anaerobic digestion and energy by MFCs. Geothermal desalination turns saltwater into freshwater, eliminating the need for fossil fuels and having less of an adverse effect on the environment. Real-time data gathering, monitoring, control, and optimization are made possible through IoT integration.

DIGESTIVE ANAEROBIC

Without the need of oxygen, a biological process called anaerobic digestion decomposes organic material to create biogas. It involves a variety of microbes cooperating to break down complicated molecules into simpler ones. The concept, associated microorganisms, and the creation and composition of biogas are all covered in this section.

The Basis and Mechanism of Anaerobic Digestion

Under anaerobic digestion, complex organic substances including carbohydrates, lipids, proteins, and cellulose are broken down via biochemical processes by bacteria that can survive under anoxic conditions. According to Mohanakrishna et al. (2016) and Pant et al. (2010), anaerobic digestion occurs in four stages: hydrolysis, acidogenesis, acetogenesis, and methanogenesis.

- The first step in the breakdown of complex organic molecules into simpler ones is called hydrolysis. Hydrolytic bacteria produce extracellular enzymes that break down big molecules into sugars, amino acids, and fatty acids.
- Acidogenesis is the process by which acidogenic bacteria convert the byproducts of hydrolysis into volatile fatty acids (VFAs), organic acids, alcohols, and gases such as carbon dioxide and hydrogen.

Acetogenic bacteria produce acetic acid, hydrogen, and carbon dioxide as a result of the conversion of VFAs and intermediate products, which is essential for the creation of methane, the main precursor.

- The process of methanogenesis, which produces biogas from acetic acid and hydrogen, is the last step of anaerobic digestion. Methanogenic archaea use the products of acetogenesis to produce methane and carbon dioxide.

Microorganisms Involved in Organic Matter Degradation

An intricate microbial community made up of numerous microbial species collaborates in anaerobic digestion to break down organic materials. According to Chen et al. (2008) and Ward et al. (2008), the major microbial groupings include hydrolytic bacteria, acidogenic bacteria, acetogenic bacteria, and methanogenic archaea. By dissolving complex substances, hydrolytic bacteria create enzymes for proteins, carbohydrates, and lipids. Acidogenic bacteria convert the substances generated during hydrolysis into VFAs, alcohols, and organic acids, allowing for flexible metabolization. For the generation of precursors for methane creation, acetogenic bacteria convert VFAs and intermediates into acetic acid, hydrogen, and carbon dioxide. For maximum action in delicate conditions, methanogenic archaea—organisms that create methane via anaerobic digestion—need a steady pH and temperature.

Production and Composition of Biogas

Biogas, a combination of gases including methane, carbon dioxide, and other gases, is created during anaerobic digestion. Various elements, including feedstock, operational circumstances, and microbial community, affect the composition. Methane makes about 50–75% of biogas, along with 25–50% carbon dioxide and trace quantities of other gases. Methane is a useful renewable energy source that helps to reduce greenhouse gas emissions. Microorganisms aid in the processes of hydrolysis, acidogenesis, acetogenesis, and methanogenesis during anaerobic digestion, which turns organic waste into biogas (Ahring, 2003; Batstone et al., 2002). For optimal production and efficient use of this renewable energy source, it is essential to comprehend the concept, procedure, microorganisms, and composition of biogas. Successful projects employ anaerobic digestion to provide power, heat, and operate wastewater treatment facilities.

Through the use of combined heat and power (CHP) systems or biogas-fired generators, electricity may be produced from biogas, improving the energy efficiency of industrial operations as well as space and water heating. Additionally, it may be used to generate direct heat, heat buildings, dry crops, and heat industrial processes, all of which help to promote energy sustainability and lessen reliance on fossil fuels. Wastewater treatment and energy generation may be combined with anaerobic digestion, which can be incorporated into wastewater treatment facilities to enhance treatment procedures and recover energy. In order to handle wastewater, the municipal of Oslo in Norway built an anaerobic digestion facility. This facility generates biogas, which is then converted to biomethane and utilized as a sustainable fuel for municipal buses, lowering greenhouse gas emissions and fostering the circular economy.

Thames Water, a supplier of water and wastewater services in the UK, creates biogas from sewage sludge and food waste using anaerobic digestion plants. Energy self-sufficiency, cost savings, and the achievement of renewable energy goals are all facilitated by the use of this biogas for power production, treatment facilities, and grid excess (Angelidaki et al., 2003; Gujer & Zehnder, 1983; Van Lier et al., 2001). In Wisconsin, the Milwaukee Metropolitan Sewerage District has created a cutting-edge anaerobic digestion system that generates biogas from food scraps, fats, oils, and grease. This plant produces extra heat for on-site operations and nearby companies in addition to renewable power for approximately 2,500 residences.

In wastewater treatment facilities, anaerobic digestion generates power and heat, among other economic and environmental advantages. By turning garbage into useful resources, this strategy lowers greenhouse gas emissions, supports energy sustainability, and promotes a circular economy. Projects using anaerobic digestion provide effective wastewater treatment and energy recovery, helping to meet renewable energy goals and advance a sustainable future.

(MFCS) MICROBIAL FUEL CELLS

Microbial Fuel Cells are a promising technology that uses microbial metabolism to turn organic material into power. As a promising decentralized and environmentally friendly energy system, they combine organic matter degradation and electricity production to provide a novel approach to sustainable energy generation and wastewater treatment (Chen et al., 2008; Mohanakrishna et al., 2016; Ward et al., 2008).

Microbial Fuel Cell Principles

Extracellular electron transfer is used in microbial fuel cells to transport electrons from the oxidation of organic materials to an electrode and generate an electric current for use in a variety of applications (Chandrasekhar et al., 2020; Pant et al., 2010).

Microbial Fuel Cell Components

Figure 1 depicts the elements of microbial fuel cells for the creation of energy and the transport of electrons.

- Anode: The anode is the electrode where organic material is oxidised. In order to improve electron transport, it is often composed of conductive material like carbon cloth or graphite and coated with a catalyst like platinum or carbon nanotubes.
- Cathode: Reduction processes take place at the cathode electrode. Typically, it is formed of a substance like platinum or graphite that catalyses the reduction of oxygen. At the cathode, oxygen serves as an electron acceptor, joining with protons and electrons to create water.
- Proton Exchange Membrane (PEM): The PEM divides the fuel cell's two compartments by sitting in between the anode and the cathode. Protons may go through the PEM without combining the anode and cathode compartments. It makes it possible to produce a proton gradient, which propels the flow of electrons from the anode to the cathode.
- Microbial population: To carry out the oxidation of organic materials and transmit electrons to the anode, microbial fuel cells depend on a diversified microbial population. These microorganisms, also known as electrogenic bacteria, have special metabolic properties and build biofilms on the anode surface. Geobacter, Shewanella, and Pseudomonas species are a few examples of electrogenic bacteria that are often utilized.
- External Circuit: By connecting the anode and cathode, the external circuit enables the flow of electrons produced during microbial metabolism. Through the circuit, the electrons power external devices or recharge batteries.

Microbial Fuel Cells' Operational Principles

By oxidising the organic stuff in the fuel, such as wastewater, microbial fuel cells produce electrons and protons. Electrogenic bacteria near the anode electrode help in the transport of electrons there. A proton gradient is produced by proton migration, which happens when protons move across the proton exchange membrane. At the cathode, oxygen is reduced, becoming water when it interacts with protons

Figure 1. Components of microbial fuel cells

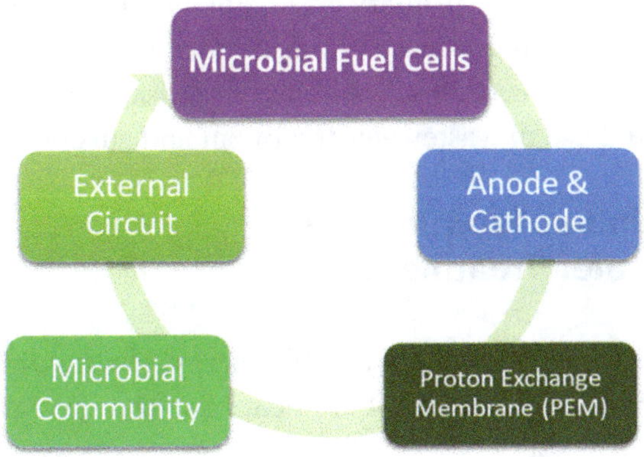

and electrons. The reduction process completes the electron transport route and produces an electric current that may power equipment or power storage.

Benefits and Uses of Microbial Fuel Cells

Microbial fuel cells (MFCs) provide renewable energy generation, low-energy sensors for sensor networks, and sustainable waste management in addition to sustainable energy production. They generate electricity from organic waste, which lessens reliance on traditional energy sources and enables long-term operation without the need for additional power sources.

Microbial fuel cells use the metabolic processes of microbes to turn organic material into energy. They function by extracellular electron transfer and have uses in sensor networks, bioenergy recovery, wastewater treatment, and the production of renewable energy. In mediator-based and mediator-less designs, MFCs provide environmentally responsible energy systems and environmental stewardship.

Energy Production through Microbiological Metabolism

Through metabolic processes, microorganisms break down chemical substances and release electrons as byproducts. Electrogenic bacteria transmit electrons directly to electrode surfaces, which is important for power production in MFCs (Liu et al., 2013; Pant et al., 2012). Organic matter oxidation, electron transfer, proton migration, electron acceptance at the cathode, and electricity generation all contribute to the energy production of MFCs. Organic material is oxidized by microorganisms in the anode chamber, releasing electrons and protons (H+). Protons go via the electrolyte or proton exchange membrane while electrons are transferred by electrogenic bacteria employing cytochromes. The electron transfer route is finished by an electron acceptor, such as molecular oxygen or another, creating water or reduced molecules.

MFC Configuration Types

MFCs may be divided into two categories according on whether a mediator is present or not:

Chemical mediators, such as ferricyanide or neutral red, are used in mediator-based systems to promote electron transport between bacteria and electrodes. These technologies increase transfer rates and broaden the selection of electrogenic microorganisms. Specialized outer membrane proteins or nanowires are used in mediator-less systems to make direct contact, which simplifies operation and lowers costs. Microbial fuel cells (MFCs) use chemical intermediaries for the transport of electrons while producing power via microbial metabolism. The two primary arrangements are mediator-based systems, which rely on chemical intermediaries, and mediator-less systems, which rely on direct electron transfer between bacteria and electrodes. Understanding these combinations is essential for improving MFC performance and investigating their potential use in wastewater treatment and sustainable energy production.

Applications of MFCs in the Production of Bioenergy and Wastewater Treatment

Especially in the fields of wastewater treatment and bioenergy generation, microbial fuel cells (MFCs) have shown considerable promise in a variety of applications. The special capacities of MFCs provide various benefits in these areas.

Water Treatment: By simultaneously treating wastewater and producing energy, MFCs provide a novel method of wastewater treatment. The MFC's microbes use the organic stuff in the wastewater as fuel, which encourages the breakdown of contaminants. MFCs provide energy recovery and nutrient removal in wastewater treatment, lowering energy demand and powering plants or grids, among other advantages of employing them in wastewater treatment. They enable microbial metabolism, transforming nutrients into less hazardous forms, and provide decentralised, sustainable, energy-efficient treatment in off-grid or distant sites.

MFCs provide a novel method for producing bioenergy by swiftly transforming organic matter into electricity. The bioenergy potential of MFCs comprises the following: By oxidizing organic matter, MFCs produce renewable electricity, making it possible for devices and batteries to be powered. They promote sustainable waste management by converting organic waste into biofuel. For off-grid populations, agricultural activities, and remote sensing systems, MFCs provide decentralized energy options.

MFCs combine energy production and wastewater treatment, encouraging resource recovery and environmental sustainability as benefits of MFC technology. They lessen their dependency on fossil fuels by using renewable resources like organic matter or wastewater. MFCs are appropriate for distant or off-grid applications because to their cheap running costs and energy recovery.

DESALINATION GEOTHERMAL

Geothermal energy is used in geothermal desalination to desalinate brackish or saltwater. This sustainable energy source uses heat that is naturally produced by the decay of materials like uranium, thorium, and potassium to extract heat from the Earth's crust. Conduction and convection processes are used to move the heat to the Earth's surface (Goosen et al., 2010; Gude, 2018; Kiaghadi et al., 2017). There are three primary categories of geothermal energy:

- High-Temperature Geothermal Energy: In areas where the Earth's crust is relatively thin, hot fluids and steam may rise closer to the surface, producing this form of geothermal energy. Geothermal resources that reach temperatures of 150 °C (302 °F) or above are ideal for steam turbine power production.
- Medium-Temperature Geothermal Energy: These resources have temperatures between 176°F and 302°F (80°C to 150°C). They may be used to direct heating tasks like district heating and space heating.

Low-temperature geothermal resources are those that have temperatures under 80°C (176°F). They are appropriate for several uses, such as desalination, water heating, and greenhouse heating.

Desalination Using Geothermal Energy

Geothermal energy may be utilized for desalination in areas with a limited supply of water, offering long-term solutions to the rising need for freshwater.

- Heat Source for Desalination: Thermal desalination and membrane-based desalination are two types of desalination technologies that may both use geothermal energy as a heat source. The heat produced may be utilized to make steam or as thermal energy for desalination.
- Thermal desalination methods: Multi-effect distillation (MED) and multi-stage flash (MSF) are two examples of thermal desalination systems that may make use of geothermal energy. These procedures include heating saltwater using geothermal energy to create steam, which is subsequently condensed to make freshwater.
- Hybrid Systems: To build hybrid systems, geothermal energy may be combined with other desalination methods. Reverse osmosis (RO) and geothermal heat, for instance, may be used to increase the desalination process' effectiveness while using less energy and enhancing system performance.

Geothermal desalination has advantages for the environment over conventional desalination techniques. It lessens reliance on fossil fuels and cuts down on greenhouse gas emissions brought on by desalination operations that need a lot of energy.

- Regional Suitability: Geothermal desalination is most suited for places with geothermal resources and water constraint, such as geothermal hot spots close to coastal areas. It offers a viable and accessible local alternative for the generation of freshwater.

For desalination applications, geothermal energy is a viable heat source, particularly when there are limited resources and water shortages. It may be incorporated with other technologies or employed in thermal procedures. Efficiency, cost-effectiveness, and environmental sustainability are all goals of research. Direct techniques and indirect approaches are the two basic categories.

Methods of Direct Geothermal Desalination

These techniques directly drive the desalination process using geothermal heat. According to these techniques, heat transfer takes place when geothermal fluid or steam comes into direct contact with the salty water (Gude, 2016; Kaczmarczyk et al., 2022). The two often used direct techniques are:

- Multi-Stage Flash (MSF) Desalination: MSF is a thermal desalination method that depends on salty water evaporating under decreased pressure. Freshwater is produced by condensing the steam that is produced by geothermal heat. The procedure involves a number of chambers or stages where the pressure is progressively reduced, resulting in the steaming of the water. Water is more easily evaporated and condensed at each cycle because each stage runs at a lower pressure and temperature. The residual brine is then separated from the freshwater that was collected during the condensation process.

Another thermal desalination method that makes use of geothermal heat is Multi-Effect Distillation (MED). Multiple evaporator units—referred to as effects—are stacked in sequence in MED. Every effect runs at a lower temperature and pressure than the one before it. Steam is produced using the geothermal heat and then transferred via the evaporator units. In each consequence, the steam warms the salt water, which causes it to evaporate. Multiple stages of evaporation and condensation are made possible by the fact that the vapour generated in one action serves as the heating source for the next. Condensation produces freshwater, which is collected as the brine is released.

Methods of Indirect Geothermal Desalination

Through the use of a heat exchanger, geothermal energy is used in these techniques to indirectly offer a heat source. Instead of coming into direct contact with the salty water, geothermal fluid or steam transfers heat to a secondary working fluid that is then employed in the desalination process. The two often used indirect techniques are:

- Reverse osmosis (RO) desalination: This membrane-based desalination technique uses a semi-permeable membrane to separate freshwater from salt water. Geothermal heat is utilized in indirect geothermal desalination to warm a secondary working fluid, such as a heat transfer fluid or a vapor compressor system. The temperature of the saline water supply before it reaches the reverse osmosis system is then raised using this heated fluid or vapor. By lowering water viscosity and promoting membrane performance, the increased temperature aids in increasing the effectiveness of the reverse osmosis procedure.
- Mechanical Vapor Compression (MVC) Desalination: This desalination method compresses vapor to raise its temperature and pressure. In indirect geothermal MVC, steam is produced from geothermal heat and mechanically compressed using a fan or a compressor. The salty water is subsequently heated by the compressed vapor, which causes it to evaporate. The leftover brine is released while the vapor is then condensed to create freshwater.

Saline water desalination processes use geothermal energy, including direct and indirect approaches. Availability of resources, required capacity, and application needs all influence the decision. In terms of efficiency, cost-effectiveness, and system complexity, each technique offers benefits and drawbacks. Enhancing performance and profitability for sustainable freshwater demand is the goal of research and development.

Combining Multiple-Effect Distillation (MED) and Reverse Osmosis (RO) with Geothermal Heat

Reverse osmosis (RO) and multiple-effect distillation (MED) desalination techniques may be efficiently combined with geothermal heat to improve overall performance. Let's look at how various techniques combine geothermal heat:

Multiple-Effect Distillation Using Geothermal Heat

Geothermal heat may be used as the main energy source in MED desalination to create steam for the evaporation process. A heat exchanger is where the geothermal fluid or steam is channeled so that it may heat the saline water supply. The MED system produces more freshwater because of the accelerated evaporation of water brought on by the greater temperature.

- Greater Efficiency: Geothermal heat offers MED a steady and dependable supply of energy. The geothermal fluid's high temperature enables quick and effective water evaporation in each effect, increasing freshwater output per unit of energy input.

- Enhanced Heat Recovery: The MED system can successfully recover and repurpose geothermal heat. The hot brine created by the condensation of steam in each effect may be utilized to heat the saline water supply in the next effects, further increasing energy efficiency.
- Lower Operating Costs: Compared to conventional fossil fuel-based energy, geothermal energy is a more affordable source of heat. Desalination can be done more cheaply by using geothermal energy in MED, which lowers operational expenses.

Reverse Osmosis (RO) With Geothermal Energy

A membrane-based desalination method called reverse osmosis may gain from the incorporation of geothermal heat. Through the preheating of the saline water supply in this instance, geothermal heat is indirectly exploited to increase the efficiency of the RO process (Aghababaie et al., 2015; Xu et al., 2015; Yaqoob et al., 2021).

- Geothermal heat increases the temperature of the saline water supply, lowers water viscosity, and increases desalination efficiency, all of which enhance membrane performance. This lowers the need for energy, leading to energy savings and increased effectiveness. Higher temperatures also deter scaling and fouling, resulting in longer membrane lifetime and less frequent maintenance.

Geothermal Desalination Advantages in Terms of Cost- and Energy-Efficiency

In comparison to conventional desalination techniques, geothermal desalination has a number of advantages in terms of energy efficiency and cost-effectiveness: geothermal energy is a renewable and environmentally friendly solution for desalination, reducing reliance on fossil fuels and promoting sustainable freshwater production. It provides geothermal power generation synergy, high energy efficiency, reduced operational costs, localized and decentralized production, and synergy with both, maximizing resource utilization and cost-effectiveness. Desalination is becoming more economically feasible in areas with a wealth of geothermal resources.

Reverse osmosis desalination and multiple-effect distillation provide energy efficiency and cost-effectiveness when combined with geothermal heat. Geothermal energy offers a dependable, renewable source of heat that cuts down on energy use, operational expenses, and dependency on fossil fuels. Geothermal desalination is a practical method for producing freshwater, particularly in areas with a lack of water and rich resources. Desalination technologies are being improved via research and development in order to be widely used and have a sustainable water future.

Case Studies of Effective Geothermal Desalination Initiatives

The Caribbean's Nevis Geothermal Desalination Plant

In order to alleviate the water shortage, the Caribbean island of Nevis has constructed a geothermal desalination plant. The project combines geothermal energy production with desalination, which uses leftover heat from the power plant. Geothermal fluid is removed from the subsurface reservoir to produce steam, which is then sent to the power generating unit to produce clean, renewable energy and desalinate water. The island's reliance on imported water is decreased by the desalination plant, which creates freshwater by repeatedly evaporating saltwater. By using renewable energy for both power generation and desalination, this successful example of integrating geothermal power generation with desalination lowers greenhouse gas emissions and highlights the advantages of waste heat for sustainable freshwater production (Kaczmarczyk et al., 2022; Kiaghadi et al., 2017).

Alaska's Akutan Geothermal-Powered Desalination Facility

Due to the shortage of freshwater resources, the island town of Akutan in Alaska has water scarcity. In order to solve this problem, a desalination facility driven by low-temperature geothermal resources was created. Reverse osmosis desalination uses waste heat while the ORC power production system produces energy. Through the removal of salt and contaminants, freshwater fit for drinking and other purposes is created. By eliminating reliance on expensive and ecologically hazardous water transport methods, the geothermal-powered desalination plant increases water security and self-sufficiency. The experiment shows that geothermal energy may be used to desalinate water in isolated areas with little access to freshwater resources.

Florida, USA: Green Cove Springs Geothermal Desalination Pilot Plant

Geothermal desalination for brackish water treatment has been successfully demonstrated at the Green Cove Springs Geothermal Desalination Pilot Plant in Florida. In the pilot plant, brackish water is heated before it enters the reverse osmosis (RO) system using a hybrid system that combines geothermal energy with RO desalination. This procedure lowers the energy needed to produce freshwater while increasing RO efficiency. The pilot plant offers important insights into energy efficacy, financial viability, and environmental sustainability while demonstrating the promise of geothermal desalination for brackish water treatment. Geothermal energy and desalination technology combination demonstrates the possibility for sustainable and green methods of freshwater production.

INTEGRATION OF IOT IN WASTEWATER BIOENERGY PROCESSES

The Internet of Things (IoT) is a network of linked systems, sensors, and gadgets that interact and share data online. With the improvement of operational effectiveness, resource optimization, and real-time monitoring and control, it has gained traction in the wastewater treatment industry. According to Boopathi, 2023a, Harikaran et al., 2023, Reddy et al., 2023, and Samikannu et al., 2023, IoT is a key factor in the transformation of energy generating systems in wastewater treatment facilities. Figure 2 depicts the integration of IOT in bioenergy from wastewater processes.

- Real-time monitoring and data collection: The Internet of Things (IoT) makes it possible to gather data in real-time from a variety of sensors placed all around the wastewater treatment facility. These sensors are capable of detecting variables including pH levels, temperature, dissolved oxygen levels, organic matter concentrations, and gas generation. Operators may learn more about how the anaerobic digestion process, microbial fuel cells, or any other bioenergy generating technologies installed into the plant are doing by regularly monitoring these metrics.

Figure 2. Integration of IoT in wastewater bioenergy processes

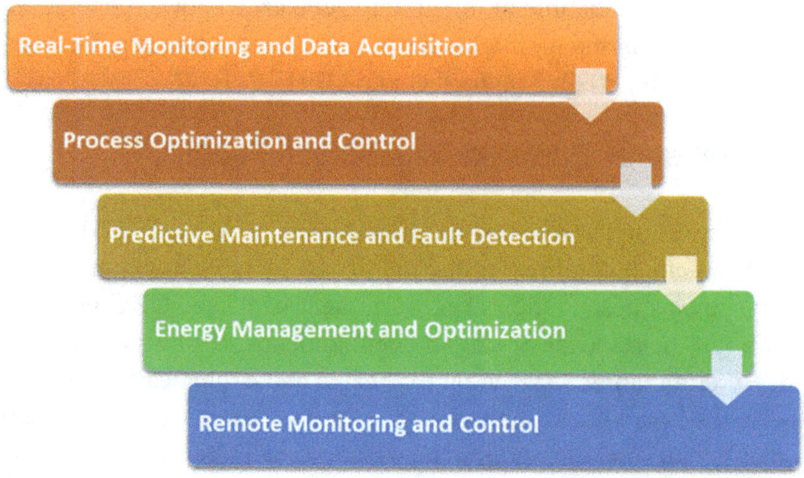

- Process Optimization and Control: The Internet of Things (IoT) makes it possible to use cutting-edge analytics and control algorithms to enhance the efficiency of bioenergy producing processes. Real-time sensor data paired with machine learning algorithms may find patterns and correlations, enabling more accurate parameter management and energy production optimization. For instance, the Internet of Things (IoT) may maximize the generation of biogas by adjusting the rate at which organic matter is fed into anaerobic digesters (Boopathi, Balasubramani, et al., 2023; Gowri et al., 2023; Janardhana et al., 2023; Sathish et al., 2023; Yupapin et al., 2023).
- Predictive Maintenance and Fault Detection: The Internet of Things (IoT) makes it easier to do predictive maintenance by continually checking the state of the equipment and identifying future faults or failures. The analysis of sensor data and use of predictive analytics algorithms enable maintenance professionals to spot early indications of equipment deterioration or anomalies. As a result, preventative maintenance is possible, downtime is reduced, and the overall effectiveness of the bioenergy producing systems is maximized.
- Energy Management and Optimization: The Internet of Things (IoT) may be integrated with energy management systems to optimize energy production and consumption in wastewater treatment facilities. To make wise choices about energy consumption, storage, and grid interaction, real-time data on energy output from bioenergy processes, energy demand from different plant activities, and energy pricing may be employed. According to Boopathi (2019, 2021, 2022a, 2022c, 2022b, 2022d), this integration provides effective energy management, cost savings, and perhaps the involvement of wastewater treatment facilities in demand response programs.
- Remote Monitoring and Control: IoT makes it possible to remotely monitor and manage devices that produce bioenergy. Remote access to real-time data and parameter control by operators enables effective administration and troubleshooting without requiring on-site physical presence. This remote accessibility improves operational flexibility, speeds up problem-solving and decision-making, and decreases reaction times.
- Data Analytics and Optimization: Advanced data analytics and optimization can be done with the vast amounts of data that IoT devices generate. In order to find trends, optimize process param-

eters, forecast energy output, and boost overall effectiveness, machine learning algorithms may analyze both historical and current data. These information may help with resource allocation decisions and continuously improve processes.

Integration of IoT into wastewater bioenergy operations has advantages for increasing sustainability, energy production, and operating cost reduction. Resource management is improved and sustainable energy is generated as a result of the shift to smart, data-driven operations.

Components and IoT in Wastewater Treatment Working Process

In order to allow data collecting, monitoring, control, and optimization, IoT in wastewater treatment needs the integration of many components and devices (Boopathi, Siva Kumar, et al., 2023; Kumara et al., 2023; Selvakumar et al., 2023; Vanitha et al., 2023). An overview of the main elements and the operation of IoT in wastewater treatment is provided below:

Devices and Sensors

- Sensors: An array of sensors are used by IoT systems to detect and gather data on variables including pH levels, temperature, dissolved oxygen, flow rates, turbidity, and chemical concentrations. The influent, primary treatment, secondary treatment, and effluent phases of the wastewater treatment process are all possible deployment locations for these sensors.
- Data Loggers: Over time, data loggers gather and store sensor data. They are essential to continuous data gathering because they make it possible to gather real-time data for analysis and management.
- Actuators: In the wastewater treatment plant, actuators are used to control and regulate a number of processes, such as altering valves, pumps, or aeration systems. They get instructions from the IoT system based on algorithms for data analysis and optimization.

Communication Facilities

- Network connection: In order to communicate data from sensors to the main control system, IoT systems depend on network connection. Both conventional (Ethernet) and wireless (Wi-Fi, cellular, or satellite) communication methods may be used to accomplish this.
- Gateway Devices: Gateway devices serve as a bridge between the central control system and the sensors. They gather information from several sensors and send it to remote or local servers for archiving and analysis.

  Cloud computing and data archiving
- Data Storage: The wastewater treatment plant's sensor data is kept in databases or cloud storage platforms. These databases make it possible to store, retrieve, and analyze historical data securely.
- Cloud Computing: Cloud platforms provide computing power and resources for machine learning, optimization techniques, and data analysis. Based on the gathered data, they allow sophisticated analytics and immediate decision-making.

Figure 3. Components and IoT in wastewater treatment working process

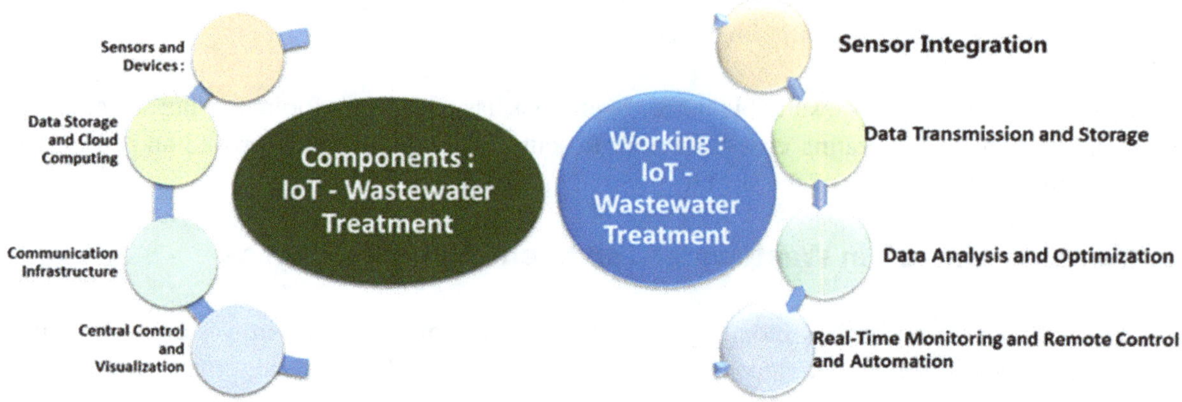

Visualisation and Central Control

- Central Control System: The central control system serves as the IoT network's brain, processing and analyzing data so that it can be utilized to make choices and manage different parts of the wastewater treatment process.
- Data Analysis and Algorithms: To gain insights, spot trends, and optimize process parameters, advanced analytics algorithms, such as machine learning and artificial intelligence approaches, are used to the obtained data.
- Visualization Tools: Graphical user interfaces (GUIs), like as dashboards, provide a simple way to show data trends, process conditions, and performance indicators. Through these interfaces, operators may keep an eye on the system, examine the data, and come to wise conclusions.

Microbial Fuel Cells With IoT Support for Real-Time Data Processing

Microorganisms are used by microbial fuel cells (MFCs) to turn organic material into power. According to Babu et al. (2023), Boopathi, Arigela, et al. (2023), and Subha et al. (2023), MFCs may improve their performance and efficiency by incorporating IoT technology. IoT may be used in MFCs in the following ways:

- Sensor Integration: The Internet of Things (IoT) makes it possible to integrate different sensors into MFCs in order to gather real-time data on variables like voltage, current, temperature, pH levels, and microbial activity. These sensors provide useful information for analysis and control by continually monitoring the MFC's functionality and state.
- Data Transmission and Storage: Through wired or wireless connectivity, the gathered sensor data is sent to a central control system or cloud platform. This information is safely kept for future analysis and is accessible from a distance for management and monitoring needs.
- Data Analysis and Optimization: IoT makes it possible to apply sophisticated analytics algorithms to the data that has been gathered. It is possible to analyze trends, find connections, and improve MFC performance using machine learning algorithms. Understanding how microorganisms be-

have, increasing the effectiveness of electricity production, and improving operational conditions may all be aided by this approach?

- Real-Time Monitoring and Alerts: IoT makes it possible to monitor MFCs in real-time and get alerts when they are performing poorly. Any deviations from the required parameters may result in warnings or notifications, enabling operators to respond quickly to resolve problems and improve the operation of the system.
- Remote Control and Automation: MFCs may be automated and controlled remotely via IoT. Based on real-time data and analysis, operators may change parameters like organic matter content or aeration rates. This option for remote control improves operational effectiveness and enables rapid reactions to optimize electricity production and microbial activity.

Microbial fuel cells for the production of bioenergy may now be better understood, optimized, and controlled in real-time thanks to the integration of IoT in MFCs.

Systems for Geothermal Desalination That Use IoT-Based Automation and Control

Geothermal heat sources are used to power the desalination process in geothermal desalination systems, turning saltwater or brackish water into freshwater. According to Anitha et al., 2023, Boopathi, Arigela, et al., 2023, and Jeevanantham et al., 2023, IoT technology may be very useful in automating and managing these systems to maximize energy use and operational efficiency. IoT may be used in geothermal desalination in the following ways:

- Sensor Integration: The Internet of Things (IoT) makes it possible to integrate sensors to track vital geothermal desalination system characteristics including temperature, pressure, flow rates, and salt levels. In order to help operators make wise choices, these sensors give real-time data on the system's performance and ambient circumstances.
- Remote Monitoring and Control: The Internet of Things (IoT) enables operators to access real-time data and control parameters from a central control system or via mobile apps. This allows for remote monitoring of geothermal desalination systems. This remote accessibility makes it easier to operate systems effectively, diagnose issues, and tweak operational settings for the best results.
- Geothermal desalination systems may use IoT-based algorithms and analytics to optimize energy use. Adjusting operational parameters, such as flow rates, temperature differences, and heat transfer rates, may be done by analyzing real-time data on energy consumption, heat exchange efficiency, and environmental variables. In addition to maximizing energy efficiency, this optimization also lowers operating expenses.
- Problem diagnosis and Maintenance: The Internet of Things (IoT) offers continuous monitoring of system parameters, which enables early problem diagnosis in geothermal desalination systems. Operators may get warnings or messages when operational conditions deviate from standard ones, allowing them to quickly plan maintenance tasks and spot possible problems. To improve maintenance schedules and save downtime, predictive maintenance algorithms may also be used.
- Integration with Energy Management Systems: Geothermal desalination systems and energy management systems may be integrated via IoT. Better coordination between energy production,

storage, and desalination processes is made possible by this integration. To maximize energy use and cost-effectiveness, demand response initiatives and real-time energy pricing may be used.

Geothermal desalination systems may increase operational effectiveness, energy optimization, remote monitoring, and maintenance by leveraging IoT-based control and automation. The sustainability and efficiency of geothermal-powered desalination processes are improved by this connection.

JOB STUDY

Control and automation in geothermal desalination systems using IoT capabilities

IoT was effectively used to geothermal desalination, energy production, and district heating at Iceland's Reykjanes Geothermal Power Plant. In addition, the facility uses waste heat from the production of power to desalinate saltwater using geothermal energy. According to many studies (Jeevanantham et al., 2023; Koshariya et al., 2023b; Reddy et al., 2023; Sahal et al., 2022; Samikannu et al., 2023), the system's performance was greatly enhanced by the incorporation of IoT technology.

IoT implementation

- Sensor Integration: The Reykjanes Power Plant's geothermal desalination system uses a variety of sensors to track vital indicators including temperature, pressure, flow rates, salt levels, and energy use. For the system to collect data in real time, these sensors are positioned in carefully chosen locations.
- Data Collection and Transmission: Through a dependable IoT network, the sensor data is gathered and delivered to a central control system. The network guarantees rapid and secure data delivery, enabling ongoing oversight and management.
- Remote Monitoring and Control: Operators have access to real-time data gathered from the sensors through the IoT-enabled central control system. Operators may make educated judgments about system parameters, energy consumption, and efficiency while remotely monitoring the geothermal desalination system's performance.
- Data Analysis and Optimization: To analyze trends, find correlations, and improve the performance of the geothermal desalination system, advanced analytics algorithms are used to the acquired data. The prediction and optimization of energy consumption, heat transfer rates, and operating parameters are made possible by machine learning approaches.
- Energy Optimization and Cost-Effectiveness: The Internet of Things-based automation and control allow for the geothermal desalination system to use energy more efficiently. Energy price data and real-time data on production and consumption are analyzed to change operational parameters to maximize energy efficiency and minimize operating expenses.

Geothermal desalination at Reykjanes Power Plant is improved by the incorporation of IoT technology, which has several advantages.

- Greater Operational Efficiency: System performance and energy consumption have been optimized by operators thanks to real-time monitoring and management of the geothermal desalination system. Improved freshwater production efficiency has resulted from changes to operating settings based on real-time data analysis.

- Energy Optimization: By using real-time data on energy use and price, the IoT-enabled system has permitted the optimization of energy usage. Costs were reduced as a consequence of this optimization, which also increased sustainability.
- Remote Monitoring and Maintenance: Geothermal desalination systems may now be monitored remotely, reducing the requirement for on-site presence and enabling effective monitoring, troubleshooting, and maintenance procedures. Early defect identification and preventative maintenance have reduced downtime and made sure everything is running smoothly.
- Enhanced Sustainability: The geothermal desalination process is now more sustainably run as a result of the inclusion of IoT technology. The facility has produced freshwater in a way that is both more ecologically friendly and economically feasible by optimizing energy use and lowering operating expenses.

IoT has the potential to improve the effectiveness, performance, and sustainability of geothermal-powered desalination operations, as shown by the Reykjanes Power Plant's successful use of IoT-enabled control and automation in its geothermal desalination system. This case study highlights the useful advantages of IoT integration in geothermal desalination systems, showing how it has a favorable effect on freshwater production and resource optimization.

CONCLUSION

The combined difficulties of wastewater treatment and energy sustainability may be addressed with the help of cutting-edge technologies including geothermal desalination, anaerobic digestion, microbial fuel cells, and IoT integration. Organic material is transformed into biogas by anaerobic digestion, a sustainable energy source that may be utilized to generate power and heat. Microbial fuel cells (MFCs) allow for the simultaneous treatment of wastewater and production of renewable energy by using microorganisms' metabolic activities to turn organic materials into electricity. By converting salty or brackish water into freshwater using heat from geothermal sources, geothermal desalination lessens the need for fossil fuels and has a less negative effect on the environment. Geothermal desalination systems and bioenergy from wastewater processes can gather, monitor, regulate, and optimize data in real-time thanks to IoT connectivity.

Case studies like the Reykjanes Geothermal Power Plant show how these technologies may be successfully implemented, leading to increased operational efficiency, energy optimization, remote monitoring, and maintenance, as well as cost-effectiveness and sustainability. A more sustainable and resilient future may be achieved by combining sustainable energy generation from wastewater with cutting-edge technology to effectively treat wastewater, produce bioenergy, and provide freshwater.

REFERENCES

Aftab, S., Shah, A., Nisar, J., Ashiq, M. N., Akhter, M. S., & Shah, A. H. (2020). Marketability Prospects of Microbial Fuel Cells for Sustainable Energy Generation. *Energy & Fuels*, *34*(8), 9108–9136. doi:10.1021/acs.energyfuels.0c01766

Aghababaie, M., Farhadian, M., Jeihanipour, A., & Biria, D. (2015). Effective factors on the performance of microbial fuel cells in wastewater treatment–a review. *Environmental Technology Reviews*, *4*(1), 71–89. doi:10.1080/09593330.2015.1077896

Ahring, B. K. (2003). Perspectives for anaerobic digestion. *Advances in Biochemical Engineering/Biotechnology*, *81*, 1–30. doi:10.1007/3-540-45839-5_1 PMID:12747559

Angelidaki, I., Ellegaard, L., & Ahring, B. K. (2003). Applications of the anaerobic digestion process. *Advances in Biochemical Engineering/Biotechnology*, *82*, 1–33. doi:10.1007/3-540-45838-7_1 PMID:12747564

Anitha, C., Komala, C. R., Vivekanand, C. V., Lalitha, S. D., Boopathi, S., & Revathi, R. (2023, February). Artificial Intelligence driven security model for Internet of Medical Things (IoMT). *Proceedings of 2023 3rd International Conference on Innovative Practices in Technology and Management, ICIPTM 2023*. 10.1109/ICIPTM57143.2023.10117713

Babu, B. S., Kamalakannan, J., Meenatchi, N., M, S. K. S., S, K., & Boopathi, S. (2023). Economic impacts and reliability evaluation of battery by adopting Electric Vehicle. *IEEE Explore*, 1–6. doi:10.1109/ICPECTS56089.2022.10046786

Batstone, D. J., Keller, J., Angelidaki, I., Kalyuzhnyi, S. V., Pavlostathis, S. G., Rozzi, A., Sanders, W. T., Siegrist, H., & Vavilin, V. A. (2002). The IWA Anaerobic Digestion Model No 1 (ADM1). *Water Science and Technology : A Journal of the International Association on Water Pollution Research*, *45*(10), 65–73. doi:10.2166/wst.2002.0292

Boopathi, S. (2019). Experimental investigation and parameter analysis of LPG refrigeration system using Taguchi method. *SN Applied Sciences*, *1*(8), 892. doi:10.1007/s42452-019-0925-2

Boopathi, S. (2021). Improving of Green Sand-Mould Quality using Taguchi Technique. *Journal of Engineering Research*. doi:10.36909/jer.14079

Boopathi, S. (2022a). An experimental investigation of Quench Polish Quench (QPQ) coating on AISI 4150 steel. *Engineering Research Express*, *4*(4), 45009. doi:10.1088/2631-8695/ac9ddd

Boopathi, S. (2022b). Cryogenically treated and untreated stainless steel grade 317 in sustainable wire electrical discharge machining process: A comparative study. *Environmental Science and Pollution Research International*, *30*(44), 1–10. doi:10.1007/s11356-022-22843-x PMID:36057706

Boopathi, S. (2022c). Experimental investigation and multi-objective optimization of cryogenic Friction-stir-welding of AA2014 and AZ31B alloys using MOORA technique. *Materials Today. Communications*, *33*, 104937. doi:10.1016/j.mtcomm.2022.104937

Boopathi, S. (2022d). Performance Improvement of Eco-Friendly Near-Dry Wire-Cut Electrical Discharge Machining Process Using Coconut Oil-Mist Dielectric Fluid. *Journal of Advanced Manufacturing Systems*. Advance online publication. doi:10.1142/S0219686723500178

Boopathi, S. (2023a). An Investigation on Friction Stir Processing of Aluminum Alloy-Boron Carbide Surface Composite. In *Materials Horizons: From Nature to Nanomaterials* (pp. 249–257). Springer. doi:10.1007/978-981-19-7146-4_14

Boopathi, S. (2023b). Deep Learning Techniques Applied for Automatic Sentence Generation. In Promoting Diversity, Equity, and Inclusion in Language Learning Environments (pp. 255–273). IGI Global. doi:10.4018/978-1-6684-3632-5.ch016

Boopathi, S. (2023c). Internet of Things-Integrated Remote Patient Monitoring System: Healthcare Application. In *Dynamics of Swarm Intelligence Health Analysis for the Next Generation* (pp. 137–161). IGI Global. doi:10.4018/978-1-6684-6894-4.ch008

Boopathi, S., Arigela, S. H., Raman, R., Indhumathi, C., Kavitha, V., & Bhatt, B. C. (2023). Prominent Rule Control-based Internet of Things: Poultry Farm Management System. *IEEE Explore*, 1–6. doi:10.1109/ICPECTS56089.2022.10047039

Boopathi, S., Balasubramani, V., & Sanjeev Kumar, R. (2023). Influences of various natural fibers on the mechanical and drilling characteristics of coir-fiber-based hybrid epoxy composites. *Engineering Research Express*, *5*(1), 15002. doi:10.1088/2631-8695/acb132

Boopathi, S., & Myilsamy, S. (2021). Material removal rate and surface roughness study on Near-dry wire electrical discharge Machining process. *Materials Today: Proceedings*, *45*(9), 8149–8156. doi:10.1016/j.matpr.2021.02.267

Boopathi, S., Siva Kumar, P. K., & Meena, R. S. J., S. I., P., S. K., & Sudhakar, M. (2023). Sustainable Developments of Modern Soil-Less Agro-Cultivation Systems. In Human Agro-Energy Optimization for Business and Industry (pp. 69–87). IGI Global. doi:10.4018/978-1-6684-4118-3.ch004

Chandrasekhar, K., Kumar, G., Venkata Mohan, S., Pandey, A., Jeon, B. H., Jang, M., & Kim, S. H. (2020). Microbial Electro-Remediation (MER) of hazardous waste in aid of sustainable energy generation and resource recovery. *Environmental Technology & Innovation*, *19*, 100997. doi:10.1016/j.eti.2020.100997

Chen, Y., Cheng, J. J., & Creamer, K. S. (2008). Inhibition of anaerobic digestion process: A review. *Bioresource Technology*, *99*(10), 4044–4064. doi:10.1016/j.biortech.2007.01.057 PMID:17399981

Goosen, M., Mahmoudi, H., & Ghaffour, N. (2010). Water Desalination using geothermal energy. *Energies*, *3*(8), 1423–1442. doi:10.3390/en3081423

Gowri, N. V., Dwivedi, J. N., Krishnaveni, K., Boopathi, S., Palaniappan, M., & Medikondu, N. R. (2023). Experimental investigation and multi-objective optimization of eco-friendly near-dry electrical discharge machining of shape memory alloy using Cu/SiC/Gr composite electrode. *Environmental Science and Pollution Research International*, *0123456789*(49), 1–19. doi:10.1007/s11356-023-26983-6 PMID:37126160

Gude, V. G. (2016). Geothermal source potential for water desalination - Current status and future perspective. *Renewable & Sustainable Energy Reviews*, *57*, 1038–1065. doi:10.1016/j.rser.2015.12.186

Gude, V. G. (2018). Geothermal Source for Water Desalination-Challenges and Opportunities. *Renewable Energy Powered Desalination Handbook: Application and Thermodynamics*, 141–176. doi:10.1016/B978-0-12-815244-7.00004-0

Gujer, W., & Zehnder, A. J. B. (1983). Conversion processes in anaerobic digestion. *Water Science and Technology*, *15*(8–9), 127–167. doi:10.2166/wst.1983.0164

Guo, Z., Sun, Y., Pan, S. Y., & Chiang, P. C. (2019). Integration of green energy and advanced energy-efficient technologies for municipal wastewater treatment plants. *International Journal of Environmental Research and Public Health*, *16*(7), 1282. doi:10.3390/ijerph16071282 PMID:30974807

Haribalaji, V., Boopathi, S., & Asif, M. M. (2021). Optimization of friction stir welding process to join dissimilar AA2014 and AA7075 aluminum alloys. *Materials Today: Proceedings*, *50*, 2227–2234. doi:10.1016/j.matpr.2021.09.499

Harikaran, M., Boopathi, S., Gokulakannan, S., & Poonguzhali, M. (2023). Study on the Source of E-Waste Management and Disposal Methods. In *Sustainable Approaches and Strategies for E-Waste Management and Utilization* (pp. 39–60). IGI Global. doi:10.4018/978-1-6684-7573-7.ch003

Janardhana, K., Anushkannan, N. K., Dinakaran, K. P., Puse, R. K., & Boopathi, S. (2023). Experimental Investigation on Microhardness, Surface Roughness, and White Layer Thickness of Dry EDM. *Engineering Research Express*, *5*(2), 025022. doi:10.1088/2631-8695/acce8f

Jeevanantham, Y. A., A, S., V, V., J, S. I., Boopathi, S., & Kumar, D. P. (2023). Implementation of Internet-of Things (IoT) in Soil Irrigation System. *IEEE Explore*, 1–5. doi:10.1109/ICPECTS56089.2022.10047185

Kaczmarczyk, M., Tomaszewska, B., & Bujakowski, W. (2022). Innovative desalination of geothermal wastewater supported by electricity generated from low-enthalpy geothermal resources. *Desalination*, *524*, 115450. doi:10.1016/j.desal.2021.115450

Karn, A. L., Pandya, S., Mehbodniya, A., Arslan, F., Sharma, D. K., Phasinam, K., Aftab, M. N., Rajan, R., Bommisetti, R. K., & Sengan, S. (2021). An integrated approach for sustainable development of wastewater treatment and management system using IoT in smart cities. *Soft Computing*, 1–17. doi:10.1007/s00500-021-06244-9

Kiaghadi, A., Sobel, R. S., & Rifai, H. S. (2017). Modeling geothermal energy efficiency from abandoned oil and gas wells to desalinate produced water. *Desalination*, *414*, 51–62. doi:10.1016/j.desal.2017.03.024

Koshariya, A. K., Kalaiyarasi, D., Jovith, A. A., Sivakami, T., Hasan, D. S., & Boopathi, S. (2023a). AI-Enabled IoT and WSN-Integrated Smart Agriculture System. In *Artificial Intelligence Tools and Technologies for Smart Farming and Agriculture Practices* (pp. 200–218). IGI Global. doi:10.4018/978-1-6684-8516-3.ch011

Koshariya, A. K., Kalaiyarasi, D., Jovith, A. A., Sivakami, T., Hasan, D. S., & Boopathi, S. (2023b). AI-Enabled IoT and WSN-Integrated Smart. *Practice, Progress, and Proficiency in Sustainability*, 200–218. doi:10.4018/978-1-6684-8516-3.ch011

Kumara, V., Mohanaprakash, T. A., Fairooz, S., Jamal, K., Babu, T., & B., S. (2023). Experimental Study on a Reliable Smart Hydroponics System. In *Human Agro-Energy Optimization for Business and Industry* (pp. 27–45). IGI Global. doi:10.4018/978-1-6684-4118-3.ch002

Liu, Y., Liu, H., Wang, C., Hou, S. X., & Yang, N. (2013). Sustainable energy recovery in wastewater treatment by microbial fuel cells: Stable power generation with nitrogen-doped graphene cathode. *Environmental Science & Technology*, *47*(23), 13889–13895. doi:10.1021/es4032216 PMID:24219223

Llácer-Iglesias, R. M., López-Jiménez, P. A., & Pérez-Sánchez, M. (2021). Hydropower technology for sustainable energy generation in wastewater systems: Learning from the experience. *Water (Basel)*, *13*(22), 3259. doi:10.3390/w13223259

Meneses-Jácome, A., Diaz-Chavez, R., Velásquez-Arredondo, H. I., Cárdenas-Chávez, D. L., Parra, R., & Ruiz-Colorado, A. A. (2016). Sustainable Energy from agro-industrial wastewaters in Latin-America. *Renewable & Sustainable Energy Reviews*, *56*, 1249–1262. doi:10.1016/j.rser.2015.12.036

Mohanakrishna, G., Srikanth, S., & Pant, D. (2016). Bioprocesses for waste and wastewater remediation for sustainable energy. In *Bioremediation and Bioeconomy* (pp. 537–565). Elsevier. doi:10.1016/B978-0-12-802830-8.00021-6

Palaniappan, M., Tirlangi, S., Mohamed, M. J. S., Moorthy, R. M. S., Valeti, S. V., & Boopathi, S. (2023). Fused Deposition Modelling of Polylactic Acid (PLA)-Based Polymer Composites. In Development, Properties, and Industrial Applications of 3D Printed Polymer Composites (pp. 66–85). IGI Global. doi:10.4018/978-1-6684-6009-2.ch005

Pant, D., Singh, A., Van Bogaert, G., Irving Olsen, S., Singh Nigam, P., Diels, L., & Vanbroekhoven, K. (2012). Bioelectrochemical systems (BES) for sustainable energy production and product recovery from organic wastes and industrial wastewaters. *RSC Advances*, *2*(4), 1248–1263. doi:10.1039/C1RA00839K

Pant, D., Van Bogaert, G., Diels, L., & Vanbroekhoven, K. (2010). A review of the substrates used in microbial fuel cells (MFCs) for sustainable energy production. *Bioresource Technology*, *101*(6), 1533–1543. doi:10.1016/j.biortech.2009.10.017 PMID:19892549

Radeef, A. Y., & Ismail, Z. Z. (2021). Bioelectrochemical treatment of actual carwash wastewater associated with sustainable energy generation in three-dimensional microbial fuel cell. *Bioelectrochemistry (Amsterdam, Netherlands)*, *142*, 107925. doi:10.1016/j.bioelechem.2021.107925 PMID:34392137

Reddy, M. A., Reddy, B. M., Mukund, C. S., Venneti, K., Preethi, D. M. D., & Boopathi, S. (2023). Social Health Protection During the COVID-Pandemic Using IoT. In *The COVID-19 Pandemic and the Digitalization of Diplomacy* (pp. 204–235). IGI Global. doi:10.4018/978-1-7998-8394-4.ch009

Sahal, B. C., R, D., A, A., Thrinath, B. V. S., Boopathi, S., J. R., & Sudhakar, M. (2022). *IoT based smart energy meter for smart grid*. Academic Press.

Samikannu, R., Koshariya, A. K., Poornima, E., Ramesh, S., Kumar, A., & Boopathi, S. (2023). Sustainable Development in Modern Aquaponics Cultivation Systems Using IoT Technologies. In *Human Agro-Energy Optimization for Business and Industry* (pp. 105–127). IGI Global. doi:10.4018/978-1-6684-4118-3.ch006

Sampath, B., Pandian, M., Deepa, D., & Subbiah, R. (2022). Operating parameters prediction of liquefied petroleum gas refrigerator using simulated annealing algorithm. *AIP Conference Proceedings*, *2460*(1), 70003. doi:10.1063/5.0095601

Sathish, T., Sunagar, P., Singh, V., Boopathi, S., Sathyamurthy, R., Al-Enizi, A. M., Pandit, B., Gupta, M., & Sehgal, S. S. (2023). Characteristics estimation of natural fibre reinforced plastic composites using deep multi-layer perceptron (MLP) technique. *Chemosphere*, *337*(June), 139346. doi:10.1016/j.chemosphere.2023.139346 PMID:37379988

Selvakumar, S., Adithe, S., Isaac, J. S., Pradhan, R., Venkatesh, V., & Sampath, B. (2023). A Study of the Printed Circuit Board (PCB) E-Waste Recycling Process. In Sustainable Approaches and Strategies for E-Waste Management and Utilization (pp. 159–184). IGI Global.

Subha, S., Inbamalar, T. M., Komala, C. R., Suresh, L. R., Boopathi, S., & Alaskar, K. (2023, February). A Remote Health Care Monitoring system using internet of medical things (IoMT). *Proceedings of 2023 3rd International Conference on Innovative Practices in Technology and Management, ICIPTM 2023*. 10.1109/ICIPTM57143.2023.10118103

Van Lier, J. B., Tilche, A., Ahring, B. K., Macarie, H., Moletta, R., Dohanyos, M., Hulshoff Pol, L. W., Lens, P., & Verstraete, W. (2001). New perspectives in anaerobic digestion. *Water Science and Technology*, *43*(1), 1–18. doi:10.2166/wst.2001.0001 PMID:11379079

Vanitha, S. K. R., & Boopathi, S. (2023). Artificial Intelligence Techniques in Water Purification and Utilization. In *Human Agro-Energy Optimization for Business and Industry* (pp. 202–218). IGI Global. doi:10.4018/978-1-6684-4118-3.ch010

Ward, A. J., Hobbs, P. J., Holliman, P. J., & Jones, D. L. (2008). Optimisation of the anaerobic digestion of agricultural resources. *Bioresource Technology*, *99*(17), 7928–7940. doi:10.1016/j.biortech.2008.02.044 PMID:18406612

Xu, B., Ge, Z., & He, Z. (2015). Sediment microbial fuel cells for wastewater treatment: Challenges and opportunities. *Environmental Science. Water Research & Technology*, *1*(3), 279–284. doi:10.1039/C5EW00020C

Yaqoob, A. A., Ibrahim, M. N. M., Umar, K., Parveen, T., Ahmad, A., Lokhat, D., & Setapar, S. H. M. (2021). A glimpse into the microbial fuel cells for wastewater treatment with energy generation. *Desalination and Water Treatment*, *214*, 379–389. doi:10.5004/dwt.2021.26737

Yupapin, P., Trabelsi, Y., Nattappan, A., & Boopathi, S. (2023). Performance Improvement of Wire-Cut Electrical Discharge Machining Process Using Cryogenically Treated Super-Conductive State of Monel-K500 Alloy. *Iranian Journal of Science and Technology. Transaction of Mechanical Engineering*, *47*(1), 267–283. doi:10.1007/s40997-022-00513-0

Chapter 12
Intelligent Charging Methods for Electric Vehicles

Aicha Degla
https://orcid.org/0000-0002-0415-9479
Centre de Développement des Energies Renouvelables, Algeria

ABSTRACT

This chapter examines electric vehicles (EVs) and their charging systems, focusing on contemporary infrastructure effectiveness. It explores hybrid vehicle architectures and recharging technologies, categorizing them into traditional and autonomous chargers. The following research examines photovoltaic and wireless charging systems, electronic components, and a simple management cycle for uninterrupted operation. This chapter proposes that extensive modelling and simulation studies utilising the MATLAB/Simulink programme are conducted to evaluate the energetic performance of an electric vehicle fitted with hybrid recharge equipment.

INTRODUCTION

The adoption of electrified transportation systems has become imperative in contemporary urban settings. The utilisation of electric vehicles, as opposed to traditional fuel-powered ones, is increasingly crucial due to the technological advancements in communication techniques, as well as the benefits of tax incentives and the lower cost of electrical energy in comparison to fossil fuels (Kumar et al., 2023). Various models and architectures with internal differences have been subjected to the implementation of this transportation system. In recent times, a majority of nations have endeavoured to enhance their transportation infrastructure(Kaiser & Barstow, 2022). The advancement of new technologies and the growing worldwide focus on mitigating air pollution have significantly contributed to the increasing use of electrification in transportation networks. Hence, the primary goal for the future entails the eradication of transport systems reliant on environmentally harmful energies. Consequently, the need to substitute such systems with alternative modes powered by clean energy sources has emerged as an important requirement in several nations (Ercan et al., 2022). This implies that contemporary transport systems rely on electrical energy, either entirely or partially, which is a form of energy that does not

DOI: 10.4018/979-8-3693-1014-4.ch012

contribute to pollution. The depletion of fossil fuel reserves and growing environmental apprehensions are driving the onset of an energy transition phase.(Holechek et al., 2022) The rate at which oil may be replaced determines the dynamics of the transport industry, which currently represents 90% of energy use in transportation. In 2050 this significant growth is expected to occur in both passenger and freight sectors, driven by an increased need for transportation. The emerging world is seeing economic growth and population expansion, accompanied by an improvement in living conditions. Reductions With regards to emissions, it is projected that there will be a reduction of approximately 25% by 2030 (International Energy Agency, 2022). The motor vehicle industry is vital to these issues. Hence, the imperative nature of conducting research and development in the realm of electric vehicle technology has become evident (M. S. Bin Ahmad et al., 2022). The storage of electrical energy is considered a crucial determinant of success within this industry, as it encompasses fundamental components (He et al., 2021). Since it is important for a vehicle powered by electricity to achieve competitiveness within the market, it is imperative that the aforementioned aspect of profitability be significantly elevated. The selection of the charging system model has a direct impact on the categorization of the vehicle model in relation to other models. The investigation and examination of these charging mechanisms have been the subject of several academic research since the earliest days of accumulators. The concept of a charger being linked to a network emerged as a result (Palit et al., 2022). The typical systems in question serve to immobilise the system responsible for carrying the accumulators, hence restricting the range of motion of these accumulators (Edition, 2018). This behaviour encompasses both benefits and faults in certain applications. Regarding the purpose investigated in this paper, electric cars that utilise grid-connected chargers are confronted with a multitude of challenges (Kumar et al., 2023). From an alternative perspective, the incorporation of wireless charging for vehicles during parking was initially introduced in reference (Palani et al., 2023), subsequently leading to complications with the frequency variable. The authors in references (Omori et al., n.d.) and (Li et al., 2019) have put out a variant that involves doing a frequency analysis of this system, including both static and dynamic issues (Mohamed et al., 2022). The efficiency of charging systems is contingent upon the specific control and energy distribution mechanisms employed. These types of control methods can be classified as indirect, as they utilise techniques like as fuzzy logic (FL), neural networks (NN), and ANFIS-based strategies(F. Ahmad et al., 2021)(Azar et al., 2022)(Cao et al., 2023), which do not rely on explicit recharging system information or knowledge. Within the given background, the objective of the research provided in this chapter is to examine the contemporary trajectory of electric cars. Our attempt to provide this method aligns with the recommendations found in the existing literature. We have extensively discussed the many topologies that are present, along with the established designs. The subsequent section encompasses an overview of established and utilised recharge systems. The current and traditional structures of the subject have been effectively utilised by our team. Subsequently, employ an intelligent power distribution system to optimise energy conservation inside the battery and effectively utilise the existing energy sources. In this chapter, the authors conducted a study on the integration of several recharging devices within the vehicle, enabling its use while in motion. The recharge system comprises photovoltaic panels affixed to the vehicle's structure, which gather solar energy for storage in the power pack. Furthermore, the wireless charging apparatus is detached from the automobile to ensure regulation, even while the vehicle's operation on the rechargeable routes. The recharge system is defined via the modelling and explanation of these systems. Furthermore, the investigation of this system is conducted utilising the Matlab/Simulink software tool to get insights into the power flux and the correlation between external elements and vehicle speed with respect to the battery state of charge parameter. Consequently, the present chapter has been structured into four distinct sections. Following the introductory section, a comprehensive overview is provided, outlining the broad categorization of electric cars. The third half of the paper elucidates the architectural framework of electric cars, providing a precise and comprehensive depiction of its constituent elements. The subsequent part provides an overview of electric car charging, presents the mathematical model for the autonomous charging system, and presents the results of simulations. Ultimately, the chapter is brought to a close with a concluding section.

AXIAL-FLUX PERMANENT-MAGNET (AFPM) TOPOLOGIES

This section provides a concise explanation of the shape and features of the axial flux permanent magnet (AFPM) topology, with the aim of facilitating comprehension of AFPMs. According to the diagram presented in Fig. 1, the configuration of the axial flux permanent magnet (AFPM) machine can be categorised into four distinct types based on the combination of the rotor and stator: single-stator single-rotor (SSSR), double-stator single-rotor (DSSR), single-stator double-rotor (SSDR), and multistage AFPM.

SSSR is the most basic structure of the AFPM, and it is composed of one rotor and one stator, as shown in Figure 1A. This topology is relatively easy to manufacture and has the thinnest structure among AFPM topologies (Atacak, 2023). Thus, it is often used for driving small fans. However, it is difficult to maintain the gap between the rotor and stator due to the strong axial attraction force. To solve this problem, it is necessary to consider using thrust bearings instead of ball bearings or using a coreless stator. DSSR is a structure composed of two stators and one rotor, as shown in Figure 1B. In this topology, the attraction force is balanced because the rotor is between the two stators. Therefore, unlike other AFPM topologies, this topology does not require mechanical measures to withstand axial attraction forces (Azar et al., 2022). However, since there are two stators through which current flows, the length in the axial direction is relatively long, making it difficult to implement the flat-type generator, which is characteristic of the AFPM. SSDR is a structure composed of one stator and two rotors, as shown in Figure 1C. Since this topology has two rotors, it has more magnets than the previous two topologies. As a result, relatively high power can be achieved compared with the topologies that only use one rotor. In addition, this topology can eliminate the yoke of the stator depending on how the poles of the permanent

Figure 1. The configuration of the Axial Flux Permanent Magnet (AFPM) topology: (A) SSSR, (B) DSSR, (C) SSDR, (D) multi-stage
Kang et al. (2022)

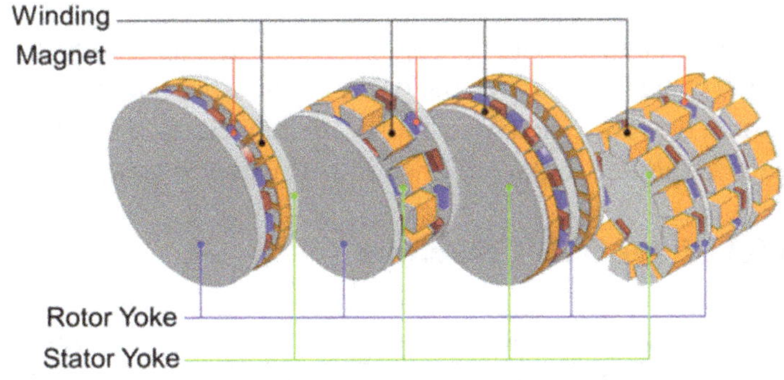

magnets attached to the two rotors are arranged, thereby reducing the loss and weight compared with other topologies (Azar et al., 2022). Multistage AFPM is a structure composed of two or more stators and rotors, as shown in Fig. 1D. This topology can produce the highest power among AFPM topologies (Atacak, 2023). However, its disadvantage is that the axial length is long, making it unsuitable for the purpose of this paper: determining the design specifications of disk-type generators. Therefore, starting from the next section, this paper does not discuss this topology.

THE DESIGN AND MATHEMATICAL MODELING OF ELECTRICAL TRACTION SYSTEMS

Theoretical Framework of the DSSR Machine

The aforementioned electric motor will be utilised in the development of the electric vehicle prototype, as mentioned in the preceding section. The electric motor under consideration is characterised by its distinct design, featuring a dual stator configuration and a singular rotor as its fundamental component. Henceforth, the motor in question shall be referred to as the Dual Stator Machine with Single Rotor (DSSR)(Yilmaz & Krein, 2013). It is crucial to illustrate the specific aspects pertaining to the internal design of this machine. This model exhibits a distinctive design characterised by the presence of two fixed stators: one located within the rotor, referred to as the winding inner rotor, and another positioned outside the rotor, known as the winding outer rotor. These two rotors have the capability to be independently stimulated or excited in tandem. The machine's internal rotor exhibits a cylindrical shape and is comprised of a permanent magnet. The potential to generate a magnetic flux can be achieved by either the outer stator, the inner rotor, or both simultaneously. Figure 2 provides a comprehensive depiction of the machine's design, illustrating the arrangement of the inner and outer stator as well as the rotor (Atacak, 2023).

Consequently, this machine has the capability to employ two distinct designs. The primary purpose of the larger stator in the initial iteration is to supply electrical power to the primary traction motor of the vehicle via a power converter. The second component (Domínguez-Navarro et al., 2019)contains the lower stator, which is intended to supply power to the Batteries system via a separate rectifier. In this context, the DC/AC machine establishes a connection between the battery pack and the outer stator component, so facilitating the traction phase. The required power is sourced by the battery pack. On the contrary, the aforementioned battery can also be charged through the internal rotor, as indicated in the corresponding diagram. Additionally, the rectifier AC/DC is linked in a manner consistent with Figure 3 (Ashok et al., 2022).

The second iteration involves the concurrent application of magnetic force to the single rotor through the utilisation of two stators. Fig. 4 illustrates the design in question and provides a magnified view of the flux movement between the inner and outer stator, which interacts with the rotor flux. It is assumed that the rotor flux remains fixed due to the presence of permanent magnets (Xie et al., 2013).

Mathematical Model of the DSSR Machine

The voltage equation is denoted by equations (1) and (2). There are six voltage equations that can be observed. The discussion pertains to the voltage of phase A in the outer stator, which is represented as V_

Figure 2. Dual stator single rotor machine design
Mahmoudi et al. (2011)

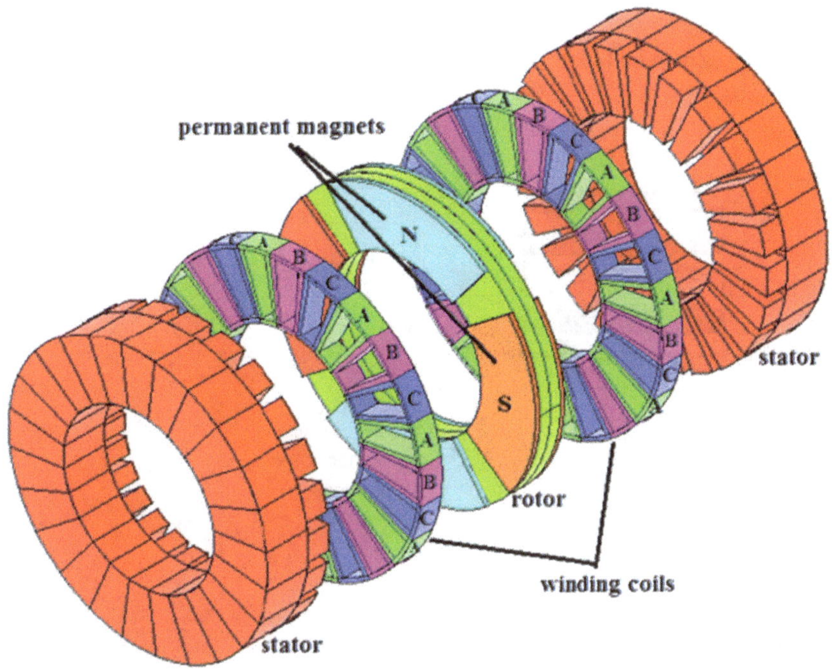

Figure 3. Dual stator single rotor machine connected to two different converters for unique traction action and a possible recharge phase

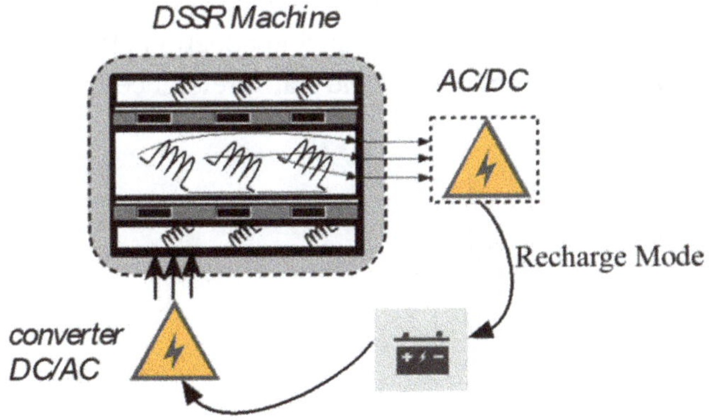

(s, a,1). The voltage of phase A of the inner stator, specifically referred to as V_ (s, a,2), is documented in reference (Zhao et al., 2020).

$$\begin{Bmatrix} V_{s,a,1} = R_{s,a,1}.I_{s,a,1} + \phi_{s,a,1} \\ V_{s,b,1} = R_{s,b,1}.I_{s,b,1} + \phi_{s,b,1} \\ V_{s,c,1} = R_{s,c,1}.I_{s,c,1} + \phi_{s,c,1} \end{Bmatrix} \quad (1)$$

Figure 4. Dual stator single rotor machine connected to two different converters for a double traction action

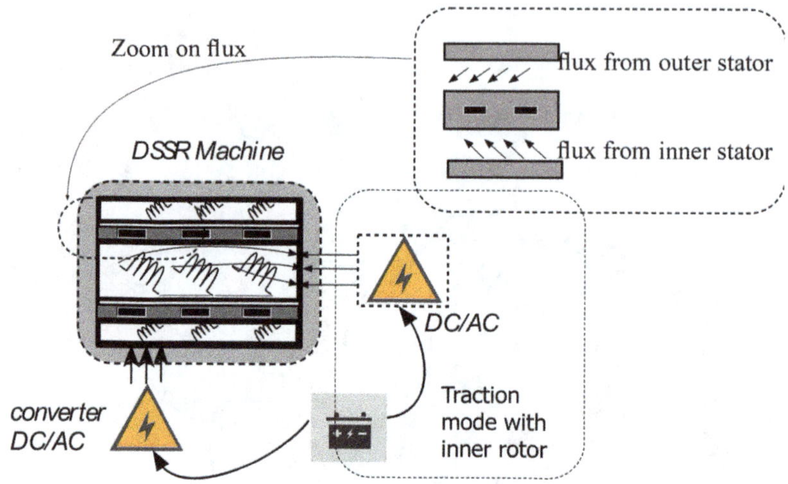

$$\begin{Bmatrix} V_{s,a,2} = R_{s,a,2}.I_{s,a,2} + \phi_{s,a,2} \\ V_{s,b,2} = R_{s,b,2}.I_{s,b,2} + \phi_{s,b,2} \\ V_{s,c,2} = R_{s,c,2}.I_{s,c,2} + \phi_{s,c,2} \end{Bmatrix} \tag{2}$$

The stator voltage signal, stator resistance, stator current, and stator flux for each phase are denoted as Vs, R, I, and ø, respectively. It is assumed that the resistance of the dual stator is uniformly distributed over its entirety. The flux equation system is introduced in equation (10). The symbol "i" serves as the phase indicator, while the symbol "j" functions as the stator reference (Nagendra et al., 2014).

$$\phi_{s,i,j} = \left(\begin{pmatrix} [L_{s,i1\to 1}] matrix 3\times 3 & \\ \begin{pmatrix} \cdot & \cdot & \cdot \\ \cdot & \cdot & \cdot \\ \cdot & \cdot & \cdot \end{pmatrix} & [L_{s,i1\to 2}] \\ [L_{s,i2\to 1}] & [L_{s,i2\to 2}] \end{pmatrix} \begin{pmatrix} \begin{pmatrix} \cdot \\ \cdot \\ \cdot \end{pmatrix} \\ [L_{s,i1}] \\ [L_{s,i2}] \end{pmatrix} \right) \tag{3}$$

The equation system presented in equation (3) is implicitly translated to the park model (4) by the equation system:

First stator (dq) model

$$\left\{\begin{aligned} V_{d,1} &= R_1 I_{d,1} - (\omega_{s,1})\phi_{q,1} + \frac{d\phi_{d,1}}{dt} \\ V_{q,1} &= R_1 I_{q,1} - (\omega_{s,1})\phi_{d,1} + \frac{d\phi_{q,1}}{dt} \end{aligned}\right\} \quad (4)$$

Second stator (dq) model

$$\left\{\begin{aligned} V_{d,1} &= R_2 I_{d,2} - (\omega_{s,2})\phi_{q,2} + \frac{d\phi_{d,2}}{dt} \\ V_{q,2} &= R_2 I_{q,2} - (\omega_{s,2})\phi_{d,2} + \frac{d\phi_{q,2}}{dt} \end{aligned}\right\}$$

Equation (5), derived from the park frame, provides a concise representation of the flux equations.

$$\left\{\begin{aligned} \phi_{d,j} &= L_{d,j} I_{d,j} + L_m \left\| I_{d,j} \right\| + \lambda_m \\ \phi_{q,j} &= L_{q,j} I_{q,j} + L_m \left\| I_{q,j} \right\| \end{aligned}\right\} \quad (5)$$

The simplification mentioned serves as the fundamental basis for the mechanical equations, namely in equation (6), which expresses the mechanical speed. It is customary for the torque generated by an electromagnet to be represented as a function of the combined flux originating from both the outer and inner stator. However, it is crucial to note that the provided torque of the electromagnet might be represented as a combination of the components from both the inner and outer stator, or solely from one of them. Nevertheless, for the purposes of this simplified model, only the flux from the main stator was considered. The given expression is represented by Equation (6). The inertia factor is denoted by the symbol J, whereas the friction coefficient is represented by the symbol B. The load torque (Tl) refers to the torque exerted on the motor during the operation of the vehicle. The expression for the applied load torque will be derived from equation (7) in the referenced model (Dheeban et al., 2019)

$$\omega_m = \frac{T_e - T_l}{(J.s + B_s)} \quad (6)$$

$$T_e = \frac{p}{\rho}\left(\frac{[\phi_{d,1} i_{q,2} + \phi_{q,1} i_{d,2}]}{[\frac{1}{\rho} + \frac{\sqrt{L_{d1}{}^2 + L^2_{q1}}}{L_m}]}\right) \quad (7)$$

Battery Model

The key component of this energy loop is the electric energy accumulator, which utilises a lithium battery variant. An essential requirement for accurately simulating and comprehensively analysing the behaviour of a battery system is the development of a mathematical model that effectively describes its electrical properties. The mathematical model presented in Fig.5 is founded on a Second-Order RC Equivalent Circuit Model (ECM) in order to describe the external electrical performance of the battery. The Electrochemical Model (ECM) consists of an internal ohmic resistance, denoted as R-0, an open circuit voltage source (OCV), and two parallel RC arrays. To effectively represent the reactivity distribution and account for the unique characteristics of the electrolyte, the use of resistance (R_1) and capacity (C_1) is employed. The battery's interface impedance is denoted by the charge transfer resistor, R_2, and the double-layer electrical capacity, C_2 (Liu et al., 2021).

Equation (8) presents the differential equations governing the extracellular matrix (ECM).

$$V_{batt/cell} = V_{oc} + R_{batt}.I + \int \frac{R_1.I - V_1}{R_1.C_1} dt + \int \frac{R_2.I - V_2}{R_2.C_2} dt \tag{8}$$

The number of parallel cells, N-pbatt, determines the voltage of the battery pack, denoted as V-batt, and series cells, N-sbatt, employed. The link in question is described by Equation (9) (Nguyen et al., 2020).

$$V_{batt} = \frac{N_{sbatt}}{N_{pbatt}} . V_{batt/cell} \tag{9}$$

The expression for the battery state of charge (SOC) can be represented as a function of time and the number of cells, in relation to the current and voltage output, as shown in equation (20) (Guo et al., 2019). It is feasible to establish the same parameter by using the battery's rated capacity, denoted as Q_n. Consequently, the equivalent equation can be expressed as equation (10).

$$SOC(t+1) = SOC(t) - \int_{t-1}^{t} \frac{1}{60} (SOC(t-1).N_{bat} - W(V_{bat}.I_{bat}))dt \tag{10}$$

Figure 5. Battery cell equivalent electrical diagram

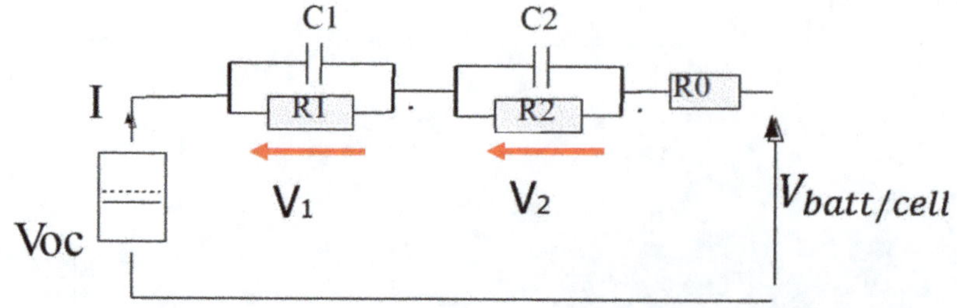

$$SOC(t+1) = SOC(t) - \int_{t}^{t+1} \frac{I(t)}{Q_n} dt \tag{11}$$

The rated capacity of a battery is determined by the extent of capacity fading. The expression for the remaining battery capacity is given in Equation (12).

$$Q_n = 3600.Q_{init}.CCF \tag{12}$$

The symbol Q-init represents the initial battery capacity in ampere-hours (Ah), whereas CCF is an abbreviation for the capacity correction factor. Additionally, it is feasible to depict the battery model as a function of the voltage charge and discharge function. Equations (13) and (14) encapsulate these forms, as referenced in

$$E_{ch}(t) = V_{bat}(t-1) + (E_{HES}(t) - E_L(t)).\eta b_{at} \tag{13}$$

$$E_{dis}(t) = V_{bat}(t-1) + (E_L(t) - E_{HES}(t)) \tag{14}$$

The variable E_ch denotes the amount of electrical energy stored in the battery during the charging process. The term "E_dis" represents the amount of electrical energy available throughout the discharging process, which is analogous to the power provided by a battery. The variable η_bat denotes the efficiency of the battery. The load demand is denoted as E_L, while E_HES represents the power supplied by the energy system, measured in watts (W). The mathematical models were implemented on the Matlab Simulink platform in order to achieve the specified battery discharge standard, as shown in Fig. 6. The discharge characteristics of a used battery are depicted in Fig. 6. This experiment examines the performance of a battery with a rated power of 1500 W and a rated capacity of 10 Ah. The battery in question has a rated voltage of 150 V and is capable of providing a discharge duration of 2 hours when subjected to a load demand of 5 A.

CONVERTER ASSOCIATION

Associated Converter With the DSSR Machine

First Version of Four Legs Inverter Associated With the DSSR Machine

As commonly understood, every component within this traction chain, such as the primary electrical motor or the battery power pack, necessitates connection to an appropriate power electronics module to ensure optimal performance of the respective element. For instance, when the diver needs to modify the velocity of the vehicle, the command will be converted into a signal that is transmitted to the electrical machine as a reference voltage and frequency. This signal is transmitted by a DC/AC converter, commonly referred to as a three-phase converter. It is imperative to note that there are two distinct designs for supplying power to the DSSR machine in the context of electrical machine design. The initial as-

Figure 6. Battery specifications

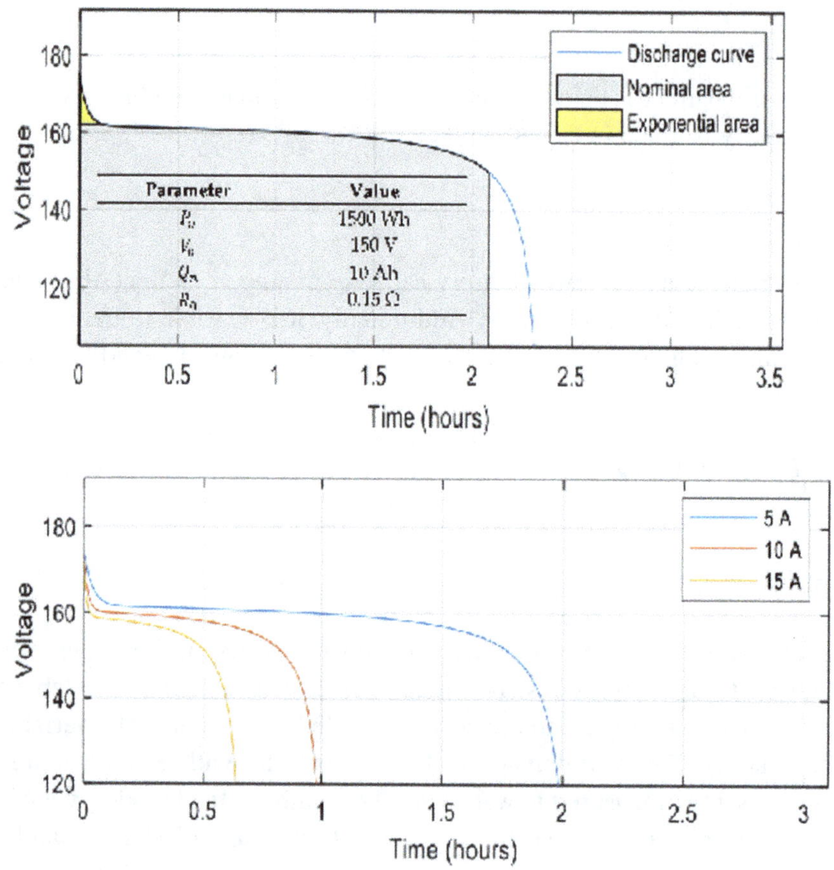

sumption is that there are two distinct three-phase converters installed, as depicted in the motor control loop diagram, or alternatively, a four-leg inverter is utilised to enable control over all phases within this unique motor design. Fig. 7 provides a comprehensive illustration of the design characteristics pertaining to both the first and second categories. The initial design displays a traditional three-phase inverter that operates using three legs, while the subsequent design illustrates an alternative variant including four legs.

When operational, this type of inverter will generate voltage equations that differ from those generated by a conventional three-phase inverter. In reality, the midpoint functions as the reference point for determining the voltages across each branch (between the two capacitors). The relationship between the output voltages and the state of the switchers can be defined using Equation (15). The four voltage phases can be represented by equation (16).

$$\begin{bmatrix} V_{aN} \\ V_{bN} \\ V_{cN} \\ V_{nN} \end{bmatrix} = \frac{V_{dc}}{2} \times \begin{bmatrix} 2S_a - 1 \\ 2S_b - 1 \\ 2S_c - 1 \\ 2S_1 - 1 \end{bmatrix} \quad (15)$$

Figure 7. Four legs inverter associated with DSSR machine

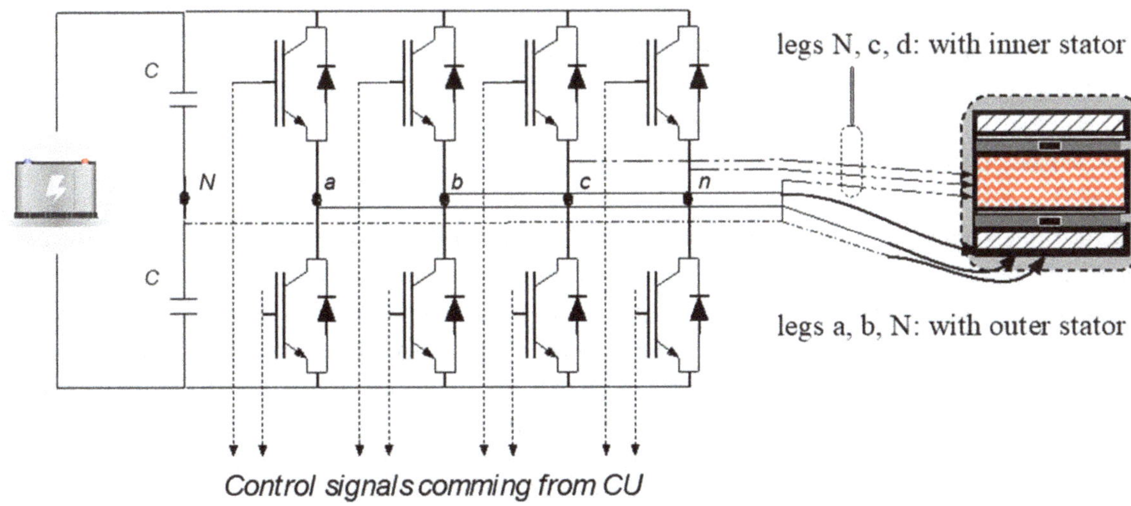

$$\begin{bmatrix} V_{an} \\ V_{bn} \\ V_{cn} \\ V_{n} \end{bmatrix} = \begin{bmatrix} 1 & 0 & 0 & -1 \\ 0 & 1 & 0 & -1 \\ 0 & 0 & 1 & -1 \\ 0 & 0 & 0 & 0 \end{bmatrix} \begin{bmatrix} V_{aN} \\ V_{bN} \\ V_{cN} \\ V_{nN} \end{bmatrix} \tag{16}$$

Second Version of the Conventional Three Legs Inverter Associated With the DSSR Machine

The inverter depicted in Figure 8 is widely utilised in the majority of three-phase AC devices. In this particular application, it is imperative to employ two out of the three available phase inverters in order to adequately supply each component with the requisite power. Each individual inverter will be associated with its respective control signals. The control signals will be computed according to the specifications given in the overarching control framework of the machine. It is crucial to note that in order to provide consistent input voltage standards, both inverters must be linked to the same battery pack. The point N represents the midpoint between the two capacities that are connected in parallel with the battery pack. The equation system denoted as (17) establishes this relationship.

$$\left\{ \begin{array}{l} V_{aN} = \frac{V_{dc}}{3}(2S_a - S_b - S_c) \\ V_{bN} = \frac{V_{dc}}{3}(2S_a - S_b - S_c) \\ V_{cN} = \frac{V_{dc}}{3}(2S_a - S_b - S_c) \end{array} \right\} \tag{17}$$

This study utilises the second iteration of the three-phase inverter and implements it within the Simulink platform to facilitate the control of the DSSR machine. The use of an IGBT transistor is observed in this particular iteration, and the many electrical parameters are concisely outlined within the application section incorporated within this report.

Association of the Converter With Battery Back

The battery pack included into the electric vehicle type need recharging whenever its state of charge reaches a low level. This can be achieved using several approaches and techniques. There are a multitude of recharge tools available to facilitate the acquisition of power within this storage system. However, a DC/DC converter is commonly seen in many applications that incorporate a battery storage system. In this particular electrical vehicle application, a buck-boost DC/DC converter is employed and integrated. The utilisation of this bidirectional DC/DC converter was implemented due to its capability to facilitate the charging of the battery during the operation of the regenerative braking system. it is feasible to recharge the battery using a dedicated equipment. The Buck-Boost (DC/DC) chopper is commonly employed in battery storage systems. Loading and unloading procedures are facilitated by their inherent reversibility. Hence, it can be observed that the voltage exhibits an increase during the discharge process, whereas it demonstrates a decrease during the charging process. The schematic diagram of the Buck-Boost chopper concept is depicted in Figure 8.

The Buck-Boost converter is characterised by two distinct modes of operation. The positioning of the IGBT, which is located internally, is a determining factor. Assuming that the two switches in question are labelled as K_10 and K_11. The correlation between the two variables can be mathematically represented by equation (18).

(18)

The converter will be linked to a battery on one end and connected to the DC bus line on the other end. V_bat represents the voltage at the battery side, while V_dc represents the voltage at the DC bus side. Hence, the correlation between the two voltages may be mathematically represented as stated in equation (19), while the resultant current can be constructed as indicated in equation (20). The variables R_bat and L_bat represent the equivalent resistance and inductance of the battery, respectively.

Figure 8. Buck/boost converter for charging the vehicle

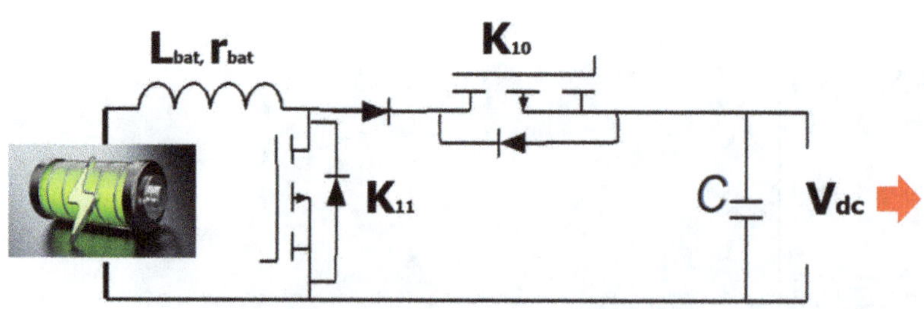

Buck- Boost converter

$$L_{bat}.\frac{di_{bat}}{dt}+R_{bat}.i_{bat}=V_{bat}-V_{dc}.(1-K_{10}) \tag{19}$$

$$I_{out}=I_{bat}.(1-K_{10}) \tag{20}$$

The parameters utilised in this application are succinctly outlined in Table .1.

The Control Architecture of the DSSR System

The variable r, commonly known as tracking error, represents the discrepancy between the desired input (w_{ref}) and the actual output (m_{act}). The tracking fault signal is delivered to the controller, which generates the q-axis command current, denoted as iq*. The use of the inverse Park's transform results in the conversion of the controller and id* (= 0) outputs into ABC and XYZ current instructions. The command currents are subsequently compared with the actual currents to generate pulse width modulation (PWM) signals. These PWM signals are then utilised to activate the semiconductor devices, thereby supplying the necessary voltages for the Six Phase motor to operate effectively. The diagram presented in Figure 9.

Proportional-Integral Controller (PI)

The proportional-integral (PI) controller produces an output signal with two components: a proportional term, which is directly proportional to the input signal, and an integral term, which is directly proportional to the integral of the input signal. The primary objective of the proportional-integral (PI) controller implemented in the system is to minimise the steady-state error while simultaneously elevating the order and type of the system by one. This is illustrated in Figure 10.

$$PI=K_p+\frac{K_i}{S} \tag{21}$$

Table 1. Buck boost converter parameters

Parameters	values
V_{bat} (V)	150
V_{dc} (V)	350
I_{out} (A)	42.85
L_{bat} (H)	0.1
C_{dc} (μF)	1
K_{10}	[0,1]
K_{11}	[0,1]

Figure 9. The DSSR control architecture

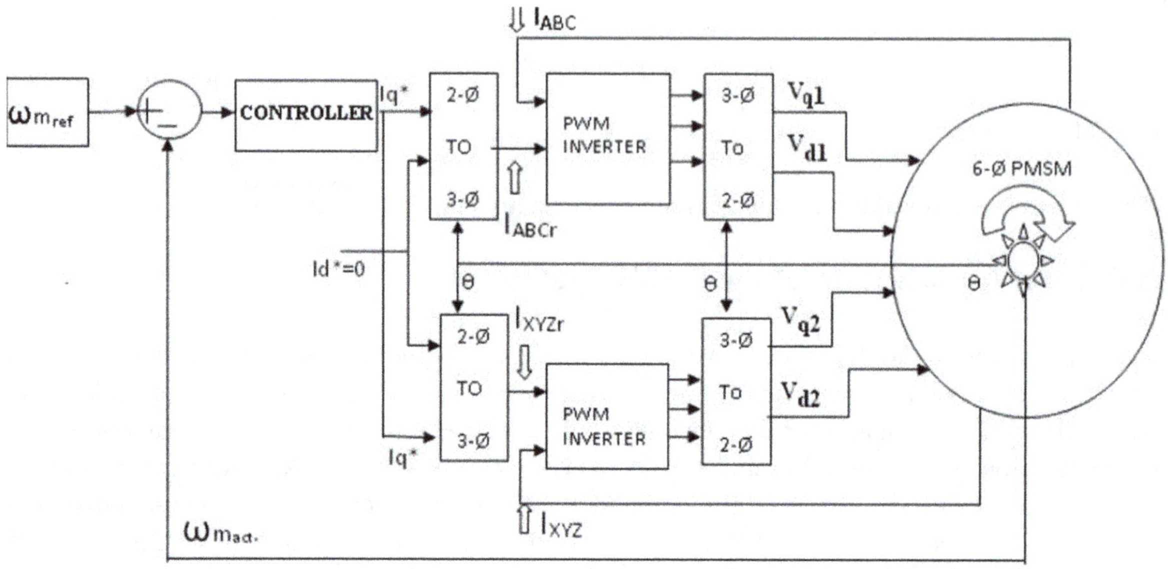

$$iq^* = K_p.\Delta\omega_r + K_i \int \Delta\omega_r dt \tag{22}$$

The intelligence quotient (IQ) is subsequently transmitted inside the system to control the operation of the Six Phase Permanent Magnet Synchronous Motor (PMSM) drive system, as depicted in the block diagram shown in Figure 9. This facilitates the adjustment of the Proportional-Integral (PI) controller using the Closed Loop Ziegler-Nichols Method. The duration of the process is extended due to the inherent nature of trial and error. In the initial stage, a random selection is made for the values of K_p and K_i. Subsequently, by the observation of the model's enhanced responsiveness, the best suitable values for K_p and Ki are determined.

The Fuzzy Logic Controller

The Fuzzy Logic Controller (FLC) is a computational system that utilises fuzzy logic principles to make decisions and control processes. It is a type of control system that incorporates the initial step in the

Figure 10. Proportional integral controller

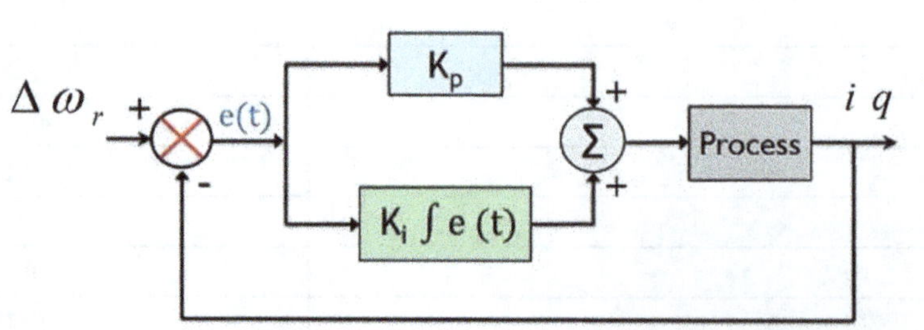

process is to provide a clear definition for the fuzzy input vector. The composition of this entity consists of two variables, namely speed error and its derivative.

$$e(t) = \omega^*_{ref} - \omega_{mact} \tag{23}$$

$$\frac{d[e(t)]}{dt} = \frac{d}{dt}(\omega^*_{ref} - \omega_{mact}) \tag{24}$$

A fuzzy set is established to represent the input and output variables. Figure 11 (a) and 11(b) depict the seven linguistic variables employed for each fuzzy input variable, while Figure 11 illustrates the fuzzy set of the output variable. In order to construct the set of rules, it is vital to utilise a look-up table that establishes the connection between the input variables, e(t) and d[e(t)/ dt], and facilitates the generation of the output variable of the fuzzy logic controller. The results obtained from the proportional-integral (PI) controller present a valuable opportunity and offer valuable information for the process of justifying rules while establishing control rules.

The graphic in Figure 12 illustrates the membership function for the output variable iq*.

Consequently, after undergoing a comprehensive set of investigations, 49 rules have been substantiated, as depicted in Table 2. This lookup table is afterwards utilised within the simulation programme.

Neural Network Controller

The neural network controller is a computational model that utilises interconnected nodes, or neurons, to process and transmit information. The neural network system is considered an intelligent control solution because to its ability to acquire knowledge from a pre-existing database, enabling it to make optimal decisions in many scenarios (Villegas-Mier et al., 2021). The efficacy of a neural network controller is mostly influenced by the scale of the database and the quantity of occurrences. The total performance

Figure 11. The membership function plots: (a) the input errors, (b) the change in error Δe

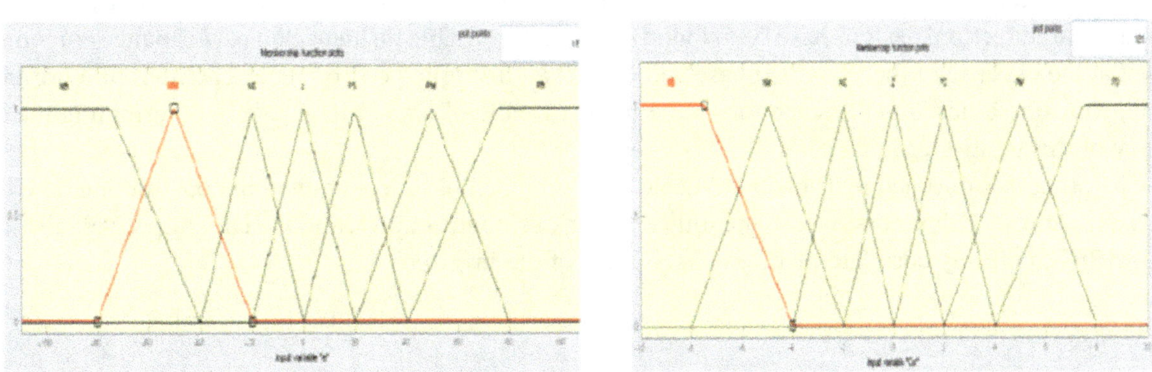

(a) Error

(b) Change in Error

*Figure 12. Membership function plot for output variable Iq**

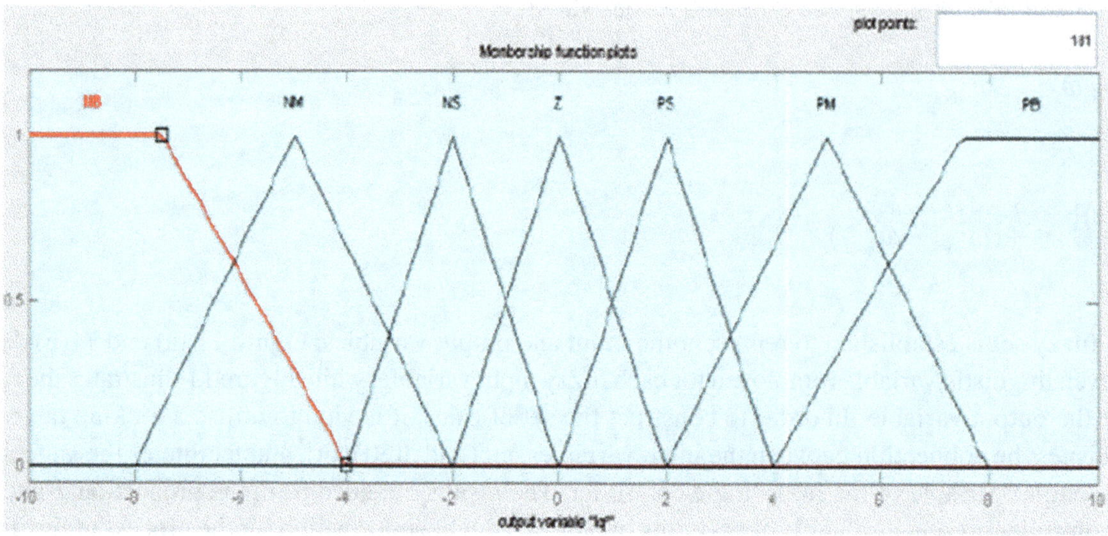

Table 2. Fuzzy roul loock up table

CE E	NB	NM	NS	Z	PS	PM	PB
NB	NB	NB	NB	NB	NM	NS	Z
NM	NB	NB	NB	NM	NS	Z	PS
NS	NB	NB	NM	NS	Z	PS	PM
Z	NB	NM	NS	Z	PS	PM	PB
PS	NM	NS	Z	PS	PM	PB	PB
PM	NS	Z	PS	PM	PB	PB	PB
PB	Z	PS	PM	PB	PB	PB	PB

of the intelligent supervisor is influenced by the neural network architecture (Messalti et al., 2017). Complex architectural designs have been found to have a favourable influence on performance outcomes, albeit at the expense of the learning phase and overall algorithmic performance. The literature presents a range of topologies, and the selection of the input variable plays a crucial role in determining the efficacy of this controller.

The processes outlined in Table 3 delineate the fundamental steps involved in constructing a neural network controller that possesses adaptability to various control applications. The output value of the NN-MPPT is closely correlated with the duty cycle of the inverter.

Table 3. NN-MPPT algorithm implementation stems

Step 1: Choose the inputs signals
Step 2: Normalize the inputs and the outputs
Step 3: Built the database
Step 4: Configure the Neural Network architecture
Step 5: Activate the learning step Repeat until learning error is minimum If learning error is minimum Generate the neural controller Else more learning
Step 6: use the generated optimal NN controller
Step 7: End

ASSOCIATION OF VARIOUS CONSTRUCTION BLOCKS FOR THE EV MATLAB MODEL

As a resume for all of these modelling steps, and in order to have a global mathematical model that can resume the global function of the electric vehicle, Figure 13 provides a clear overview of the relationship between the main components of this model. The mechanical model, which represents all vehicle parameters and is dependent on external and internal parameters, is linked to the electrical machine by coupling the load torque to the load demand on the DSSR machine. This machine will be attached to the two three-phase inverters, which are connected to the buck/boost converter in order to make charging the battery pack possible in certain regenerative circumstances. On the basis of this architecture, it is possible to monitor all electrical motor parameters and provide information regarding the vehicle's speed based on a designated acceleration form. All electronic converters and inverters are connected to the control unit, which regroups within all control blocks such as the DSSR machine control application, the recharge control loop, and the proposed solution for power management.

The achieved results and discussion of the presented contribution are presented and discussed in the next chapter, and it is vital to note that the electrical specification of the main of employed blocks are supplied in this chapter as well.

SIMULATION AND RESULTS

This final section will test and assess the proposed model of the revealed pure electric car prototype, which is powered by a unique electrical motor. In reality, the entire model of the electric car is constructed and run on the Matlab Simulink platform. Each bloc component, such as the mechanical bloc system, which makes up the primary forces acting on the vehicle, also includes the dual stator electric motor and related converters. The battery pack was installed such that it would serve as the vehicle's primary energy source. There were links between the groups, and many findings were shared and discussed.

Figure 13. Overall blocks for budling the global EV model

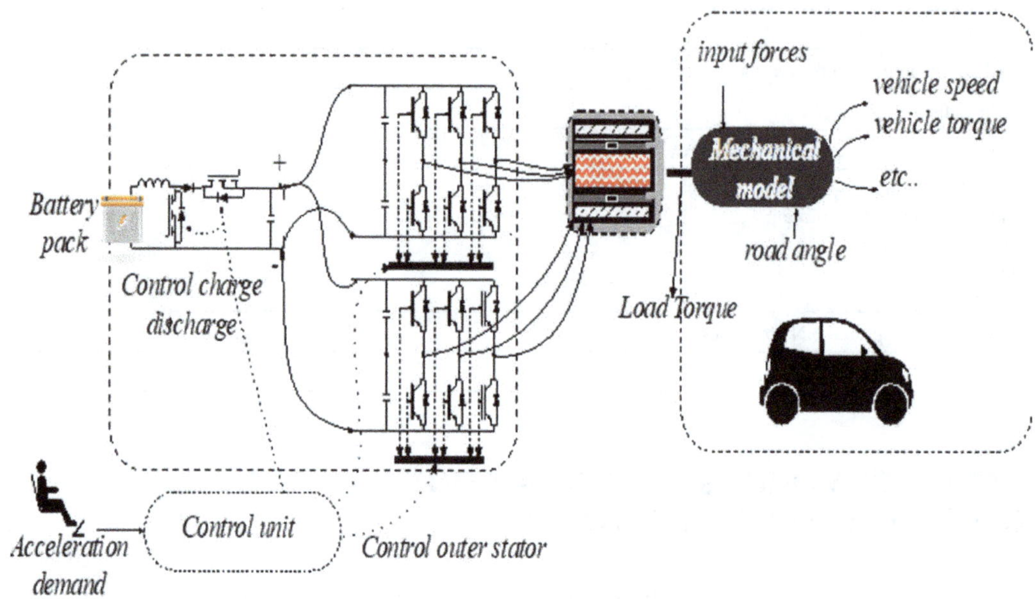

Simulation Conditions

The simulation conditions at this stage must be mentioned. But first, we'd like to state that all of these simulation results were generated using Matlab Simulink platform version 2020, which runs on an I7 CPU processor with 16 GB of RAM. Table 4 recaps the vehicle and road specifications needed to run the simulation phase. In reality, the mass of the vehicle, air density, and wind speed are all defined in this table. Furthermore, all external factors are real.

Detailed Descriptions of the Vehicle's Dynamics and Battery Capacity

These parameters were settled on after extensive study, and have been chosen in this way to reduce the complexity of the vehicle's exterior while still minimising the load torque on the electrical machine. The technical details of the battery pack system are laid bare in table 5. In fact, we have based our model on a genuine battery, such as the one used in TESLA vehicles, and it has a capacity of 85 kWh. The internal resistance of this battery is 0.15, and its nominal voltage is 210 V. The battery's initial charge level will be referenced throughout the rest of the experiment.

The details of this dual stator machine are listed in Table 6. The truth is that this machine can be fed by either a single inner or outer stator, or by a dual stator, as has already been revealed. In reality, the maximum output power is 65 kW, with 50 kW available from the outer stator and 30 kW from each of the dual stators. That's 12.5 kW of input from the inner stator. Similarly, in two-stator mode, the maximum rated speed of the motor is 4000 rpm. The specified torque. However, the given torque depends which stator is used and if the dual exists. The voltage input for each of these stators is 500 V.

The IGBT semiconductor family was chosen for this application's electronic converter due to its ability to achieve a commutation frequency of 20 kHz using only six components. The three-phase inverter

Table 5. Simulated system specifications

Vehicle Specification		Battery System Parameters	
Vehicle Mass	500 kg	Voltage	210V
Air density	1.225 kg/m3	Rated autonomy	20.9 [kW/h] / 100 km
Wind speed	0 m/s	Rated capacity	85 kWh
Road Slop	0	Inner resistance R_0	0.15 Ω
The coefficient of rolling resistance	0	Initial state of charge	
Aerodynamic drag coefficient	0.3		

Table 6. Special machines specifications

	Inner stator	Outer stator	Dual Alimentation
Maximum speed	8000 rpm	8000 rpm	8000 rpm
Maximum Outputted power	50000 W	12500 W	60000 W
Maximum current	100 A	25 A	-
Voltage output	500 V	500V	500 V

accepts input voltage up to 660 V in the DC Bus. Therefore, a buck boost converter is installed between the two inverters and the battery pack to transform the battery's output voltage into the appropriate input voltage for the inverters. Table 7 summarises the buck-boost converter parameters. The power management control loop is in charge of operating both K_{10} and K_{11} switches. In reality, we have relied on the conventional power management control loop, which is dependent on the ON/OFF toggle of the Buck/boost converter. Figure 8 depicts the operating principle, which is based on the diagram provided before. This application did not involve the battery recharging operation.

Global EV Control Loop

As may be seen in Figure 14, the global blocks on MATLAB Simulink platform were implemented as shown. With knowledge of the actual vehicle speed and the acceleration ratio provided by the driver or

Table 7. Buck-boost parameters

Parameters	values	Units
V_{bat}	250	V
V_{dc}	660	V
I_{out}	200	A
L_{bat}	0.1	H
C_{dc}	1	μF
K_{10}	[0,1]	-
K_{11}	[0,1]	-

the given driving cycle, the electrical motor reference speed may be determined. The two three-phase inverters will be activated and deactivated based on commutation signals calculated by the motor control algorithm upon activation by the control loop. These electronic power converters, having received the appropriate input voltage from Chopper, will transmit the power from the DC battery to the machine. Since the chopper can function as either a boost or a buck, the method used to recharge it determines which of its two switches will be activated.

After include all of these factors in the developed electric vehicle model, the resulting outputs for a segment of the new European driving cycle (NEDC) were displayed, including the vehicle's speed and the key blocs it uses. The effectiveness of the EV control package can be verified by zooming in on a particular area of acceleration, which can be seen as proof of the machine's efficiency. Figure 15 depicts the provided acceleration shape. At 50 seconds, the acceleration ratio reduces to 15%, then begins to increase to 35% of the maximum acceleration ratio at the end of the simulation length of 150 seconds.

According to the acceleration demand, the electric motor will assist the vehicle in reaching 65 kilometres per hour at the end of the simulation phase. Figure 16 illustrates the speed distribution of electric vehicles. The maximum speed attained for the given acceleration form is 65 km/h. This is valid for the simulation time of 150 seconds.

According to this acceleration form, the reference speed is calculated as shown in Figure 17. In fact, the electric motor reaches 7500 rpm in the stable region at the end of the acceleration period. Figure 17depicts the dual stator response when the proposed motor control topology is implemented. It is evident that the electrical motor response is flawless, as there is no perturbation or excess when the acceleration form is altered in the vicinity of 50 seconds.

Figure 18 depicts the shape of the injected power into the dual stator machine. Here, the two stators cooperated to provide the necessary electricity to the electric vehicle. In Figure 18, we present the two operating regions for each stator; the vehicle will begin with the outer stator at modest acceleration ra-

Figure 14. Overall EV control on MATLAB Simulink

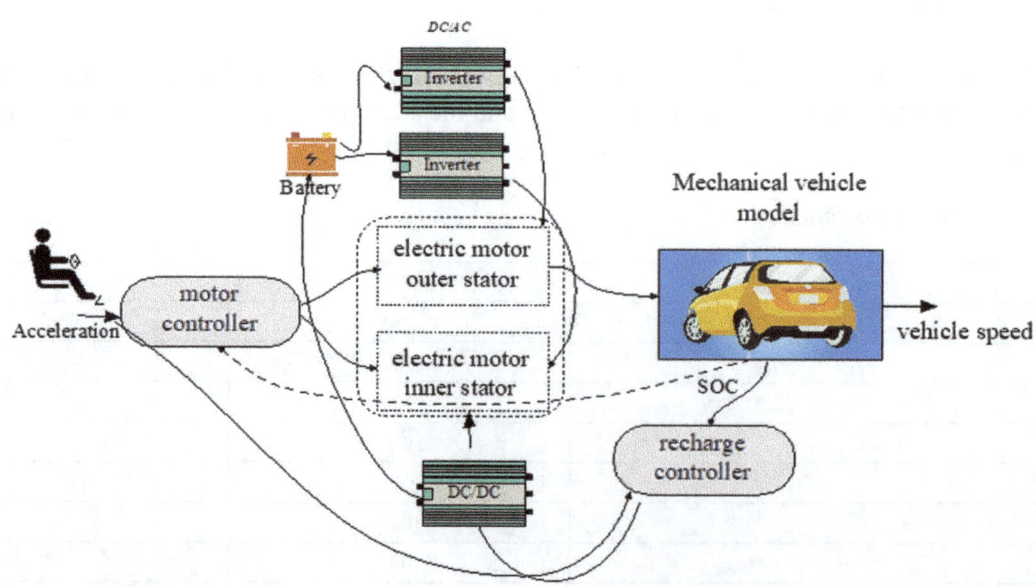

Figure 15. Form of acceleration

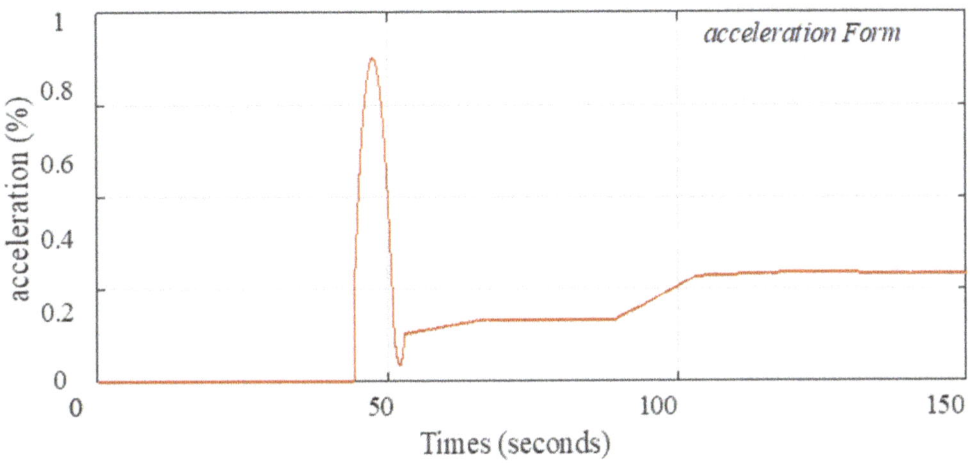

Figure 16. Vehicle speed

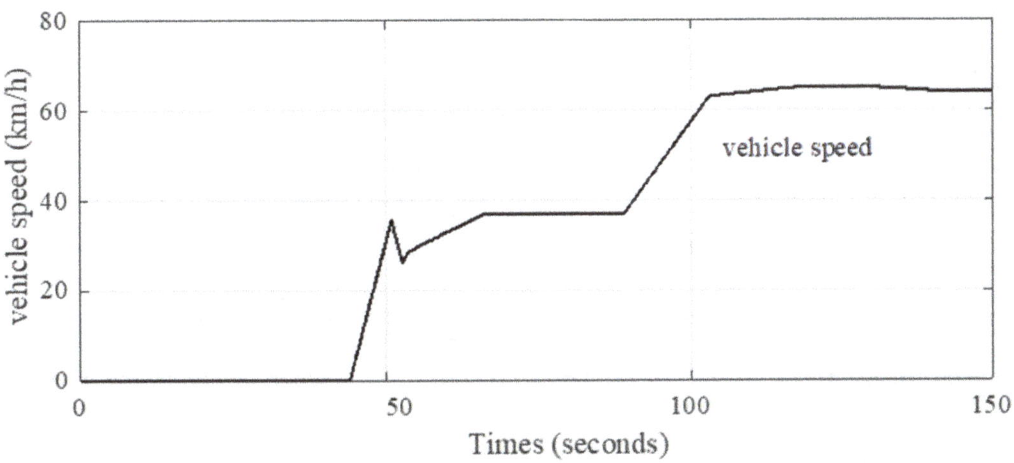

tios, and when the acceleration demand increases, the second stator will assist in increasing the electric vehicle's power.

The interior and outer stator currents are represented in Figure 19. Evidently, the two stators interact to provide electricity to the vehicle. Nonetheless, it is evident that the maximum current originates from the inner stator, which is capable of supplying 50 amps as the maximum rated current for the maximum acceleration form. In this instance, the outer stator current has peaked at 30 A after 46 seconds of testing. At the medium acceleration region between 80 and 110 seconds, the same margin of current is reached. In fact, the obtained results demonstrate that the motor control loop is operating flawlessly and demonstrate the advantage of this machine in the acceleration slope change. With this two-stator alimentation, there is no impact on the vehicle's pace. There is no delay in the vehicle's response when the reference pace is altered.

Figure 17. Dual stator machine real and reference speed

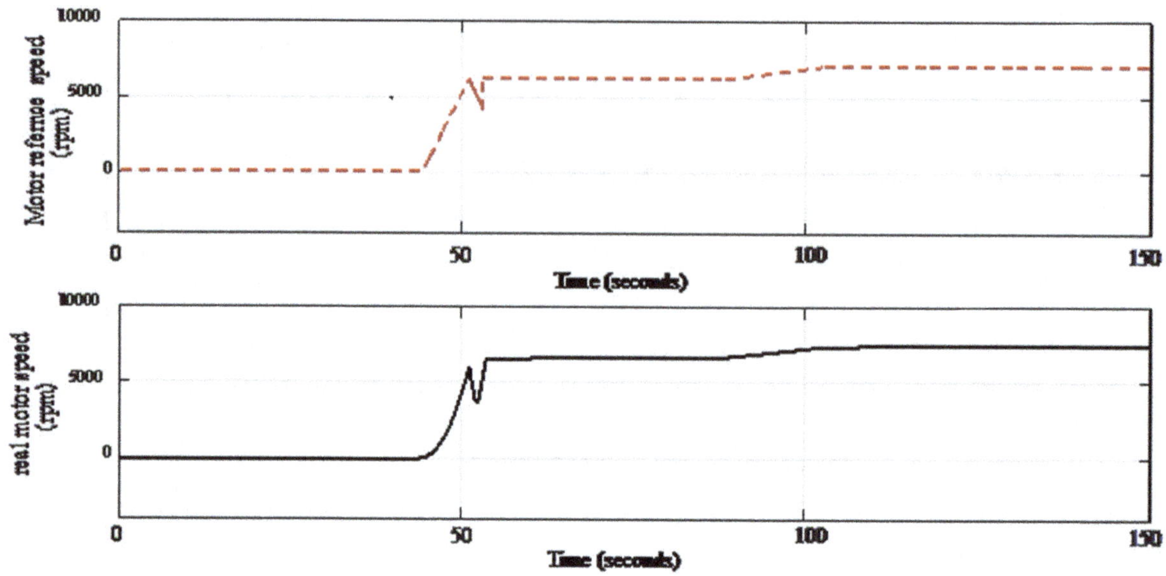

Figure 18. Dual stator machine power curve

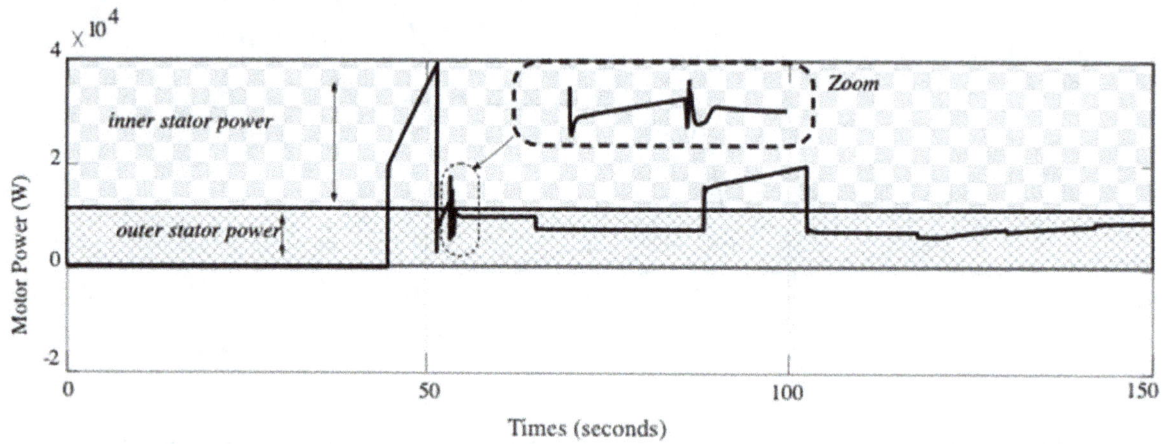

In reality, the recharge principle is founded on Figure 20, and the goal is to relate the recharge phase to the acceleration form and battery state of charge. If the state of charge is greater than 80% and the acceleration is high, no recharge is required; however, if the state of charge falls below 40%, the vehicle must be recharged; otherwise, the high acceleration mode cannot be applied. Therefore, the constructed power management loop depends on the acceleration ratio and the battery's state of charge. If the state of charge is low, the acceleration ratio cannot exceed the limit shown in the acceleration limit curve.

Observation: The various PI parameters in the dual stator machine control loop have been determined through trial-and-error testing. Due to the system's complexity and high nonlinearity, it is difficult to calculate the PI parameters, so all regulators' parameters were set explicitly. The evolution of the battery power system's state of charge is depicted in Figure 20. In reality, the initial State of charge value

Figure 19. Dual stator machine currents

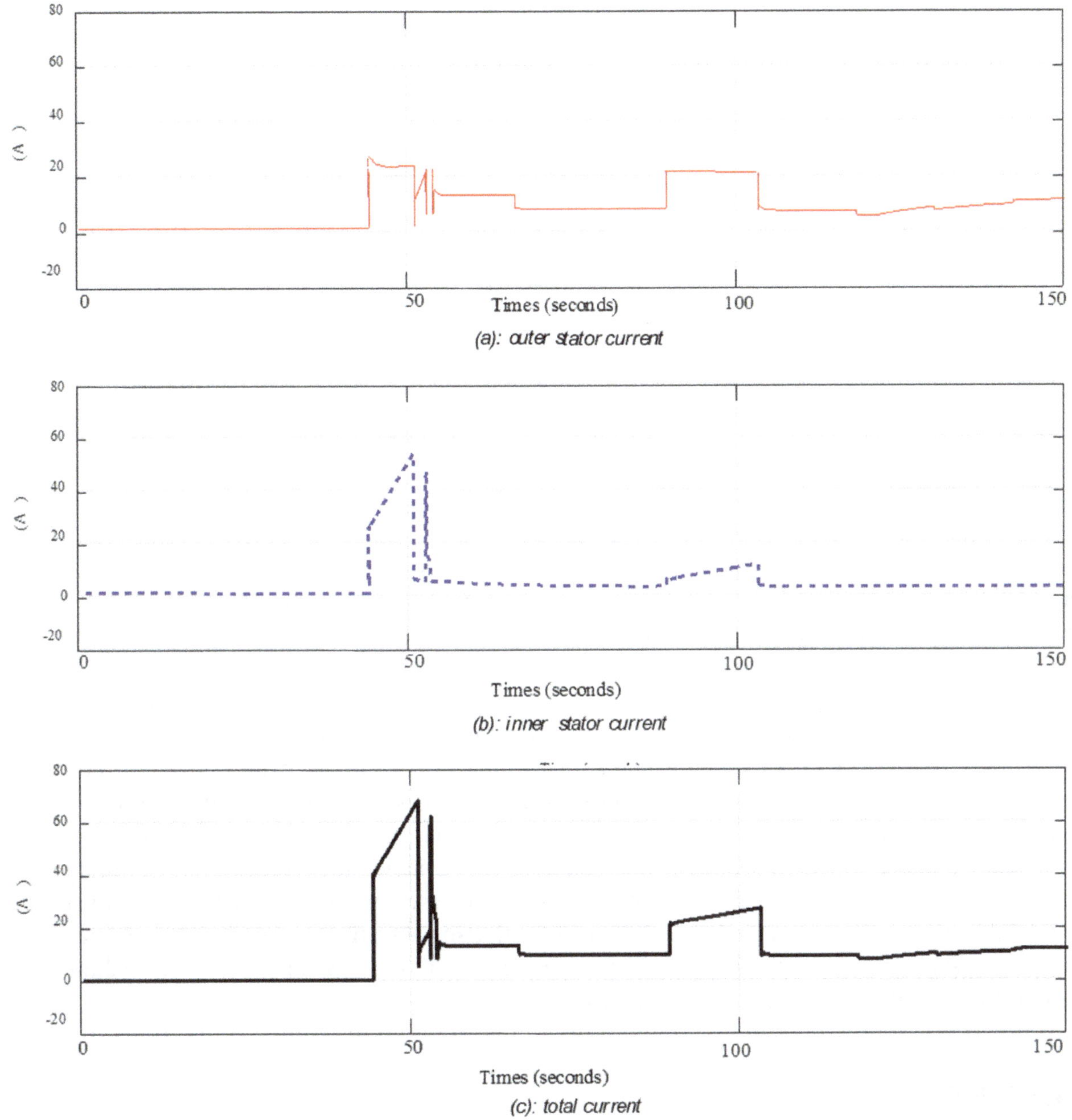

was set to 80%, and Figure 20 depicts the ultimate performance of the battery state of charge based on this acceleration form.

This state of charge curve does not include the regeneration case because this mode was deactivated during the simulation phase. As an extension of this work, it would be fascinating to test the influence of this special machine in the regenerative case, where the dual stator can contribute to charging the battery, thereby accelerating the recharge action.

Figure 20. State of charge of the battery

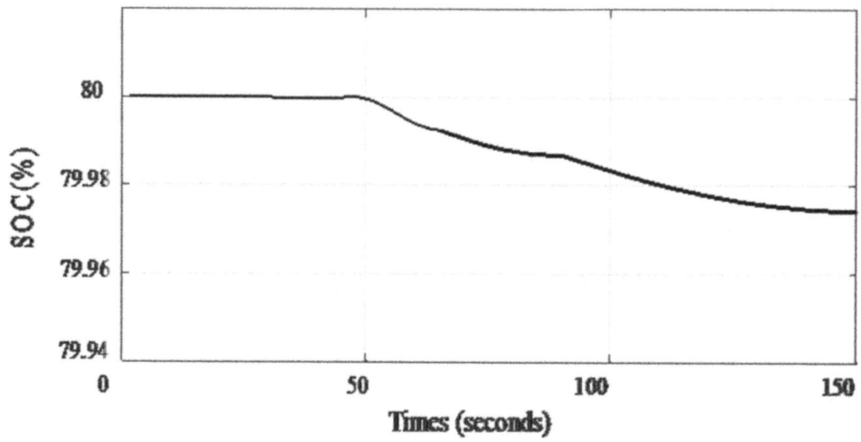

CONCLUSION

In this chapter, we have examined the current state of the electrified transportation system as it pertains to electric vehicles. In fact, we have endeavoured to present the various architectures and models mentioned in the literature, such as electric models. Specifically, this chapter demonstrates the utilised replenishment systems as well as their various internal architectures. During the course of this investigation, we separated these types of chargers into two categories: a group of modern or advanced chargers, solar chargers, and Inductive Power Transfer. By concentrating on the second category, the remainder of this work provides an in-depth examination of these charge systems and exposes detailed modelling for these various blocks. To ascertain their performance under specific operating conditions, an operational simulation is applied. The simulation result for the recharge system and energy is then displayed by a detailed mathematical model. This charging system is installed in an electric vehicle to increase the autonomy of the vehicle. Each recharge block, including the photovoltaic recharge system and the wireless recharge tool, was modelled, and their respective mathematical expressions are provided. As the battery state of charge performances became more crucial, the vehicle's overall profitability was demonstrated by the enhancement of its energetic performance.

REFERENCES

Ahmad, F., Khalid, M., & Panigrahi, B. K. (2021). Development in energy storage system for electric transportation: A comprehensive review. *Journal of Energy Storage*, *43*(September), 103153. doi:10.1016/j.est.2021.103153

Ashok, B., Kannan, C., Usman, K. M., Vignesh, R., Deepak, C., Ramesh, R., Narendhra, T. M. V., & Kavitha, C. (2022). Transition to Electric Mobility in India: Barriers Exploration and Pathways to Powertrain Shift through MCDM Approach. *Journal of The Institution of Engineers (India): Series C*, *103*(5), 1251–1277. doi:10.1007/s40032-022-00852-6

Atacak, İ. (2023). An Ensemble Approach Based on Fuzzy Logic Using Machine Learning Classifiers for Android Malware Detection. *Applied Sciences (Basel, Switzerland)*, *13*(3), 1484. Advance online publication. doi:10.3390/app13031484

Azar, A. T., Abed, A. M., Abdulmajeed, F. A., Hameed, I. A., Kamal, N. A., Jawad, A. J. M., Abbas, A. H., Rashed, Z. A., Hashim, Z. S., Sahib, M. A., Ibraheem, I. K., & Thabit, R. (2022). A New Nonlinear Controller for the Maximum Power Point Tracking of Photovoltaic Systems in Micro Grid Applications Based on Modified Anti-Disturbance Compensation. *Sustainability (Basel)*, *14*(17), 10511. Advance online publication. doi:10.3390/su141710511

Bin Ahmad, M. S., Pesyridis, A., Sphicas, P., Mahmoudzadeh Andwari, A., Gharehghani, A., & Vaglieco, B. M. (2022). Electric Vehicle Modelling for Future Technology and Market Penetration Analysis. *Frontiers of Mechanical Engineering*, *8*(July), 1–18. doi:10.3389/fmech.2022.896547

Cao, Y., Yao, M., & Sun, X. (2023). An Overview of Modelling and Energy Management Strategies for Hybrid Electric Vehicles. *Applied Sciences (Basel, Switzerland)*, *13*(10), 5947. Advance online publication. doi:10.3390/app13105947

Dheeban, S. S., Muthu Selvan, N. B., & Senthil Kumar, C. (2019). Design of standalone pv system. *International Journal of Scientific and Technology Research*, *8*(11), 684–688.

Domínguez-Navarro, J. A., Dufo-López, R., Yusta-Loyo, J. M., Artal-Sevil, J. S., & Bernal-Agustín, J. L. (2019). Design of an electric vehicle fast-charging station with integration of renewable energy and storage systems. *International Journal of Electrical Power and Energy Systems, 105*(March), 46–58. doi:10.1016/j.ijepes.2018.08.001

Edition, T. (2018). *Control Systems for Drilling Well Control Equipment and Control Systems for Diverter Equipment*. Academic Press.

Ercan, T., Onat, N. C., Keya, N., Tatari, O., Eluru, N., & Kucukvar, M. (2022). Autonomous electric vehicles can reduce carbon emissions and air pollution in cities. *Transportation Research Part D, Transport and Environment*, *112*(September), 103472. doi:10.1016/j.trd.2022.103472

Guo, J., He, H., & Sun, C. (2019). ARIMA-based road gradient and vehicle velocity prediction for hybrid electric vehicle energy management. *IEEE Transactions on Vehicular Technology*, *68*(6), 5309–5320. doi:10.1109/TVT.2019.2912893

He, W., King, M., Luo, X., Dooner, M., Li, D., & Wang, J. (2021). Technologies and economics of electric energy storages in power systems: Review and perspective. *Advances in Applied Energy*, *4*(May), 100060. doi:10.1016/j.adapen.2021.100060

Holechek, J. L., Geli, H. M. E., Sawalhah, M. N., & Valdez, R. (2022). A Global Assessment: Can Renewable Energy Replace Fossil Fuels by 2050? *Sustainability (Basel)*, *14*(8), 1–22. doi:10.3390/su14084792

International Energy Agency. (2022). *International Energy Agency (IEA) World Energy Outlook 2022*. https://www.iea.org/reports/world-energy-outlook-2022

Kaiser, N., & Barstow, C. K. (2022). Rural Transportation Infrastructure in Low-and Middle-Income Countries: A Review of Impacts, Implications, and Interventions. *Sustainability (Basel), 14*(4), 2149. Advance online publication. doi:10.3390/su14042149

Kang, J. B., Lee, J. Y., & Lee, J. H. (2022). Guidelines for Determining the Initial Shape and Specifications of High-Speed AFPM Based on Recent Research. *Energies, 15*(16), 6099. Advance online publication. doi:10.3390/en15166099

Kumar, M., Panda, K. P., Naayagi, R. T., Thakur, R., & Panda, G. (2023). *Applied sciences Comprehensive Review of Electric Vehicle Technology and Its Impacts : Detailed Investigation of Charging Infrastructure, Power Management, and Control Techniques*. Academic Press.

Li, Z., Shao, T., Zheng, T. Q., Li, H., & Huang, B. (2019). Interference source analysis and EMC design for All-SiC power module in EV charger. *Microelectronics and Reliability, 100–101*(September), 113458. Advance online publication. doi:10.1016/j.microrel.2019.113458

Liu, H. C., Wu, S. M., Wang, Z. L., & Li, X. Y. (2021). A New Method for Quality Function Deployment with Extended Prospect Theory under Hesitant Linguistic Environment. *IEEE Transactions on Engineering Management, 68*(2), 442–451. doi:10.1109/TEM.2018.2864103

Mahmoudi, A., Rahim, N. A., & Hew, W. P. (2011). Axial-flux permanent-magnet machine modeling, design, simulation and analysis. *Scientific Research and Essays, 6*(12), 2525–2549.

Messalti, S., Harrag, A., & Loukriz, A. (2017). A new variable step size neural networks MPPT controller: Review, simulation and hardware implementation. *Renewable and Sustainable Energy Reviews, 68*(August), 221–233. doi:10.1016/j.rser.2016.09.131

Mohamed, N., Aymen, F., Alqarni, M., Turky, R. A., Alamri, B., Ali, Z. M., & Abdel Aleem, S. H. E. (2022). A new wireless charging system for electric vehicles using two receiver coils. *Ain Shams Engineering Journal, 13*(2), 101569. Advance online publication. doi:10.1016/j.asej.2021.08.012

Nagendra, G. R., Chen, L., Covic, G. A., & Boys, J. T. (2014). Detection of EVs on IPT highways. *Conference Proceedings - IEEE Applied Power Electronics Conference and Exposition - APEC*, 1604–1611. 10.1109/APEC.2014.6803521

Nguyen, B. H., Trovão, J. P. F., German, R., & Bouscayrol, A. (2020). Real-time energy management of parallel hybrid electric vehicles using linear quadratic regulation. *Energies, 13*(21), 1–19. doi:10.3390/en13215538

Palani, G., Sengamalai, U., Vishnuram, P., & Nastasi, B. (2023). Challenges and Barriers of Wireless Charging Technologies for Electric Vehicles. *Energies, 16*(5), 2138. Advance online publication. doi:10.3390/en16052138

Palit, T., Bari, A. B. M. M., & Karmaker, C. L. (2022). An integrated Principal Component Analysis and Interpretive Structural Modeling approach for electric vehicle adoption decisions in sustainable transportation systems. *Decision Analytics Journal, 4*(August), 100119. doi:10.1016/j.dajour.2022.100119

Villegas-Mier, C. G., Rodriguez-Resendiz, J., Álvarez-Alvarado, J. M., Rodriguez-Resendiz, H., Herrera-Navarro, A. M., & Rodríguez-Abreo, O. (2021). Artificial neural networks in mppt algorithms for optimization of photovoltaic power systems: A review. *Micromachines*, *12*(10), 1–19. doi:10.3390/mi12101260 PMID:34683311

Xie, L., Shi, Y., Thomas Hou, Y., & Lou, W. (2013). W Ireless P Ower T Ransfer and a Pplications To S Ensor N Etworks. *IEEE Wireless Communications*, *20*(4), 140–145.

Yilmaz, M., & Krein, P. T. (2013). Review of battery charger topologies, charging power levels, and infrastructure for plug-in electric and hybrid vehicles. *IEEE Transactions on Power Electronics*, *28*(5), 2151–2169. doi:10.1109/TPEL.2012.2212917

Zhao, C., Zu, B., Xu, Y., Wang, Z., Zhou, J., & Liu, L. (2020). Design and analysis of an engine-start control strategy for a single-shaft parallel hybrid electric vehicle. *Energy*, *202*, 117621. doi:10.1016/j.energy.2020.117621

Chapter 13
Energy Pattern During Unforeseen Spells of COVID-19

Ashok Asthana
Sarala Birla University, India

Swati Srivastava
KR Mangalam University, India

ABSTRACT

The clean energy curve is anticipated to avoid flattening with the help of investment stimulation, COVID-19 alleviation conditions, innovation, and infrastructure opportunities in low-carbon and digital technology. Over the coming years, the global renewable energy picture may ultimately resume its long-term trend towards the production of green electricity. The experience of the COVID-19 pandemic period may persuade countries to speed up their attempts to implement renewable energy policies before another global economic shock is caused by a different illness or an unanticipated weather calamity. Overall, there is optimism that the development of renewable energy will proceed.

1. INTRODUCTION

Many people believe that as time goes on, technology will develop, the climate will get better, and people's lives will be better all around the world. There are major setbacks, such as in 2020 when the worldwide pandemic COVID-19 first emerged. This virus initially surfaced in China during the end of 2019 or the beginning of 2020 (Eroğlu, 2020). Many didn't realize how serious it was or how rapidly it would spread. It quickly grew to become a highly contagious, deadly virus that might distress anybody. On January 30, 2020, the World Health Organization (WHO) declared the 2019 Coronavirus disease (COVID-19) pandemic a public health emergency of global concern after infection cases outnumbered those of the 2003, SARS epidemic and spread to all 34 provinces of China (BBC News, 2020). Although there is no precise confirmation, it is thought to have been initiated in late December 2019 at a seafood wholesale market in Wuhan, Hubei Province; the number of cases skyrocketed rapidly within and outside Wuhan, spreading extensively throughout the world (Birol, 2020; Eroğlu, 2020). No country was ready

DOI: 10.4018/979-8-3693-1014-4.ch013

for this sad occurrence. There is currently a serious health problem, and there is only so much that can be done to minimize it. There are mask-wearing faces everywhere, and most individuals are required to keep a six-foot distance from one another in most institutions. COVID-19 resulted in a large number of fatalities and health problems. This has had a significant influence on economies all around the world, whether they are developed or poor. The COVID-19 epidemic is a serious issue that has to be addressed and taken into account on all levels.

The influence that this epidemic has had on the world will live with everyone forever, even though many would like to forget 2020 from their life. This pandemic has affected everyone on a mental, emotional, physical, and financial level from all over the globe; the virus is oblivious to a person's race, ethnicity, or money. The American economy has been severely impacted by the coronavirus. The largest economic shock to hit the US since the 2008 financial crisis has been brought on by the coronavirus. Currently, both the number of infections and fatalities are highest in the United States (Worldometer, 2020). As of the end of December 2020, there have been over 82 million coronavirus cases, 1.8 million deaths throughout the world that have been reported related to the Coronavirus, and approximately 59 million people have recovered from the virus (Worldometer, 2020). The overall number of infections in the United States is likely to approach up to 20 million, with almost 350 thousand deaths (Worldometer, 2020).

The global economy, financial markets, and sustainable economic development have all been negatively impacted by COVID-19. Among the effects of the illness are large revenue losses, a surge in unemployment, and disruptions in the industrial, service, and transportation sectors. Various nations are now implementing mitigating efforts to lessen the effects of this epidemic, but it has become evident that none of the governments or the world's healthcare systems were ready to handle such a sad occurrence. The whole world underrated the dangers of the rapid spread of COVID-19 and reacted to the problem mostly in a reactionary manner. In order to preserve lives and safeguard economic development, strong international initiatives are needed as disease outbreaks are unlikely to cease very soon. Due to decreased productivity, fatalities, company closures, disruptions in commerce, and the devastation of the tourism sector, there has already been a substantial economic impact on different parts of the world. COVID-19 may serve as a "wake-up" call for world leaders to step up collaboration on pandemic preparation and give the funding required for global cooperation. The projected financial and health consequences of infectious disease outbreaks are well known, yet the world hasn't made enough investments in preparedness and prevention measures to reduce the likelihood of major epidemics.

Humanity has been severely impacted by COVID-19, with developing countries bearing the brunt of the harm. In order to fulfil the 17 Sustainable Development Goals (SDGs), the UN and top financial individuals and communities were urged to work together before COVID-19. providing the poorest and most vulnerable with technical help (Barbier & Burgess, 2020). The pandemic is predicted to harm 12 of the 17 SDGs, as shown in Figure 1. This will occur at a crucial time for some long-term economic objectives since 673 million people still use open defecation, 821 million people are anemic, 785 million people lack even the most basic infrastructure for accessing clean water, and 736 million people continue to live in extreme poverty. 87% of the 840 million people without access to electricity live in rural areas, while almost 3 billion people lack access to safe cooking technologies and fuels (Barbier & Burgess, 2020).

The fact that COVID-19 first struck China, a significant player in the economy, had a significant influence was another. Since the outbreak, it has been clear that the disease's effects stretch beyond those who have died from it and others who would be unable to work owing to high risk factors because of it. Since there have been production halts and a slowdown in China. Companies worldwide that depend

Figure 1. The Impact of COVID-19 on the UN sustainable development goals
Adopted from Barbier and Burgess (2020)

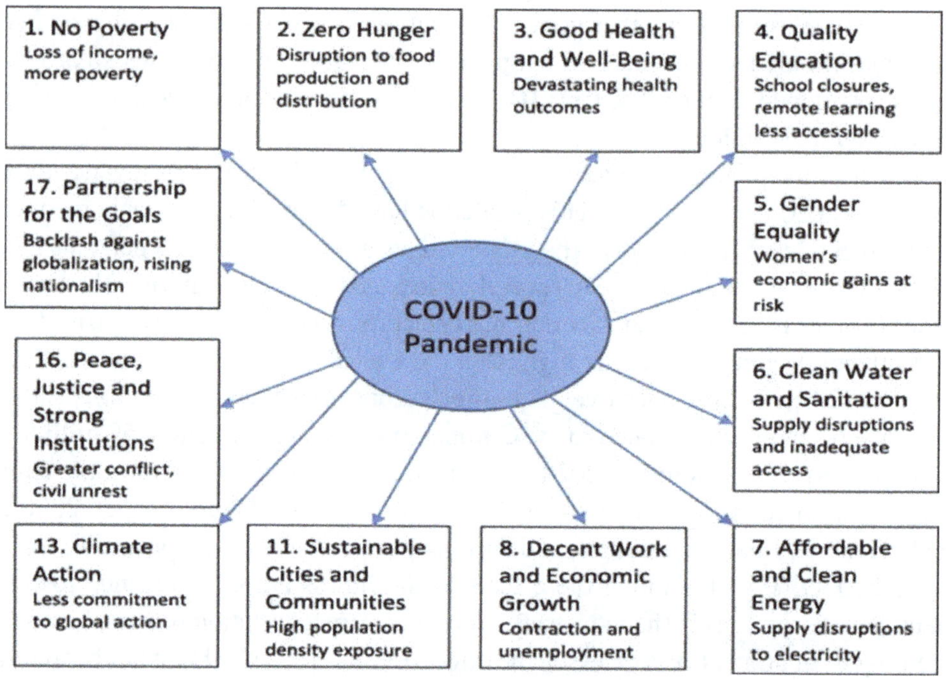

on Chinese imports are starting to encounter output restrictions, regardless of size or location, which has hindered the operation of global supply chains. Transport between nations is restricted, if not outright forbidden, which has further slowed down global economic growth. Most importantly, consumer, organization, and corporate anxiety has distorted regular buying habits and produced weird market phenomena. Global stock indexes have fallen as a result of the changes, which have also been reflected in global financial markets (Barbier & Burgess, 2020).

The continent's economy is interconnected through transnational flows of goods, services, information, people, financial capital, foreign investment, international banking, and exchange rates. All of these items are also channeling for the spread of economic shocks or economic turbulence since they are connected by shared ideas. Limitations on transportation have also had an effect on economic productivity and activity. Organizations, corporations, and financial markets all appear to be impacted. The increased risk may indicate that one or more important market players have adopted unprofitable investment positions in light of the current circumstances, further undermining confidence in financial products and markets. As evidenced by the variation in cases per capita in several countries, independent of their economic standing, it is predicted that different nations and territories will have diverse recovery pathways, with the shape of those pathways being influenced by the interaction between their experience in containing and managing the spread of COVID-19 and the underlying socio-economic features of each country or territory. The study hypothesizes that trade-exposed nations may recover from the COVID-19 outbreak more slowly than less trade-exposed ones if it persists (McKibbin & Fernando, 2020). Simply put, the impact on global trade increases as the pandemic persists. However, some economists believe that a realistically sluggish distribution of vaccines throughout emerging countries may prevent a return to pre-pandemic levels of activity. Recent work on coronavirus immunizations has brightened the economic outlook. Repeated lockdowns across Europe to stop a disease outbreak may impede economic recovery, even among advanced nations (World Economy, 2020).

2. EFFECT OF COVID-19 ON THE GLOBAL ECONOMY

Since the Great Depression and the 2008 financial crisis, the globe has never witnessed such extreme economic implications. The COVID-19 virus damaged the economic boom of the preceding 4-5 years. (Bachman, 2020). It has also induced future concern and unpredictability. G-20 economic growth has been greatly influenced by COVID-19. It was repaired to a 4.2% contraction presenting the COVID-19 pandemic, as seen in Table 1, as opposed to the positive increase anticipated before COVID-19. (The Economist, 2020).

The International Monetary Fund (IMF) predicted that the global GDP will decrease by 4.9 percent in 2020, the largest decrease ever, in its June World Economic Outlook report (Boatwright & Wynne, 2020). In the second quarter of this year, the United Kingdom saw the biggest annual GDP decrease of any major country, with a decline of 59.8% in the USA, or around 32%. Of the emerging economies, India's GDP contracted at a pace of 69.4 percent annually (Boatwright & Wynne, 2020). Even while many countries still have a long way to go and there is still a lot of suffering in the world, the world may

Table 1. Pre and post COVID-19 GDP growth forecast for G-20 nations

Country	Post COVID-19 Projection (%)	Pre COVID-19 Projection (%)
Argentina	-9	-2
Australia	-4.2	2
Brazil	-5.5	2.4
Canada	-4.3	1.8
China	1	5.9
France	-8.8	1
Germany	-6.1	0.9
India (2020/21 fiscal year)	-4	6
Indonesia	1	5.1
Italy	-10.8	0.4
Japan	-5.2	0.4
South Korea	-2.1	2.2
Mexico	-9	1.1
Russia	-5.2	1.6
Saudi Arabia	-3.2	1
South Africa	-5.6	1.4
Turkey	-5.4	3.8
UK	-8.7	1.1
US	-4	1.7
Global (market exchange rates)	-4.2	2.3

be at a point where this crisis will eventually come to an end and a sustainable recovery may reappear (Capital Group, 2020). Projected 2021 GDP growth by International Monitory Fund is expected to go and it shows in Figure 2.

2.1 Direct Impact on Production

The shutdown in regions of the world United States and others have already had implications for global industrial production (Yardeni et al., 2020). The most impacted industries are those that depend on workers whose jobs cannot be carried out remotely. The sectors most affected are those that rely on workers whose duties cannot be practiced remotely. The majority of US firms anticipate financial consequences from the epidemic. (Capital Group, 2020). Some major industrial enterprises have halted operations and are probably weighing how many people they would lay off to restrict the virus' transmission as well as for financial reasons. The manufacturing industry in the United States, which employs around 13 million people, is projected to be badly disrupted by this outbreak for two reasons. First, many industrial tasks are on-site and cannot be performed remotely. Second, weaker economic development has lowered local and foreign demand for industrial goods. (Capital Group,2020). Not to mention that being on-site increases the chance of exposure and demands a 14-day quarantine for anybody who becomes exposed, significantly slowing production for firms that are already facing difficulties. According to some academics, nearly 20 million Americans, or 15% of all jobs, might be performed regularly remotely. (Fottrell, 2020).

Figure 2. Projected GDP for 2020 and 2021
Adopted from reference, Capital Group (2020)

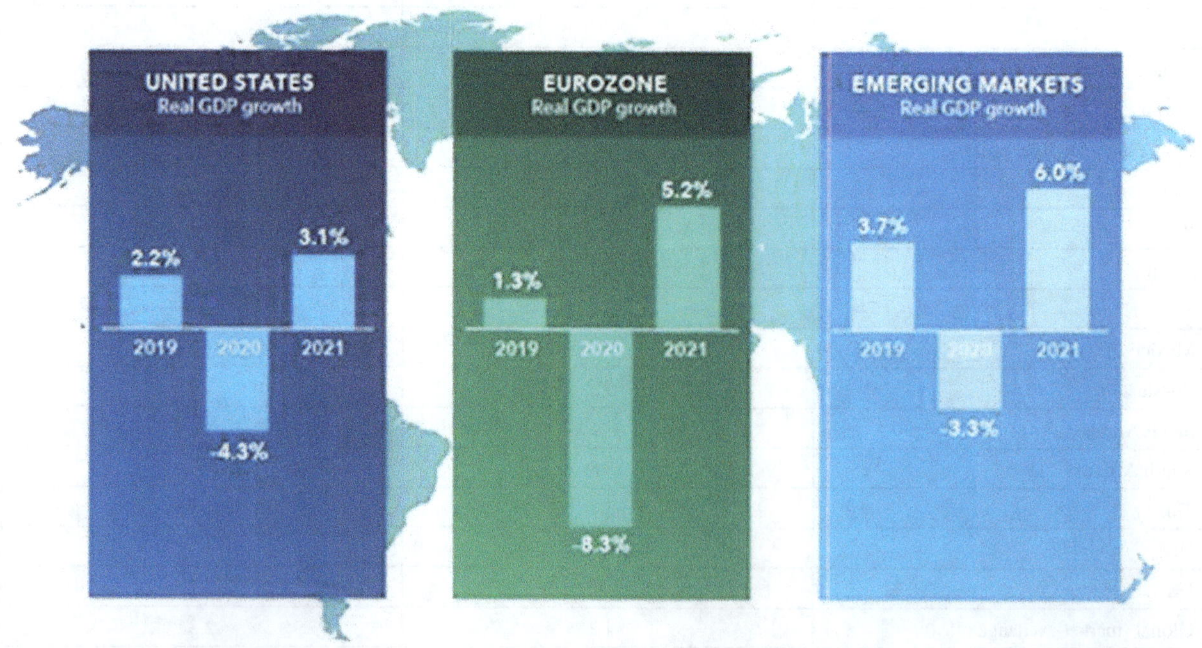

2.2 Supply Chain and Market Disruption

Many industrial companies are dependent on imported intermediate materials from China and other diseased nations. Many businesses also rely on sales in other nations to help them reach their financial objectives. Production and profitability of certain international enterprises have been harmed, notably in manufacturing and raw materials used in manufacturing, as a result of the slowdown in economic activity and shipping restrictions in affected nations. The speed at which the outbreak subsides might affect how much of an impact there is for businesses who depend on intermediate items from the impacted areas and are unable to readily switch sourcing. Small and medium-sized businesses may struggle more to survive the disruption. Businesses involved in tourism and travel are suffering losses that are probably not recoupable.

2.3 Financial Impact on Firms/Organizations and Financial Markets

Some businesses, especially those with little money, may experience stress as a result of temporary interruptions in inputs or output. Financial market traders may or may not accurately predict or comprehend which companies may be susceptible. The consequent increase in risk might indicate that one or more significant participants in the financial system have adopted investing strategies that are unprofitable in the current environment, further eroding confidence in financial instruments and markets. A substantial financial market disruption as investors grow concerned about counterparty risk is a potential (likely low likelihood) scenario. A major decrease in equities and corporate bond markets is a considerably more possible scenario, with investors choosing to keep government securities (especially US treasuries) because to the uncertainty the epidemic has caused.

3. HISTORICAL GROWTH OF RENEWABLE ENERGY

Energy is a crucial component in the growth of the economy. The relationship between energy and economic growth is widely known. Along with land and capital, some economists view energy as a fourth component of production. A developing industrial sector needs energy inputs like coal and electric electricity. The easy access to affordable energy promotes industrial development and can be referred to as a source of economic growth. On the other hand, growth in the development of energy-intensive industries would result in higher energy output and consumption per person. Due to the comfort needs that must be met by energy-intensive gadgets like cars, air conditioners, and other electrical appliances, rising living standards are associated with rising energy consumption, which suggests that energy supply may impede economic progress. It could have an impact on a variety of industrial and agricultural operations. Almost all of the country's energy requirements for heating, cooking, and lighting were satisfied by wood up until the mid-1800s (U.S. Energy Information Administration [EIA], 2019). Since the late 1800s till the present, fossil fuels including coal, oil, and natural gas have been the primary energy sources. Wood and hydropower were the most widely used renewable energy sources.

Although harnessing the power of nature for transportation, lighting, heating, and other uses is nothing new, renewable energy is typically seen as a relatively recent invention. grain mills and ships that rely on wind strength to cross the oceans. Fires burned far into the evening thanks to the sun, which also kept things warm during the day. But over the past 500 years or more, humans have become more dependent on dirtier, less costly energy sources, such coal and fracked gas. As innovation drives down costs and begins to fulfil the promise of a clean energy future, renewable energy is expanding. The country's electrical grid is combining record-breaking American solar and wind energy output without sacrificing reliability. Renewable energy sources now make up more than one-eighth of U.S. generation because of ever-innovative and more affordable methods of capturing and storing wind and solar energy. In both large and small scales, from enormous offshore wind farms to rooftop solar panels on homes that can sell electricity back to the grid, the usage of renewable energy is increasing. In other circumstances, entire rural towns use renewable energy for heating and lighting. A cost-effective source of fresh energy, renewable energy protects power markets and customers from market instability, promotes economic stability, and encourages sustainable growth. Many nations and regions are aware of the extent to which the energy transition may result in beneficial consequences because renewable additions accounted for the majority of the new capacity last year. Even while the trend is encouraging, more must be done to move global energy towards sustainable growth and climate mitigation, both of which have substantial economic advantages. We are reminded at this trying time of the value of strengthening our economies' resilience. Enabling policies are required in what must be a decade of work to boost investments and hasten the adoption of renewable energy.

3.1 Global Growth of Renewable Energy

In comparison to the 179 GW added in 2018, the renewable energy industry added 176 GW of generating capacity globally in 2019. The International Renewable Energy Agency (IRENA) has released fresh data showing that new renewable energy accounted for 72% of all power expansion last year. According to IRENA's annual Renewable Capacity Statistics 2020, the output of renewable energy increased by 7.6 percent in 2017; Asia led the way with 54 percent of all additions (International Renewable Energy Agency (IRENA), 2020). Even if the growth of renewable energy slowed last year, it nevertheless outperformed the growth of fossil fuels overall by a ratio of 2.6, confirming the dominance of renewables in power expansion that was originally established in 2012. Using solar and wind energy, renewable energy capacity increased by 90% in 2019 (International Renewable Energy Agency [IRENA], 2020).

Renewables contributed for at least 70% of the growth in capacity globally in 2019 with the exception of the Middle East and Africa, which accounted for 52% and 26% of net additions, respectively. Due to the advances, the proportion of renewable energy in the world's total power capacity increased from 33.3 percent at the end of 2018 to 34.7 percent. With net additions of non-renewable capacity in Asia, the Middle East, and Africa and net decommissioning in Europe and North America, 2019 saw the continuation of long-term trends. In 2019, solar installed 98 GW, with 60% of that capacity located in Asia. Nearly 60 GW of additional wind energy were added, with China (26 GW) and the United States (9 GW) leading the way. Currently, the two technologies produce 623 GW and 586 GW, respectively, or over half of the world's renewable energy capacity. Geothermal energy, hydropower, bioenergy, and marine energy all had moderate year-over-year growth of 12 GW, 6 GW, 700 MW, and 500 MW, respectively. Despite growing considerably more slowly than in 2018, Asia continued to account for more over half of new installations. Growth increased year over year in both Europe and North America. Half of the 4 GW of new renewable energy capacity installed in Africa in 2018 will be added in 2019 (International Renewable Energy Agency (IRENA), 2020). The global installed renewable energy trend is shown in Figure 3.

Figure 3. Global installed renewable energy trend
Adopted from International Renewable Energy Agency (IRENA) (2020)

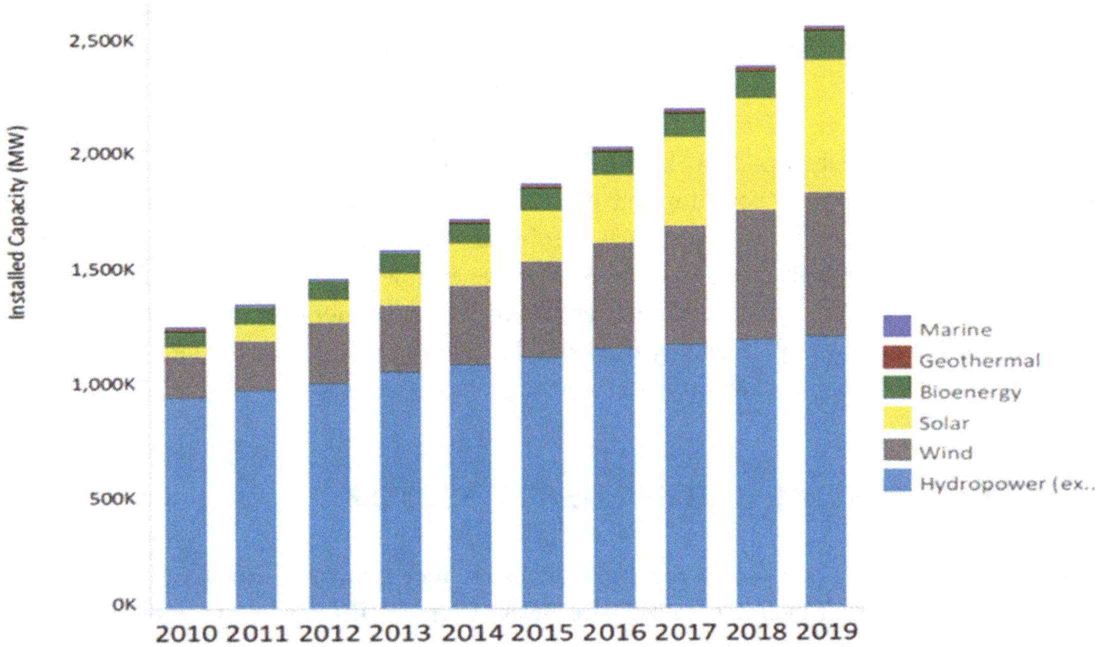

Highlights by Technology:

- Hydropower: 2019 saw abnormally slow growth, which may have been caused by certain major projects failing to finish on schedule. The majority of the growth was contributed by China and Brazil, who each added more over 4 GW.
- Wind energy: In 2019, wind performance was especially strong, increasing by about 60 GW. With increases of 26 GW and 9 GW, respectively, China and the US maintained their leadership.
- Solar energy: With a 56 GW rise, Asia continues to lead the world in the growth of solar capacity, but this was less than in 2018. The United States, Australia, Spain, Ukraine, and Germany all experienced significant rises.
- Bioenergy: In 2019, the capability for bioenergy expansion remained low. Half of the extra capacity (+3.3 GW) came from China. Turkey, Japan, Italy, and Germany all experienced growth.
- Geothermal energy: More than in 2018, geothermal power capacity increased by 682 MW in 2019. Once more, Turkey took the lead with an increase of 232 MW, followed by Kenya (+160 MW) and Indonesia (+185 MW).
- Off-grid electricity: In 2019, off-grid capacity increased by 160 MW (+2%), reaching 8.6 GW. Off-grid solar PV climbed by 112 MW, hydropower by 31 MW, and biofuel by just 17 MW in 2019, respectively.

3.2 US Growth of Renewable Energy

In 2019, the combined percentage share of these renewable energy sources was higher than the combined proportion of wood and hydropower, reflecting both a growth in the volumes and percentage shares of these sources in the overall U.S. energy consumption. US energy consumption by energy source in 2019 is shown in Figure 4. In 2019, the United States use of geothermal, solar, wind, and biofuels was more than three times higher from 2000(U.S. Energy Information Administration [EIA], 2019). About 11.4% of the nation's total energy consumption, or 11.5 quadrillion British thermal units (Btu; 1 quadrillion is the number 1 followed by 15 zeros), was met by renewable energy in 2019. In 2019, the United States' electric power sector absorbed over 56% of all renewable energy consumed there, while 17% of all electricity created there came from renewable sources (U.S. Energy Information Administration [EIA], 2019). The breakdown of US energy use since 1776 in Figure 5 shows the increase in renewable energy's share of total energy consumption during the previous ten years.

4. EFFECTS OF COVID-19 ON THE ENERGY SECTOR

One of the most devastating energy and economic shocks in recent history is the COVID-19 pandemic. Long-term effects will result from the energy shift away from fossil fuels, along with major interruptions to everyday life, travel, business, and transportation. The age of energy derived from fossil fuels has reached its end as a result of COVID-19. Global energy consumption began to drastically decrease once the lockdown measures were put in place, reaching lows not seen in the preceding 70 years (Majorro, 2020). The International Energy Agency (IEA) predicts that worldwide energy demand will mini-

Figure 4. US energy consumption by energy source
Adopted from U.S. Energy Information Administration [EIA] (2019)

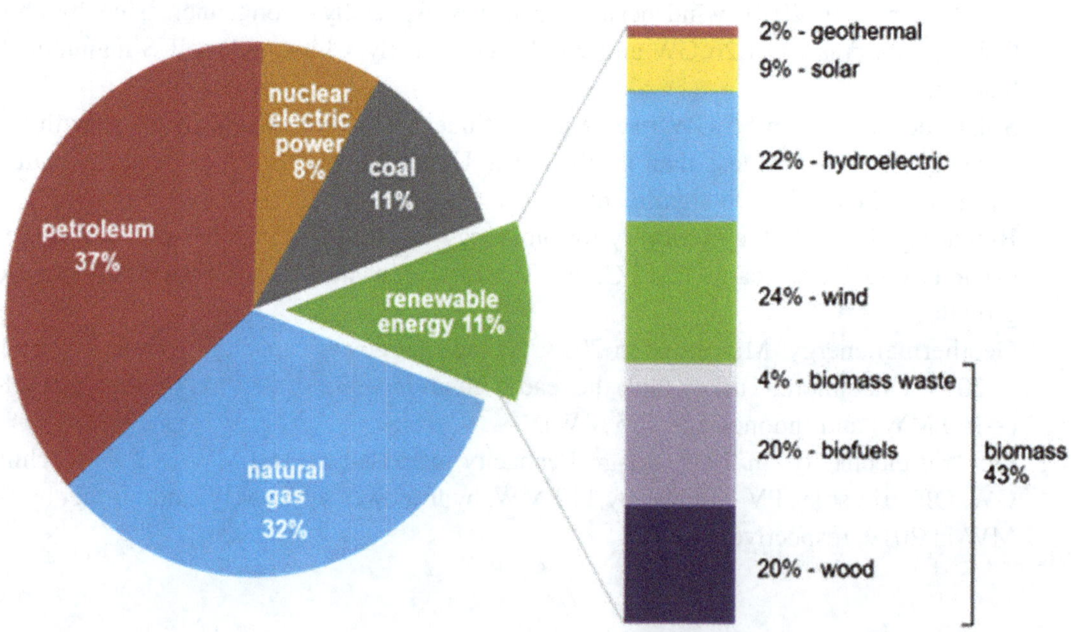

mize by 6% in 2020, while energy-related emissions would fall by 8% (IEA, 2020). S&P Global Platts Analytics predicts that after six years of increase, global oil consumption would decline by 8.7 million barrels per day (down 8.4%) from the pre-COVID-19 prediction. Particularly in poorer countries, the decreased demand for coal is thought to be the equivalent of 4.4 million less barrels of oil per day (a 5.7% decline). Due to its widespread use as a heating fuel, natural gas consumption was less adversely affected by lockdowns, falling by just 2.1 million barrels per day (down 3.0%) of oil equivalent from the pre-COVID-19 estimate. A 3.6% decrease in renewable energy output, or less than 1.0 million barrels per day of oil equivalent, is projected to occur in 2020 as a result of postponing the installation of renewable energy in important markets (Kramarchuk et al., 2020).

Global renewable energy installation is anticipated to reach a record high in 2020, according to the International Energy Agency (IEA), in stark contrast to the decreases brought on by the coronavirus pandemic in the fossil fuel sectors (Ambrose, 2020). Approximately 90% of new energy generation in 2020 will come from renewable sources, with coal and gas accounting for the remaining 10%, according to IEA statistics. In 2025, green electricity will surpass coal, which has held the top spot for the last 50 years, as the main source of energy (Ambrose, 2020).

4.1 COVID-19 Effect on Energy Security

The COVID-19 interruptions highlighted weak points in the energy grid. Thus, the energy transition's short-term prognosis is in jeopardy. The global energy system has become unstable due to the extraordinary pace and magnitude of the decline in energy consumption, as well as the concomitant price volatility and geopolitical ramifications. Extreme price volatility has financial repercussions for nations and affects the livelihoods of millions of people in the energy sector, according to the economic features of the energy transition. Additionally, it affects renewable energy technologies' ability to compete and lessens incentives for energy efficiency. For the energy transition to recover more quickly, resilience not just in markets and infrastructure, but also in policy and cooperative mechanisms will be essential. There are indications of recovery as the demand for energy appears to be progressively increasing. The economy might not make up the lost ground for some time, though. The purchase of cars and home appliances could be delayed, infrastructure development might be put on hold, and it might take longer for non-essential industrial activity to resume given the uncertain economic picture.

A successful energy transition is necessary for economic growth and social development, in addition to creating a healthy and vibrant ecology for future generations. Disruptions are the new normal in a world that is becoming more turbulent, thus energy transition roadmaps need to include resilience against these disruptions. Long-term system resilience in the energy transition entails a robust enabling environment that allows the system to recover from unexpected or external disturbances. This involves the availability of a committed political environment, stable financial markets and investment opportunities, a continual stream of innovations, infrastructure modernization, the development of human capital for emerging energy systems, etc.

The Energy Transition Index (ETI) of the World Economic Forum measures how well each nation's energy system performs in terms of economic development, environmental sustainability, and energy availability and security. The research for this year's index was just finished, and it provides strong support for the idea that the energy transition is progressing gradually. The complexities and difficulties of the energy transition are illustrated by the fact that, during the past six years, only a small number of nations have been able to achieve consistent annual progress on the ETI. Global growth has been made possible thanks to rising levels of legislation, political commitment, financial investment, and technological advancement.

When the worldwide tragedy of COVID-19 is eventually put behind us, there may be some important lessons learnt. In terms of magnitude, cascading consequences across social and economic institutions, severity for vulnerable people, and the requirement for a prompt, coordinated response, the energy transition problem is comparable to a pandemic. In order to prevent more human suffering, stakeholders must be aware that the fight against COVID-19 must not hinder efforts to solve the crucial socioeconomic conundrum of the world's energy transition.

4.2 COVID-19 Effect on Renewable Energy

4.2.1 Raw Material Supply Constraints

China is a major source of raw materials and renewable components for nations like the United States and Australia. In 2012, the United States started buying panels from other Asian countries including South Korea and Vietnam because of tariffs. The procedure, however, has been drawn out and difficult. Global Data Energy claims that even firms outside of China that produce solar energy rely primarily on imports from China for raw materials like aluminium frames and solar PV glass (McElroy, 2020). As China slowly brings back production and manufacturing in a limited capacity, prices for renewable materials and components are expected to increase before ultimately declining again. Developers, EPCs, and subcontractors are seeking lengthy lead times on orders or, if necessary, asking to postpone current projects in order to better handle any bottlenecks or slowdowns in the supply chain. Project timelines and shipments are being delayed since it is unknown if they will be there to accept delivery receipts or even to install items. One advantage of the supply chain disruption is that it drives the renewable energy sector to look for a more diverse supply chain. Alternative energy sources might stabilise such crucial renewable energy technology in the future, resulting in more independence and more affordable pricing.

4.2.2 Market Uncertainty

Every part of renewable energy is impacted by COVID-19, from the installers to the supply chain. However, the emergence of corporate demand from the quarantine will offer more immediate assurance over the direction of the renewable energy sector. Unexpectedly, investors in the renewable energy industry have cash on hand to get through the present downturn. Due to the current global uncertainty, subcontractors who rely on these investments to finance their operations are experiencing cash flow issues. Despite the current difficulties, there are no long-term alternatives that can produce extra energy as affordably and ethically as renewable energy. Future renewable energy projects won't ask us when they'll pay off, but rather if money can be secured for them and borrowing rates remain historically low to encourage price locking. It could go more quickly than most people think (McElroy, 2020). Large energy purchasers have already signed 9.33 GW worth of renewable energy contracts, according to a February 2020 announcement by the Renewable Energy Purchasers Alliance (REPA). With big giants like Facebook, Google, AT&T, and Microsoft at the top of the list, completing these projects as soon as possible will be in their best interest in attaining a zero-carbon energy future.

4.2.3 Environmental Impact

Nature is impacted by all forms of power in some manner. According to the majority of measurements, fossil fuels like coal, oil, and natural gas are much more harmful than renewable energy sources. The range of impact includes anything from animal and habitat loss to air and water pollution, water consumption, land use, and global warming emissions (Omer, 2008).

The final coal-based power reactors in Europe may very well soon switch to gas because the coronavirus has weakened the big oil producers' market sway and driven natural gas prices down to the $2 level rather than geopolitics (Wire, 2020). The decrease in fossil fuel prices is especially concerning in developing nations, where a cheap electrical power source seems essential given their dire economic circumstances as of COVID-19. Due to their great sensitivity to energy prices, these countries' governments will be forced to choose less expensive conventional energy sources rather than renewable ones, which would be disastrous for global climate policy. This unpleasant situation could have been avoided if the banks had advocated for extremely low interest rates to combat the economic stagnation that threatens the implementation of expensive renewable energy projects that would stop the energy market from moving even more in favour of fossil fuel-based power generation.

Several nations were in the process of switching to renewable and sustainable energy sources prior to this strange incident. As a consequence of important discoveries, noteworthy regulatory frameworks, and cost reductions in technology, renewable energy has become considerably more cost-effective. Recently, solar and wind energy have become more economical, and it is predicted that soon, renewable energy sources will surpass fossil fuels (Hosseini & Wahid, 2016). When oil prices reached $60 per barrel, fossil fuel companies started investing in carbon mitigation strategies; however, as a result of COVID-19 issues and the subsequent collapse in oil prices, their investments in renewable energy have stalled. The Calgary-based Suncor Energy Inc. announced the cancellation of two cogeneration units that would have lowered GHG emissions and a wind farm in northern and southern Alberta, totaling $1.5 billion (or 26%) of its capital budget for 2020. However, as the bulk of investment in renewable energy comes from sources other than the fossil fuel sectors, the data suggests that both solar and wind energy continued to rise during earlier drops in oil prices (Healing, 2020).

4.2.4 Unemployment

In the United States, employment in the renewable energy sector increased in 2019 and is anticipated to increase further in 2020. Before COVID-19, all 50 states and the District of Columbia had about 3.4 million Americans working in clean energy-related fields such renewable energy, energy efficiency, grid modernization, clean automobiles, and fuels. More Americans work in real estate, banking, or agriculture, and there are three times as many Americans employed in the fossil fuel industry, according to E2's Clean Jobs America research (Karlin, 2020). The market for sustainable energy was booming prior to Pre-Covid. The total number of jobs related to renewable energy in 2019 was 524,785, which was more than half the total number of jobs in the electric power generation industry and more than double the total number of jobs related to the oil, natural gas, and coal sectors put together. Additionally, 84,301 individuals were engaged in the energy storage industry in 2019, with 74,452 of those working in battery storage and pumped hydropower technology (Ahmed, 2020).

However, the global COVID-19 outbreak has halted the expansion of the renewable energy workforce. The workforce in the renewable energy sector is being adversely affected by supply chain interruptions, social isolation policies, project delays and cancellations, and other pandemic-related problems (Karlin, 2020). Surprisingly, March 2020 jobless claims completely eliminated all job growth in the renewable industry from 2019, and this trend is anticipated to get worse as the month goes on (Karlin, 2020). A 13 percent decrease from pre-COVID-19 employment levels, 446,000 clean energy employees were still unemployed as of December 2020, according to the government unemployment report (BusinessLeaders, 2020).

The renewable energy industry has been especially reluctant to revive, even if the country's general job recovery has halted during the past few months (S&P Global Market Intelligence, 2020). Most new building projects have been put on hold indefinitely until the United States increases its preventative measures. There are several renewable energy developments among them. Although 2020 was anticipated to be a banner year for renewable energy, the current outbreak has caused the energy sector to be plagued by ambiguous skepticism (McElroy, 2020). The job scenario in the field of renewable energy could only last a short while. Even if there are now many delays in the renewables industry, contractual work still needs to be finished. The industry will hopefully resume where it left off as soon as is reasonably practicable and go on. Many businesses are aware of the tax credits that will expire at the end of the year. When construction can continue regularly, there will likely be a tremendous push to finish these projects since no one can predict with certainty if the ITC will receive an extension at the higher rate until the end of the year (McElroy, 2020).

4.2.5 Market Outlook

The global scenario for renewable and sustainable energy, which has exploded and expanded quickly in recent decades, has been seriously threatened by the coronavirus. The COVID-19 outbreak has wreaked havoc on companies, supply chains, and industrial facilities for renewable energy, impeding the switch to renewables. Even long-standing rules governing renewable energy are being questioned, especially those that tax businesses that have suffered because of the crisis. Financial strain will very certainly affect several governments, and new renewable energy projects will almost certainly be postponed. When austerity measures are put in place, manufacturers' efforts to develop and install equipment for renewable energy technologies will come to an end. For instance, Morgan Stanley wants to decrease the amount of solar photovoltaics installed in the US (Sarad, Alexander & Donovan, 2020). Due to the fact that a large number of solar panels, connectors, modules, and cells are produced in China and East Asia, two regions that have seen rapid growth, supply chain disruptions might delay the completion of renewable energy projects that are now under construction across the world, particularly solar power plants (Fox-Penner, 2020).

Despite obstacles, investors are becoming increasingly interested in renewable energy because of the growing recognition of the necessity to confront the climate calamity by lowering carbon emissions. The value of shares in solar enterprises has more than doubled since December 2019, and shares in renewable equipment makers and project developers have surpassed most major stock market indices (Yale Environment 360, 2020). One of the world's largest markets for renewable energy is the United States, and according to the IEA, if Joe Biden's proposals are put into practice, deployment might quicken even more. COVID-19 has heightened interest in climate policy in Europe and influenced the elections in the United States. Even if stimulus programmes, particularly in China and emerging nations, give job and direct economic assistance measures a higher priority than green development, the projection for global growth for renewables likely still holds. Decreases in direct subsidies and tax credits, as witnessed in China, Europe, and the US, as well as a murkier outlook for long-term prices are major risks to the business. However, despite having lower returns and more market risks, renewable energy sources may now compete at grid-parity prices thanks to persistently strong investor demand and declining costs. As a result, larger renewables businesses with strong balance sheets and vertical integration to limit merchant risks may be better positioned and may strive to consolidate the market (Kramarchuk et al., 2020). The sector's resiliency and excellent prospects are reflected in investors' ongoing high appetite (Yale Environment 360, 2020).

By 2025, renewables are anticipated to surpass coal as the primary source of energy generation worldwide, ending coal's five-decade rule. By then, renewable energy sources are expected to provide one-third of the world's energy (Yale Environment 360, 2020). According to IEA data, solar power capacity has expanded 18 times since 2010, while wind power capacity has increased four times(Earth.org, 2020). Future power markets are predicted to be ruled by solar energy in full. By 2020, just 47% of the green electricity that was produced in 2010 came from hydropower. However, the majority of energy emissions are produced by fuel burning in transportation, industry, and heating, which makes up just around one-fifth of total energy usage (Yale Environment 360, 2020). The International Energy Agency projects that new renewable capacity will increase by a record 200 gigatons in 2020, with China and the United States leading the way as developers scramble to take advantage of incentive programmed that are about to expire (Yale Environment 360, 2020).

The IEA predicts even faster growth in 2021, when India and the European Union will be the driving force (Yale Environment 360, 2020). But the IEA cautioned that under the existing policies, growth might somewhat slowdown in 2022. Renewable energy sources can withstand the COVID-19 issue, but not political unpredictability. Governments can address these problems to hasten the transition to renewable energy and aid in a sustainable recovery. For instance, if the anticipated clean power policies of the next U.S. government are put into practice, solar PV and wind energy might be deployed in the U.S. considerably more quickly.

Renewable energy sources, including wind and solar, are booming and outpacing predictions (Team LightBox, 2020). Proving to be affordable and reliable sources of energy, they were poised for huge growth in 2020. However, the effects of COVID-19 have provided new challenges and impacted renewable energy industry forecasts. Like most industries, disruptions of supply chains and delays in construction projects due to lockdowns and social distancing are some of the major reasons that growth slowed down. Despite these setbacks, renewable energy consumption has increased by 40% in 2020 compared to 2019, and analysts say renewable energy is critical for economic recovery (Mojarro, 2020). As a result of the drop in demand, the share of power generated by renewable sources, notably wind and solar, dramatically increased to new highs across a number of countries. Figure 6 demonstrates that in less than ten weeks, the United States increased its use of renewable energy by about 40%, while India increased it by 45%. Spain, Germany, and Italy all received new rankings for the grid integration of variable renewable energy. Investors are turning to this industry to aid in the recovery of the economy (Mojarro, 2020). A recent report from UC Berkeley shows how the U.S. can achieve carbon-free electricity by 2035, and how the nation would benefit. Millions of new jobs and trillions of dollars in the economy are a few major advantages

Figure 5. Renewable energy demand after the COVID-19 lockdown
Adopted from Mojarro (2020)

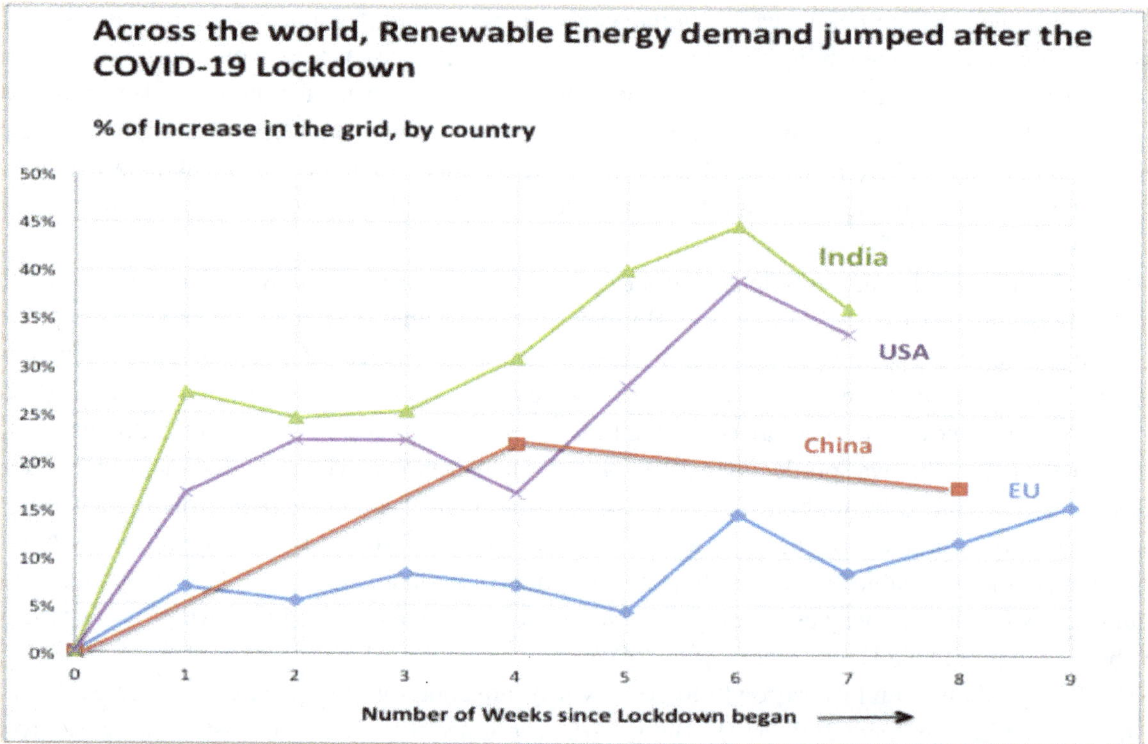

4.2.6 Renewable Energy Growth Drivers

The current situation for the electricity sector is either unforeseen or situational, despite the fact that the pandemic is. A combination of previous policies, rules, incentives, and technologies ingrained in the power sectors of many forward-thinking nations has led to the continued rise of renewable energy in the grid.

There are several factors behind the increase in renewable energy during this crisis:

1. Renewable has been aided by political favors and low financing rates shortly, providing beneficial circumstances for wind and PV projects that need big initial expenditures(International Energy Agency [IEA], 2020). Through market regulation, renewable energy sources are given preference in many nations. The cheapest source is given preference for the initial batch of energy sent to the network, favoring less expensive and cleaner sources.
2. Continuous innovation is also a discussion. Renewable energy has become the most affordable form of energy. According to IRENA, the cost of solar has dropped by 82% in the previous ten years (Sorenson, 2020). Bloomsburg (2020) states that renewable energy is now the cheapest energy source in two-thirds of the world.
3. Due to the fact that many renewable energy projects have long-term fixed-price contracts, they operate as a "haven" for some institutional investors who are dealing with the current economic recession.
4. The Covid 19 issue has so far prevented governments in important nations from abandoning or withdrawing already announced programmes, ensuring investors that compliance will endure de-

spite economic turmoil. In addition, long-term net-zero goals in the two largest renewable energy markets, the European Union and China, provide investors with long-term visibility. (International Energy Agency [IEA], 2020).

5. Another option for investing could be taken into account. Investors now favour renewable energy as a potential source for future power plants. Over the past two decades, the capacity of renewable energy has steadily expanded, and it now makes up 72% of all new power capacity (IRENA, 2020).
6. In both emerging and mature economies, stimulus packages have kept large utilities and, to a lesser extent, small businesses from investing in renewable projects (i.e., independent power producers [IPPs]). These relief measures have played a crucial role in helping them increase their cash flow and finance projects planned for the second half of this year. (International Energy Agency [IEA], 2020).

5. CONCLUSION

The pandemic struck at a time when investor concerns about the effects of fossil fuels on carbon emissions and environmental regulations, rising debt in the fossil fuel sector, declining renewable energy costs, persistently low oil prices, and one of the fastest-growing industries were already reducing capital investment in the fossil fuel sector. The COVID-19 epidemic, however, appears to have caused the enormous, two-decade-long rise in renewable energy to plateau in 2020. It is still too early to determine how severely the pandemic would affect the world's renewable energy sources, but it does not imply they will remain flat for very long. During the epidemic, governments, companies, and households "kept the lights on," but using less expensive and environmentally friendly energy. The clean energy curve is anticipated to avoid flattening with the help of investment stimulation, COVID-19 alleviation conditions, innovation, and infrastructure opportunities in low-carbon and digital technology. Over the coming years, the global renewable energy picture may ultimately resume its long-term trend towards the production of green electricity. The experience of the COVID-19 pandemic period may persuade countries to speed up their attempts to implement renewable energy policies before another global economic shock is caused by a different illness or an unanticipated weather calamity. Overall, there is optimism that the development of renewable energy will proceed.

REFERENCES

Ahmed, M. (2020). COVID-19 Hinders Progress in U.S. Renewable Energy Job Growth. *American Council of Renewable Energy*. https://acore.org/COVID-19-hinders-progress-in-u-s-renewable-energy-job-growth/

Ambrose, J. (2020). Investors fear there'll be no bright post-Covid dawn for oil majors. *Observer Business Agenda*. https://www.theguardian.com/business/2020/oct/25/investors-fear-therell-be-no-bright-post-covid-dawn-for-oil-majors

Bachman, D. (2020). The economic impact of COVID-19 (novel coronavirus). *Deloitte Insight*. https://www2.deloitte.com/us/en/insights/economy/COVID-19/economic-impact-COVID-19.html

Barbier, E. B., & Burgess, J. C. (2020). Sustainability and development after COVID-19. *World Development, 135*, 105082. doi:10.1016/j.worlddev.2020.105082 PMID:32834381

BBC News. (2020). *Worldwide cases overtake the 2003 Sars outbreak.* https://www.bbc.com/news/world-51322733

Birol, F. (2020). Put clean energy at the heart of stimulus plans to counter the coronavirus crisis. In *IEA*. https://www.iea.org/commentaries/put-clean-energy-at-the-heart-of-stimulus-plans-to-counter-the-coronavirus-crisis

Bloomsburg report. (2020). *Scale-up of Solar and Wind Puts Existing Coal, Gas at Risk.* https://about.bnef.com/blog/scale-up-of-solar-and-wind-puts-existing-coal-gas-at-risk/

Boatwright, A., & Wynne, M. A. (2020). Record Global GDP Contraction Indicative of COVID-19's Cross-Country Effect. *Dallas Federal Economics, 30*, 1–3.

Business leaders. (2020). *Clean Energy & COVID-19 Crisis* (Vol. 2). https://e2.org/reports/clean-jobs-covid-economic-crisis-april-2020/

Capital Group. (2020). 2021 Outlook : Turning points on the road to recovery. *Capital Group*. https://www.capitalgroup.com/advisor/insights/articles/2021-outlook.html

Earth.org. (2020). *Renewable Energy Hits Record Growth in 2020- Report.* https://earth.org/renewable-energy-hits-record-growth-in-2020/#:~:text=Solar

Eroğlu, H. (2020). Effects of COVID-19 outbreak on the environment and renewable energy sector. *Environment, Development and Sustainability*. Advance online publication. doi:10.1007/s10668-020-00837-4 PMID:32837274

Fottrell, Q. (2020). Will COVID-19 push more employees to work remotely after the pandemic? This economist says yes. *Market Watch*. https://www.marketwatch.com/story/does-COVID-19-have-the-power-to-revolutionize-work-this-economist-says-yes-2020-12-11

Fox-Penner, P. (2020). Will the COVID-19 Pandemic Slow The Global Shift to Renewable Energy? *The Brink, Boston University*. https://www.bu.edu/articles/2020/will-the-COVID-19-pandemic-slow-the-global-shift-to-renewable-energy/

Healing, D. (2020). Big Oil's interest in renewable energy investments is expected to waver, the report says. *The Globe and Mail*. https://www.theglobeandmail.com/business/article-big-oils-interest-in-renewable-energy-investments-expected-to-waver/

Hosseini, S. E., & Wahid, M. A. (2016). Hydrogen production from renewable and sustainable energy resources: Promising green energy carrier for clean development. *Renewable & Sustainable Energy Reviews, 57*, 850–866. doi:10.1016/j.rser.2015.12.112

IEA. (2020). *Exploring the impacts of the COVID-19 pandemic on global energy markets, energy resilience, and climate change*. IEA.

International Energy Agency. (2020). *Renewables 2020 – Analysis and Forecast to 2025*. https://www.iea.org/reports/renewables-2020

International Renewable Energy Agency (IRENA). (2020). *Renewable Capacity Statistics 2020*. https://www.irena.org/publications/2020/Mar/Renewable-Capacity-Statistics-2020

IRENA. (2020). *Renewables accounted for Almost Three-Quarters of New Capacity in 2019*. IRENA.

Karlin, M. (2020). Slow Job Growth in November Leaves 446, 000 Clean Energy Workers Unemployed as COVID Cases, Shutdowns Surge. *American Council on Renewable Energy*. https://goodmenproject.com/featured-content/slow-job-growth-in-november-leaves-446000-clean-energy-workers-unemployed-as-covid-case

Kramarchuk, R., Klein, D., Brunetti, B., Joseph, I., Schiavo, M., Georges, P., Redmond, S., Anankina, E., Prabhu, A., Lu, G., Huang, D., Amiot, M., & Roache, S. (2020). *How is COVID-19 Impacting the Energy Transition?* Available at https://www.spglobal.com/en/research-insights/featured/how-is-COVID-19-impacting-the-energy-transition

Majorro, N. (2020). COVID-19 is a game-changer for renewable energy. Here's why. *World Economic Forum*. https://www.weforum.org/agenda/2020/06/COVID-19-is-a-game-changer-for-renewable-energy/

McElroy, B. (2020). *How COVID-19 is impacting renewable energy*. Available at https://www.renewableenergyworld.com/2020/04/29/how-COVID-19-is-impacting-renewable-energy/

McKibbin, W. J., & Fernando, R. (2020). The Global Macroeconomic Impacts of COVID-19: Seven Scenarios. SSRN *Electronic Journal*, 1–43. doi:10.2139/ssrn.3547729

Mojarro, N. (2020). *COVID-19 is a game-changer for renewable energy. Here's why*. Available at https://www.weforum.org/agenda/2020/06/COVID-19-is-a-game-changer-for-renewable-energy

Omer, A. M. (2008). Energy, environment, and sustainable development. *Renewable & Sustainable Energy Reviews*, *12*(9), 2265–2300. doi:10.1016/j.rser.2007.05.001

Sarad, N. J., Alexander, M. J., & Donovan, L. P. (2020). COVID-19 Impact on US Renewable Energy Projects Article By. *The National Law Review*, *10*(363), 1–5.

Sorenson, B. (2020). Renewable Power Generation Costs in 2019. *IRENA*. https://irena.org/publications/2020/Jun/Renewable-Power-Costs-in-2019

S&P Global Market Intelligence. (2020). *As the market expands, renewable energy unemployment barely budges*. https://www.spglobal.com/marketintelligence/en/news-insights/latest-news-headlines/as-market-expands-renewable-energy-unemployment-barely-budges-60666681

Team LightBox. (2020). *The Effects of COVID-19 on the Renewable Energy Industry, and How this Industry is Important to Economic Recovery*. https://www.digmap.com/blog/the-effects-of-COVID-19-on-the-renewable-energy-industry-and-how-this-industry-is-important-to-economic-recovery/

The Economist. (2020). *COVID-19 to send almost all G20 countries into a recession - Economist Intelligence Unit*. https://www.eiu.com/n/COVID-19-to-send-almost-all-g20-countries-into-a-recession/

U.S. Energy Information Administration. (2019). *Renewable energy explained: Portfolio standards*. https://www.eia.gov/energyexplained/renewable-sources/portfolio-standards.php

Wire, F. (2020). Oil and Coronavirus Shocks Add Pressure for MEA Sovereigns. *FitchRatings.* https://www.fitchratings.com/research/sovereigns/oil-coronavirus-shocks-add-pressure-for-mea-sovereigns-10-03-2020

World Economy. (2020). *5 charts show what the global economy looks like heading into 2021.* https://www.cnbc.com/2020/12/28/5-charts-show-covid-impact-on-the-global-economy-in-2020.html

Worldometer. (2020). *COVID-19 coronavirus pandemic.* https://www.worldometers.info/coronavirus/

Yale Environment 360. (2020). https://e360.yale.edu/digest/despite-covid-driven-economic-slowdown-renewable-energy-to-hit-new-record-in-2020

Yardeni, E., Johnson, D., & Quintana, M. (2020). Global Economic Briefing: Industrial Production. *Global Economic Briefing: Industrial Production.* www.yardeni.com

LIST OF ABBREVIATIONS

BBC: British Broadcasting Corporation
COVID: Coronavirus disease
EIA: Energy Information Administration
ETI: Energy Transition Index
GW: Giga Watt
IEA: International Energy Agency
REPA: Renewable Energy Purchasers Alliance
S&P: Standard and Poor's
SARS: Severe Acute Respiratory Syndrome
SDG: Sustainable Development Goals
WHO: World Health Organization

Compilation of References

Abbasi, M. H., Abdullah, B., Ahmad, M. W., Rostami, A., & Cullen, J. (2021). Heat transition in the European building sector: Overview of the heat decarbonisation practices through heat pump technology. *Sustainable Energy Technologies and Assessments*, *48*, 101630. doi:10.1016/j.seta.2021.101630

Abdelhady, S. (2021). Performance and cost evaluation of solar dish power plant: Sensitivity analysis of levelized cost of electricity (LCOE) and net present value (NPV). *Renewable Energy*, *168*, 332–342. doi:10.1016/j.renene.2020.12.074

Abreu, V. H. S. D., Coralli, A., Proença, L. F. C., Santos, A. S. (2023). An Overview of Life Cycle Assessment of Renewable Hydrogen. *ENSUS2023 - XI Encontro de Sustentabilidade Em Projeto*, 224–240. doi:10.29183/2596-237x.ensus2023.v11.n2.p224-240

Adamou, P., Bellomi, S., Hafeez, S., Harkou, E., Al-Salem, S., Villa, A., Dimitratos, N., Manos, G., & Constantinou, A. (2023). Recent progress for hydrogen production from ammonia and hydrous hydrazine decomposition: A review on heterogeneous catalysts. *Catalysis Today*, *423*, 114022. doi:10.1016/j.cattod.2023.01.029

Aftab, S., Shah, A., Nisar, J., Ashiq, M. N., Akhter, M. S., & Shah, A. H. (2020). Marketability Prospects of Microbial Fuel Cells for Sustainable Energy Generation. *Energy & Fuels*, *34*(8), 9108–9136. doi:10.1021/acs.energyfuels.0c01766

Aggarwal, S., & Kumar, N. (2021). Core components of blockchain. In *Advances in Computers* (Vol. 121, pp. 193–209). Elsevier, doi:10.1016/bs.adcom.2020.08.010

Aghababaie, M., Farhadian, M., Jeihanipour, A., & Biria, D. (2015). Effective factors on the performance of microbial fuel cells in wastewater treatment–a review. *Environmental Technology Reviews*, *4*(1), 71–89. doi:10.1080/09593330.2015.1077896

Ahirwar, H. (2021). Life Cycle Conversion Efficiency Analysis of Roof-Top Solar Photovoltaic System. *International Journal of Innovative Research in Electrical, Electronics. Instrumentation and Control Engineering*, *9*(5), 2321–5526. doi:10.17148/IJIREEICE.2021.9558

Ahmad, F., Khalid, M., & Panigrahi, B. K. (2021). Development in energy storage system for electric transportation: A comprehensive review. *Journal of Energy Storage*, *43*(September), 103153. doi:10.1016/j.est.2021.103153

Ahmed, M. (2020). COVID-19 Hinders Progress in U.S. Renewable Energy Job Growth. *American Council of Renewable Energy*. https://acore.org/COVID-19-hinders-progress-in-u-s-renewable-energy-job-growth/

Ahring, B. K. (2003). Perspectives for anaerobic digestion. *Advances in Biochemical Engineering/Biotechnology*, *81*, 1–30. doi:10.1007/3-540-45839-5_1 PMID:12747559

Ait Bachir, L. (2021). *Algerian Green Hydrogen Production Opportunities and challenges in light of a sustainable energy system*. Academic Press.

Ajanovic, A., Sayer, M., & Haas, R. (2022). The economics and the environmental benignity of different colors of hydrogen. *International Journal of Hydrogen Energy*, *47*(57), 24136–24154. doi:10.1016/j.ijhydene.2022.02.094

Akhtar, M. (2022). Industry 4.0 Technologies Impact on Supply Chain Sustainability. In T. Bányai, Á. Bányai, & I. Kaczmar (Eds.), *Supply Chain—Recent Advances and New Perspectives in the Industry 4.0 Era*. IntechOpen. doi:10.5772/intechopen.102978

Akhtar, M. S., Dickson, R., & Liu, J. J. (2021). Life Cycle Assessment of Inland Green Hydrogen Supply Chain Networks with Current Challenges and Future Prospects. *ACS Sustainable Chemistry & Engineering*, *9*(50), 17152–17163. doi:10.1021/acssuschemeng.1c06769

AlFaris, F., Juaidi, A., & Manzano-Agugliaro, F. (2017). Intelligent homes' technologies to optimize the energy performance for the net zero energy home. *Energy and Building*, *153*, 262–274. doi:10.1016/j.enbuild.2017.07.089

Alhamid, M. I., Daud, Y., Surachman, A., Sugiyono, A., Aditya, H. B., & Mahlia, T. M. I. (2016). Potential of geothermal energy for electricity generation in Indonesia: A review. *Renewable & Sustainable Energy Reviews*, *53*, 733–740. doi:10.1016/j.rser.2015.09.032

Almutairi, K., Hosseini Dehshiri, S. J., Hosseini Dehshiri, S. S., Hoa, A. X., Arockia Dhanraj, J., Mostafaeipour, A., Issakhov, A., & Techato, K. (2022). Blockchain Technology Application Challenges in Renewable Energy Supply Chain Management. *Environmental Science and Pollution Research International*, *30*(28), 72041–72058. doi:10.1007/s11356-021-18311-7 PMID:34989989

Alsaleh, M., & Abdul-Rahim, A. S. (2021). Do global competitiveness factors effects the industry sustainability practices? Evidence from European hydropower industry. *Journal of Cleaner Production*, *310*, 127492. doi:10.1016/j.jclepro.2021.127492

Alsaleh, M., & Abdul-Rahim, A. S. (2021). The nexus between worldwide governance indicators and hydropower sustainable growth in EU 28 region. *International Journal of Environmental Research*, *15*(6), 1001–1015. doi:10.1007/s41742-021-00366-6

Alsaleh, M., & Abdul-Rahim, A. S. (2022). Moving toward sustainable environment: The effects of hydropower industry on water quality in EU economies. *Energy & Environment*, *33*(7), 1304–1325. doi:10.1177/0958305X211039452

Alsharif, M. H., Kim, J., & Kim, J. H. (2018). Opportunities and challenges of solar and wind energy in South Korea: A review. *Sustainability (Basel)*, *10*(6), 1822. doi:10.3390/su10061822

Al-Tameemi, M. A., & Chukin, V. V. (2016). Global water cycle and solar activity variations. *Journal of Atmospheric and Solar-Terrestrial Physics*, *142*, 55–59. doi:10.1016/j.jastp.2016.02.023

Althiyabi, A. (2023). Developing a Progressive Regulatory Framework to Enable Innovation in Neom's Green Hydrogen Ecosystem. *Migration Letters : An International Journal of Migration Studies*, *20*(8), 1102–1123.

Alzoubi, A. (2021). Renewable Green hydrogen energy impact on sustainability performance. *International Journal of Computations, Information and Manufacturing*, *1*(1).

Ambrose, J. (2020). Investors fear there'll be no bright post-Covid dawn for oil majors. *Observer Business Agenda*. https://www.theguardian.com/business/2020/oct/25/investors-fear-therell-be-no-bright-post-covid-dawn-for-oil-majors

Anand, S., Fernandes, B. G., & Guerrero, J. (2012). Distributed control to ensure proportional load sharing and improve voltage regulation in low-voltage DC microgrids. *IEEE Transactions on Power Electronics*, *28*(4), 1900–1913. doi:10.1109/TPEL.2012.2215055

Anatolievna, K. E., & Vladimirovna, B. V. (2020). *The main features of blockchain technologies classification*. doi:10.5281/ZENODO.4291075

Andreev, R. A., Andreeva, P. A., Krotov, L. N., & Krotova, E. L. (2018). Review of Blockchain Technology: Types of Blockchain and Their Application. *Intellekt. Sist. Proizv., 16*(1), 11. doi:10.22213/2410-9304-2018-1-11-14

Angelidaki, I., Ellegaard, L., & Ahring, B. K. (2003). Applications of the anaerobic digestion process. *Advances in Biochemical Engineering/Biotechnology, 82*, 1–33. doi:10.1007/3-540-45838-7_1 PMID:12747564

Angrish, A., Craver, B., Hasan, M., & Starly, B. (2018). A Case Study for Blockchain in Manufacturing: "FabRec": A Prototype for Peer-to-Peer Network of Manufacturing Nodes. *Procedia Manufacturing, 26*, 1180–1192. doi:10.1016/j.promfg.2018.07.154

Anitha, C., Komala, C. R., Vivekanand, C. V., Lalitha, S. D., Boopathi, S., & Revathi, R. (2023, February). Artificial Intelligence driven security model for Internet of Medical Things (IoMT). *Proceedings of 2023 3rd International Conference on Innovative Practices in Technology and Management, ICIPTM 2023.* 10.1109/ICIPTM57143.2023.10117713

Aravani, V. P., Sun, H., Yang, Z., Liu, G., Wang, W., Anagnostopoulos, G., Syriopoulos, G., Charisiou, N. D., Goula, M. A., Kornaros, M., & Papadakis, V. G. (2022). Agricultural and livestock sector's residues in Greece & China: Comparative qualitative and quantitative characterization for assessing their potential for biogas production. *Renewable & Sustainable Energy Reviews, 154*, 111821. doi:10.1016/j.rser.2021.111821

Arboleya, P., Mayet, C., Mohamed, B., Aguado, J. A., & de La Torre, S. (2020). A review of railway feeding infrastructures: Mathematical models for planning and operation. *eTransportation, 5*, 100063. doi:10.1016/j.etran.2020.100063

Arcos, J. M. M., & Santos, D. M. F. (2023). The Hydrogen Color Spectrum: Techno-Economic Analysis of the Available Technologies for Hydrogen Production. *Gases, 3*(1), 25–46. doi:10.3390/gases3010002

Arent, D. J., Green, P., Abdullah, Z., Barnes, T., Bauer, S., Bernstein, A., Berry, D., Berry, J., Burrell, T., Carpenter, B., Cochran, J., Cortright, R., Curry-Nkansah, M., Denholm, P., Gevorian, V., Himmel, M., Livingood, B., Keyser, M., King, J., ... Turchi, C. (2022). Challenges and opportunities in decarbonizing the US energy system. *Renewable & Sustainable Energy Reviews, 169*, 112939. doi:10.1016/j.rser.2022.112939

Armaroli, N., & Balzani, V. (2011). The Hydrogen Issue. *ChemSusChem, 4*(1), 21–36. doi:10.1002/cssc.201000182 PMID:21226208

Ashok, B., Kannan, C., Usman, K. M., Vignesh, R., Deepak, C., Ramesh, R., Narendhra, T. M. V., & Kavitha, C. (2022). Transition to Electric Mobility in India: Barriers Exploration and Pathways to Powertrain Shift through MCDM Approach. *Journal of The Institution of Engineers (India): Series C, 103*(5), 1251–1277. doi:10.1007/s40032-022-00852-6

Assad, M. a. (2021). *Design and performance optimization of renewable energy systems.* Academic Press.

Ásványi, K., Juhász-Dóra, K., Jászberényi, M., & Michalkó, G. (2017). Literature review of renewable energy in the tourism industry. *Journal of Environmental Management & Tourism, 8*(2 (18)), 476.

Atacak, İ. (2023). An Ensemble Approach Based on Fuzzy Logic Using Machine Learning Classifiers for Android Malware Detection. *Applied Sciences (Basel, Switzerland), 13*(3), 1484. Advance online publication. doi:10.3390/app13031484

Azadnia, A. H., McDaid, C., Andwari, A. M., & Hosseini, S. E. (2023). Green hydrogen supply chain risk analysis: A european hard-to-abate sectors perspective. *Renewable & Sustainable Energy Reviews, 182*, 113371. doi:10.1016/j.rser.2023.113371

Azar, A. T., Abed, A. M., Abdulmajeed, F. A., Hameed, I. A., Kamal, N. A., Jawad, A. J. M., Abbas, A. H., Rashed, Z. A., Hashim, Z. S., Sahib, M. A., Ibraheem, I. K., & Thabit, R. (2022). A New Nonlinear Controller for the Maximum Power Point Tracking of Photovoltaic Systems in Micro Grid Applications Based on Modified Anti-Disturbance Compensation. *Sustainability (Basel), 14*(17), 10511. Advance online publication. doi:10.3390/su141710511

Azzi, R., Chamoun, R. K., & Sokhn, M. (2019). The power of a blockchain-based supply chain. *Computers & Industrial Engineering, 135*, 582–592. doi:10.1016/j.cie.2019.06.042

Babu, B. S., Kamalakannan, J., Meenatchi, N., M, S. K. S., S, K., & Boopathi, S. (2023). Economic impacts and reliability evaluation of battery by adopting Electric Vehicle. *IEEE Explore*, 1–6. doi:10.1109/ICPECTS56089.2022.10046786

Babul, A. K., Rajput, S. K., Singh, H., & Yadaw, R. C. (2021). Energy and Economic Analysis of Grid-Type Roof-Top Photovoltaic (GRPV) System. *Lecture Notes in Electrical Engineering, 710*, 399–408. doi:10.1007/978-981-15-8815-0_34

Bachman, D. (2020). The economic impact of COVID-19 (novel coronavirus). *Deloitte Insight*. https://www2.deloitte.com/us/en/insights/economy/COVID-19/economic-impact-COVID-19.html

Bade, S. O., Tomomewo, O. S., Meenakshisundaram, A., Ferron, P., & Oni, B. A. (2024). Economic, social, and regulatory challenges of green hydrogen production and utilization in the US: A review. *International Journal of Hydrogen Energy, 49*, 314–335. doi:10.1016/j.ijhydene.2023.08.157

Bagher, A. M., Fatemeh, G., Saman, M., & Leili, M. M. (2015). Advantages and disadvantages of biogas energy. *Bulletin of Advanced Scientific Research, 1*(5), 132–135.

Balat, M. (2008). Possible methods for hydrogen production. *Energy Sources. Part A, Recovery, Utilization, and Environmental Effects, 31*(1), 39–50. doi:10.1080/15567030701468068

Balat, M. (2008). Potential importance of hydrogen as a future solution to environmental and transportation problems. *International Journal of Hydrogen Energy, 33*(15), 4013–4029. doi:10.1016/j.ijhydene.2008.05.047

Ball, M., & Wietschel, M. (2009). The future of hydrogen–opportunities and challenges. *International Journal of Hydrogen Energy, 34*(2), 615–627. doi:10.1016/j.ijhydene.2008.11.014

Barbier, E. B., & Burgess, J. C. (2020). Sustainability and development after COVID-19. *World Development, 135*, 105082. doi:10.1016/j.worlddev.2020.105082 PMID:32834381

Barde, S., Barde, A., Barde, S., & Barde, A. (2023). *Blockchain Technology: Worldwide Supplier of Digital Services and Technology – Tactics, Methods, Limitations, and Performance*. Https://Services.Igi-Global.Com/Resolvedoi/Resolve.Aspx?Doi=10.4018/978-1-6684-7808-0.Ch012

Barja-Martinez, S., Aragüés-Peñalba, M., Munné-Collado, Í., Lloret-Gallego, P., Bullich-Massagué, E., & Villafafila-Robles, R. (2021). Artificial intelligence techniques for enabling Big Data services in distribution networks: A review. *Renewable & Sustainable Energy Reviews, 150*, 111459. doi:10.1016/j.rser.2021.111459

Baron, N., (2021). *Railway terminal regulation*. Academic Press.

Baroutaji, A., Arjunan, A., Robinson, J., Abdelkareem, M. A., & Olabi, A. G. (2023). Additive Manufacturing for Proton Exchange Membrane (PEM) Hydrogen Technologies: Merits, Challenges, and Prospects. *International Journal of Hydrogen Energy*.

Barron, A. R. (2023). *2.10 Storage of Hydrogen for Use as a Fuel*. LibreTexts Chemistry.

Batstone, D. J., Keller, J., Angelidaki, I., Kalyuzhnyi, S. V., Pavlostathis, S. G., Rozzi, A., Sanders, W. T., Siegrist, H., & Vavilin, V. A. (2002). The IWA Anaerobic Digestion Model No 1 (ADM1). *Water Science and Technology : A Journal of the International Association on Water Pollution Research, 45*(10), 65–73. doi:10.2166/wst.2002.0292

Bazaee, G., Hassani, M., & Shahmansouri, A. (2020). Identifying blockchain technology maturity's levels In the oil and gas industry with a grounded theory approach. *Petroleum Business Review*. doi:10.22050/pbr.2020.256181.1138

BBC News. (2020). *Worldwide cases overtake the 2003 Sars outbreak*. https://www.bbc.com/news/world-51322733

Berga, L. (2008). Solving dam engineering challenges. *HydroWorld. com.*

Berga, L. (2016). The role of hydropower in climate change mitigation and adaptation: A review. *Engineering (Beijing)*, *2*(3), 313–318. doi:10.1016/J.ENG.2016.03.004

Bergmann, A., Colombo, S., & Hanley, N. (2008). Rural versus urban preferences for renewable energy developments. *Ecological Economics*, *65*(3), 616–625. doi:10.1016/j.ecolecon.2007.08.011

Bertuccioli, L., Chan, A., Hart, D., Lehner, F., Madden, B., & Standen, E. (2014). Development of water electrolysis in the European union. Final report. *Fuel Cells and Hydrogen Joint Undertaking*.

Bhandari, K. P., Collier, J. M., Ellingson, R. J., & Apul, D. S. (2015). Energy payback time and energy return on energy invested of solar photovoltaic systems: A systematic review and meta-analysis. *Renewable & Sustainable Energy Reviews*, *47*, 133–141. doi:10.1016/j.rser.2015.02.057

Bhaskar, A., Assadi, M., & Somehsaraei, H. N. (2020). Decarbonization of the Iron and Steel Industry with Direct Reduction of Iron Ore with Green Hydrogen. *Energies*, *13*(3), 758. doi:10.3390/en13030758

Bilgen, S., Kaygusuz, K., & Sari, A. (2004). Renewable energy for a clean and sustainable future. *Energy Sources. Part A, Recovery, Utilization, and Environmental Effects*, *26*(12), 1119–1129. https://www.irena.org/

Bin Ahmad, M. S., Pesyridis, A., Sphicas, P., Mahmoudzadeh Andwari, A., Gharehghani, A., & Vaglieco, B. M. (2022). Electric Vehicle Modelling for Future Technology and Market Penetration Analysis. *Frontiers of Mechanical Engineering*, *8*(July), 1–18. doi:10.3389/fmech.2022.896547

Birol, F. (2020). Put clean energy at the heart of stimulus plans to counter the coronavirus crisis. In *IEA*. https://www.iea.org/commentaries/put-clean-energy-at-the-heart-of-stimulus-plans-to-counter-the-coronavirus-crisis

Blaine, F. (2021, August 12). *Touted as clean, 'blue' hydrogen may be worse than gas or coal | Cornell Chronicle*. https://news.cornell.edu/stories/2021/08/touted-clean-blue-hydrogen-may-be-worse-gas-or-coal

Bloomsburg report. (2020). *Scale-up of Solar and Wind Puts Existing Coal, Gas at Risk*. https://about.bnef.com/blog/scale-up-of-solar-and-wind-puts-existing-coal-gas-at-risk/

Boatwright, A., & Wynne, M. A. (2020). Record Global GDP Contraction Indicative of COVID-19's Cross-Country Effect. *Dallas Federal Economics*, *30*, 1–3.

Bobbert, Y., & Ozkanli, N. (2020). LockChain Technology as One Source of Truth for Cyber, Information Security and Privacy. In K. Arai, S. Kapoor, & R. Bhatia (Eds.), *Intelligent Computing* (Vol. 1230, pp. 336–347). Springer International Publishing. doi:10.1007/978-3-030-52243-8_24

Bockris, J. O., & Klerer, J. (1976). Energy: The Solar Hydrogen Alternative. *Journal of the Electrochemical Society*, *123*(8), 284C. doi:10.1149/1.2133054

Bockris, J. O., & Klerer, J. (1978). *Solar-Hydrogen Energy Systems An Authoritative Review of Water-Splitting Systems by Solar Beam and Solar Heat: Hydrogen Production, Storage and Utilisation*. Pergamon Press.

Bollmann, J., Pitchaimuthu, S., & Kühnel, M. F. (2023). Challenges of Industrial-Scale Testing Infrastructure for Green Hydrogen Technologies. *Energies*, *16*(8), 3604. doi:10.3390/en16083604

Boopathi, S. (2021). Improving of Green Sand-Mould Quality using Taguchi Technique. *Journal of Engineering Research*. doi:10.36909/jer.14079

Boopathi, S. (2023b). Deep Learning Techniques Applied for Automatic Sentence Generation. In Promoting Diversity, Equity, and Inclusion in Language Learning Environments (pp. 255–273). IGI Global. doi:10.4018/978-1-6684-3632-5.ch016

Boopathi, S., Arigela, S. H., Raman, R., Indhumathi, C., Kavitha, V., & Bhatt, B. C. (2023). Prominent Rule Control-based Internet of Things: Poultry Farm Management System. *IEEE Explore*, 1–6. doi:10.1109/ICPECTS56089.2022.10047039

Boopathi, S., Siva Kumar, P. K., & Meena, R. S. J., S. I., P., S. K., & Sudhakar, M. (2023). Sustainable Developments of Modern Soil-Less Agro-Cultivation Systems. In Human Agro-Energy Optimization for Business and Industry (pp. 69–87). IGI Global. doi:10.4018/978-1-6684-4118-3.ch004

Boopathi, S. (2019). Experimental investigation and parameter analysis of LPG refrigeration system using Taguchi method. *SN Applied Sciences*, *1*(8), 892. doi:10.1007/s42452-019-0925-2

Boopathi, S. (2022a). An experimental investigation of Quench Polish Quench (QPQ) coating on AISI 4150 steel. *Engineering Research Express*, *4*(4), 45009. doi:10.1088/2631-8695/ac9ddd

Boopathi, S. (2022b). Cryogenically treated and untreated stainless steel grade 317 in sustainable wire electrical discharge machining process: A comparative study. *Environmental Science and Pollution Research International*, *30*(44), 1–10. doi:10.1007/s11356-022-22843-x PMID:36057706

Boopathi, S. (2022c). Experimental investigation and multi-objective optimization of cryogenic Friction-stir-welding of AA2014 and AZ31B alloys using MOORA technique. *Materials Today. Communications*, *33*, 104937. doi:10.1016/j.mtcomm.2022.104937

Boopathi, S. (2022d). Performance Improvement of Eco-Friendly Near-Dry Wire-Cut Electrical Discharge Machining Process Using Coconut Oil-Mist Dielectric Fluid. *Journal of Advanced Manufacturing Systems*. Advance online publication. doi:10.1142/S0219686723500178

Boopathi, S. (2023a). An Investigation on Friction Stir Processing of Aluminum Alloy-Boron Carbide Surface Composite. In *Materials Horizons: From Nature to Nanomaterials* (pp. 249–257). Springer. doi:10.1007/978-981-19-7146-4_14

Boopathi, S. (2023c). Internet of Things-Integrated Remote Patient Monitoring System: Healthcare Application. In *Dynamics of Swarm Intelligence Health Analysis for the Next Generation* (pp. 137–161). IGI Global. doi:10.4018/978-1-6684-6894-4.ch008

Boopathi, S., Balasubramani, V., & Sanjeev Kumar, R. (2023). Influences of various natural fibers on the mechanical and drilling characteristics of coir-fiber-based hybrid epoxy composites. *Engineering Research Express*, *5*(1), 15002. doi:10.1088/2631-8695/acb132

Boopathi, S., & Myilsamy, S. (2021). Material removal rate and surface roughness study on Near-dry wire electrical discharge Machining process. *Materials Today: Proceedings*, *45*(9), 8149–8156. doi:10.1016/j.matpr.2021.02.267

Bouckaert, S., Pales, A. F., McGlade, C., Remme, U., Wanner, B., Varro, L., D'Ambrosio, D., & Spencer, T. (2021). *Net zero by 2050: A roadmap for the global energy sector*. Academic Press.

Boulmrharj, S., Khaidar, M., Bakhouya, M., Ouladsine, R., Siniti, M., & Zine-dine, K. (2020). Performance Assessment of a Hybrid System with Hydrogen Storage and Fuel Cell for Cogeneration in Buildings. *Sustainability (Basel), 12*(12), 4832. doi:10.3390/su12124832

Bouraima, M. B., Alimo, P. K., Agyeman, S., Stuno, P. D., Lartey-Young, G., Ehebrecht, D., & Qiu, Y. (2023). Africa's railway renaissance and sustainability: Current knowledge, challenges, and prospects. *Journal of Transport Geography, 106*, 103487. doi:10.1016/j.jtrangeo.2022.103487

Bridgeland, R., Chapman, A., McLellan, B., Sofronis, P., & Fujii, Y. (2022). Challenges toward achieving a successful hydrogen economy in the US: Potential end-use and infrastructure analysis to the year 2100. *Cleaner Production Letters, 3*, 100012. doi:10.1016/j.clpl.2022.100012

Business leaders. (2020). *Clean Energy & COVID-19 Crisis* (Vol. 2). https://e2.org/reports/clean-jobs-covid-economic-crisis-april-2020/

Buttler, A., & Spliethoff, H. (2017, September). Current Status of Water Electrolysis for Energy Storage, Grid Balancing and Sector Coupling via Power-to-Gas and Power-to-Liquids: A Review. *Renewable & Sustainable Energy Reviews, 82*, 2440–2454. doi:10.1016/j.rser.2017.09.003

Byun, J., Hong, I., Kang, B., & Park, S. (2011). A smart energy distribution and management system for renewable energy distribution and context-aware services based on user patterns and load forecasting. *IEEE Transactions on Consumer Electronics, 57*(2), 436–444. doi:10.1109/TCE.2011.5955177

Cabrane, Z., & Lee, S. H. (2022). Control and Management of Railway System Connected to Microgrid Stations. *IEEE Access : Practical Innovations, Open Solutions, 10*, 40445–40455. doi:10.1109/ACCESS.2022.3165832

Cai, J., Weng, C., Zhang, R., Li, Q., Zhang, T., & Shi, Z. (2023). Comparative Analysis on the Dynamic Operation Performance of Photovoltaic/Thermal Powered Proton Exchange Membrane Water Electrolysis Cogeneration System (PV/T-PEMWE) under Different Connection Modes. *Renewable Energy, 219*(P2), 119566. doi:10.1016/j.renene.2023.119566

CanadianSolar. (n.d.). *CS6R-405MS Datasheet*. Retrieved September 7, 2023 (https://static.csisolar.com/wp-content/uploads/2020/06/27131418/CS-Datasheet-HiKu6_CS6R-MS_v2.0_EN.pdf)

Cao, B., Wang, Z., Zhang, L., Feng, D., Peng, M., & Zhang, L. (2021). *Blockchain Systems, Technologies and Applications: A Methodology Perspective* (arXiv:2105.03572). arXiv. http://arxiv.org/abs/2105.03572

Cao, Y., Yao, M., & Sun, X. (2023). An Overview of Modelling and Energy Management Strategies for Hybrid Electric Vehicles. *Applied Sciences (Basel, Switzerland), 13*(10), 5947. Advance online publication. doi:10.3390/app13105947

Capital Group. (2020). 2021 Outlook : Turning points on the road to recovery. *Capital Group*. https://www.capitalgroup.com/advisor/insights/articles/2021-outlook.html

Capuano, M., Sorrentino, M., & Agelin-Chaab, M. (2023). Design and analysis of a hybrid space heating system based on HT-PEM fuel cell and an air source heat pump with a novel heat recovery strategy. *Applied Thermal Engineering, 231*, 120947. doi:10.1016/j.applthermaleng.2023.120947

Casson, M. (2004). The future of the UK railway system: Michael Brooke's vision. *International Business Review, 13*(2), 181–214. doi:10.1016/j.ibusrev.2003.10.002

Castro-Santos, L. e., Filgueira-Vizoso, A., Costoya, X., Arguilé-Pérez, B., & Ribeiro, A. S. (2024). Economic viability of floating wave power farms considering the energy generated in the near future. *Renewable Energy, 222*, 119947. doi:10.1016/j.renene.2024.119947

Cayirci, E., & Rong, C. (2018). Blocknetwork for Big Information Fusion and NoSQL Knowledgebase Management. *2018 IEEE Seventh International Conference on Communications and Electronics (ICCE)*, 379–384. 10.1109/CCE.2018.8465729

Cen, S. H., Li, K., Liu, Q. W., & Jiang, Y. L. (2021). Solar energy-based hydrogen production and post-firing in a biomass fueled gas turbine for power generation enhancement and carbon dioxide emission reduction. *Energy Conversion and Management*, *233*, 113941. doi:10.1016/j.enconman.2021.113941

Chabachi, S. e., Necaibia, A., Abdelkhalek, O., Bouraiou, A., Ziane, A., & Hamouda, M. (2022). Performance analysis of an experimental and simulated grid connected photovoltaic system in southwest Algeria. *International Journal of Energy and Environmental Engineering*, *13*(2), 831–851. doi:10.1007/s40095-022-00474-9

Chae, M. J., Kim, J. H., Moon, B., Park, S., & Lee, Y. S. (2022). The present condition and outlook for hydrogen-natural gas blending technology. *Korean Journal of Chemical Engineering*, *39*(2), 251–262. doi:10.1007/s11814-021-0960-8

Chai, M. M. (2015). Evaluation of Global Solar Radiation Models for Inclined Surfaces. *Proceedings of the 2015 International Conference on Electrical, Automation and Mechanical Engineering,* (pp. 472-475). Academic Press.

Chakrabarty, P., Alam, K. C. A., Paul, S. K., & Saha, S. C. (2024). Hydrogen production by electrolysis: A sustainable pathway. In *Hydrogen energy conversion and management* (pp. 81–102). Elsevier. doi:10.1016/B978-0-443-15329-7.00013-2

Chakwesha, R. T., & Musti, K. S. (2023). Designing a Futuristic Solar Smart Bench for Smart Cities: Towards Developing the Futuristic Urban Infrastructure. In Intersecting Health, Livability, and Human Behavior in Urban Environments (pp. 42-58). IGI Global. doi:10.4018/978-1-6684-6924-8.ch003

Chandrasekhar, K., Kumar, G., Venkata Mohan, S., Pandey, A., Jeon, B. H., Jang, M., & Kim, S. H. (2020). Microbial Electro-Remediation (MER) of hazardous waste in aid of sustainable energy generation and resource recovery. *Environmental Technology & Innovation*, *19*, 100997. doi:10.1016/j.eti.2020.100997

Chang, H.-S. (2018). *Blockchain network and method of operation thereof.* Academic Press.

Chauhan, P. R., Kaushik, S. C., & Tyagi, S. K. (2022). A review on thermal performance enhancement of green cooling system using different adsorbent/refrigerant pairs. *Energy Conversion and Management-X, 14*. https://doi.org/ARTN 10022510.1016/j.ecmx.2022.100225

Cheng, L. X., Guo, Z. X., & Xia, G. D. (2023). A Review on Research and Technology Development of Green Hydrogen Energy Systems with Thermal Management and Heat Recovery. *Heat Transfer Engineering*. Advance online publication. doi:10.1080/01457632.2023.2191435

Chen, Y., Cheng, J. J., & Creamer, K. S. (2008). Inhibition of anaerobic digestion process: A review. *Bioresource Technology*, *99*(10), 4044–4064. doi:10.1016/j.biortech.2007.01.057 PMID:17399981

Chen, Z. S., Yang, L. P., Rodríguez, R. M., Zhu, Z., Pedrycz, W., & Skibniewski, M. J. (2022). BIM-aided large-scale group decision support: Optimization of the retrofit strategy for existing buildings. *Applied Soft Computing*, *131*, 109790. doi:10.1016/j.asoc.2022.109790

Chen, Z., Yiliang, X., Hongxia, Z., Yujie, G., & Xiongwen, Z. (2023). Optimal design and performance assessment for a solar powered electricity, heating and hydrogen integrated energy system. *Energy*, *262*, 125453. doi:10.1016/j.energy.2022.125453

Chhetri, R., Wangchuk, K., & Sastry, M. K. S. (2014). Home Energy Audit: A Case Study of Phuentsholing, Bhutan. *IEEE International Conf. on Communication Systems and Network Technologies (CSNT)*, 1022-1025. 10.1109/CSNT.2014.208

Chia, W. Y., Chew, K. W., Le, C. F., Lam, S. S., Chee, C. S. C., Ooi, M. S. L., & Show, P. L. (2020). Sustainable utilization of biowaste compost for renewable energy and soil amendments. *Environmental Pollution*, *267*, 115662. doi:10.1016/j.envpol.2020.115662 PMID:33254731

Chi, J., & Yu, H. (2018). Water electrolysis based on renewable energy for hydrogen production. *Chinese Journal of Catalysis*, *39*(3), 390–394. doi:10.1016/S1872-2067(17)62949-8

Cho, G. J., Lee, C., & Kim, J. (2022). Equivalent load estimation method for determining railway power requirements under extended feeding conditions using a simplex algorithm. *International Journal of Electrical Power & Energy Systems*, *137*, 107770. doi:10.1016/j.ijepes.2021.107770

Choi, W., & Kang, S. (2023). Greenhouse gas reduction and economic cost of technologies using green hydrogen in the steel industry. *Journal of Environmental Management*, *335*, 117569. doi:10.1016/j.jenvman.2023.117569 PMID:36842354

Chuadhry, M. A., Bhatti, M. G., & Shah, R. A. (2023). Impact of Blockchain Technology in Modern Banking Sector to Exterminate the Financial Scams: Impact of Blockchain Technology in Modern Banking Sector to Exterminate the Financial Scams. *Sukkur IBA Journal of Computing and Mathematical Sciences*, *6*(2), 27–38. doi:10.30537/sjcms.v6i2.1170

Çinici, O. K., Karaca, B., & Acir, A. (2023). Comparison Study in Terms of The Results of Photovoltaic Solar Energy Systems Designed with PVsyst and MATLAB Simulink Software. *Politeknik Dergisi*, *26*(4), 1635–1649. doi:10.2339/politeknik.1266464

Clark, D. R., Klein, S. A., & Beckman, W. A. (1983). Algorithm for evaluating hourly radiation utilizability function. ASME. *Journal of Solar Energy Engineering*, *105*(3), 281–287. doi:10.1115/1.3266379

Clark, D. R., Klein, S. A., & Beckman, W. A. (1984). A method for estimating the performance of photovoltaic systems. *Solar Energy*, *33*(6), 551–555. doi:10.1016/0038-092X(84)90010-0

Corbo, P., Migliardini, F., & Veneri, O. (2011). Fuel Cells for Automotive Applications. In P. Corbo, F. Migliardini, & O. Veneri (Eds.), *Hydrogen Fuel Cells for Road Vehicles* (pp. 71–102). Springer. doi:10.1007/978-0-85729-136-3_3

Crowl, D. A., & Jo, Y.-D. (2007). The hazards and risks of hydrogen. *Journal of Loss Prevention in the Process Industries*, *20*(2), 158–164. doi:10.1016/j.jlp.2007.02.002

Custom, N. R. (2022). *Hydrogen gas turbine offers promise of clean electricity*. Retrieved 29/09/2023 from https://www.nature.com/articles/d42473-022-00211-0

Dagdougui, H., Ouammi, A., Dessaint, L., & Sacile, R. (2016). Global energy management system for cooperative networked residential green buildings. *IET Renewable Power Generation*, *10*(8), 1237–1244. doi:10.1049/iet-rpg.2015.0282

Daş, E., Ungan, H., Bayrakçeken Yurtcan, A., & Fıçıcılar, B. (2020). Other possible fuels and possible use of blended fuels in fuel cells. In R. G. Akay & A. Bayrakçeken Yurtcan (Eds.), *Direct Liquid Fuel Cells Fundamentals, Advances and Future* (pp. 249–273). Academic Press. doi:10.1016/B978-0-12-818624-4.00012-1

Dawood, F., Anda, M., & Shafiullah, G. M. (2020). Hydrogen production for energy: An overview. *International Journal of Hydrogen Energy*, *45*(7), 3847–3869. doi:10.1016/j.ijhydene.2019.12.059

De Clercq, D., Wen, Z., Gottfried, O., Schmidt, F., & Fei, F. (2017). A review of global strategies promoting the conversion of food waste to bioenergy via anaerobic digestion. *Renewable & Sustainable Energy Reviews*, *79*, 204–221. doi:10.1016/j.rser.2017.05.047

del Puerto-Flores, D., & Scherpen, J. M. (2011). Power-Based Controller for a Bidirectional AC-DC Power Converter. *IFAC Proceedings Volumes, 44*(1), 9751-9756.

Delimatsis, P. (2018). *When Disruptive Meets Streamline: International Standardization in Blockchain* (*SSRN* Scholarly Paper 3184824). doi:10.2139/ssrn.3208237

Demirbas, A. J. E. E. S. T. (2000). Recent advances in biomass conversion technologies. *Energy Education Science and Technology*, *6*, 19–41.

Dheeban, S. S., Muthu Selvan, N. B., & Senthil Kumar, C. (2019). Design of standalone pv system. *International Journal of Scientific and Technology Research*, *8*(11), 684–688.

Dinçer, İ. (2001). Environmental Issues: II-Potential Solutions. *Energy Sources*, *23*(1), 83–92. doi:10.1080/00908310151092218

Dixon, B. J., Tang, J., & Zhang, J. H. (2013). The evolution of molecular hydrogen: A noteworthy potential therapy with clinical significance. *Medical Gas Research*, *3*(1), 10. doi:10.1186/2045-9912-3-10 PMID:23680032

Dodds, P. E., Staffell, L., Hawkes, A. D., Li, F., Grunewald, P., McDowall, W., & Ekins, P. (2015). Hydrogen and fuel cell technologies for heating: A review. *International Journal of Hydrogen Energy*, *40*(5), 2065–2083. doi:10.1016/j.ijhydene.2014.11.059

Domínguez-Navarro, J. A., Dufo-López, R., Yusta-Loyo, J. M., Artal-Sevil, J. S., & Bernal-Agustín, J. L. (2019). Design of an electric vehicle fast-charging station with integration of renewable energy and storage systems. *International Journal of Electrical Power and Energy Systems, 105*(March), 46–58. doi:10.1016/j.ijepes.2018.08.001

Doug, F. (2019, September 16). *How is cutting carbon like winning a world war? Applying unconventional thinking to an increasingly urgent problem.* Explore UCalgary. https://explore.ucalgary.ca/climate-change-carbon-capture-reducing-CO2

Dougherty, W., Kartha, S., Rajan, C., Lazarus, M., Bailie, A., Runkle, B., & Fencl, A. (2009). Greenhouse gas reduction benefits and costs of a large-scale transition to hydrogen in the USA. *Energy Policy*, *37*(1), 56–67. doi:10.1016/j.enpol.2008.06.039

Dou, Y., Sun, L., Ren, J., & Dong, L. (2017). Opportunities and Future Challenges in Hydrogen Economy for Sustainable Development. In A. Scipioni, A. Manzardo, & J. Ren (Eds.), *Hydrogen Economy Supply Chain, Life Cycle Analysis and Energy Transition for Sustainability* (pp. 277–305). Academic Press. doi:10.1016/B978-0-12-811132-1.00010-9

Duca, D., Maceratesi, V., Fabrizi, S., & Toscano, G. (2022). Valorising agricultural residues through pelletisation. *Processes (Basel, Switzerland)*, *10*(2), 232. doi:10.3390/pr10020232

Duca, D., & Toscano, G. (2022). Biomass energy resources: Feedstock quality and bioenergy sustainability. *Resources*, *11*(6), 57. doi:10.3390/resources11060057

Duffie, J. W. (1991). Solar Engineering of Thermal Process. John Wiley & Sons Inc.

Dunn, S. (2002). Hydrogen futures: Toward a sustainable energy system. *International Journal of Hydrogen Energy*, *27*(3), 235–264. doi:10.1016/S0360-3199(01)00131-8

Durakovic, G., del Granado, P. C., & Tomasgard, A. (2023). Are green and blue hydrogen competitive or complementary? Insights from a decarbonized European power system analysis. *Energy*, *282*, 128282. doi:10.1016/j.energy.2023.128282

Earth.org. (2020). *Renewable Energy Hits Record Growth in 2020- Report.* https://earth.org/renewable-energy-hits-record-growth-in-2020/#:~:text=Solar

Ebaid, M. S. Y., Hammad, M., & Alghamdi, T. (2015). THERMO economic analysis OF PV and hydrogen gas turbine hybrid power plant of 100 MW power output. *International Journal of Hydrogen Energy*, *40*(36), 12120–12143. doi:10.1016/j.ijhydene.2015.07.077

Economic Times. (2023). *Indian Railways set to float tender for 35 hydrogen trains.* https://economictimes.indiatimes.com/industry/transportation/railways/indian- railways-could-float-tender-for-35-hydrogen-trains/articleshow/98226195.cms

Edition, T. (2018). *Control Systems for Drilling Well Control Equipment and Control Systems for Diverter Equipment.* Academic Press.

EG&G Technical Services. (2004). Technology Overview. In Fuel Cell Handbook (7th ed., pp. 1-32). U.S. Department of Energy Office of Fossil Energy National Energy Technology Laboratory.

Ekins, P., & Shah, N. (2017). Extended Summary. In I. Staffell & P. E. Dodds (Eds.), The role of hydrogen and fuel cells in future energy systems. H2FC SUPERGEN.

Ellamla, H. R., Staffell, I., Bujlo, P., Pollet, B. G., & Pasupathi, S. (2015). Current status of fuel cell based combined heat and power systems for residential sector. *Journal of Power Sources, 293*, 312–328. doi:10.1016/j.jpowsour.2015.05.050

Elnour, M., Fadli, F., Himeur, Y., Petri, I., Rezgui, Y., Meskin, N., & Ahmad, A. M. (2022). Performance and energy optimization of building automation and management systems: Towards smart sustainable carbon-neutral sports facilities. *Renewable & Sustainable Energy Reviews, 162*, 112401. doi:10.1016/j.rser.2022.112401

ElSayed, M., Foda, A., & Mohamed, M. (2022). Autonomous drone charging station planning through solar energy harnessing for zero-emission operations. *Sustainable Cities and Society, 86*, 104122. doi:10.1016/j.scs.2022.104122

Elshafei, A. M., & Mansour, R. (2023). Green Hydrogen as a Potential Solution for Reducing Carbon Emissions: A Review. *Journal of Energy Research and Reviews, 13*(2), 1–10. doi:10.9734/jenrr/2023/v13i2257

Energy Observer. (2023). *Quel potentiel pour l'hydrogène naturel?* Energy Observer. https://www.energy-observer.org/fr/ressources/hydrogene-naturel

Energy World, E. T. (2022). *Is India ready for the switch to Green Hydrogen in passenger trains?* https://energy.economictimes.indiatimes.com/news/renewable/is-india-ready-for-the-switch-to-green-hydrogen-in-passenger-trains/96535240

Ercan, T., Onat, N. C., Keya, N., Tatari, O., Eluru, N., & Kucukvar, M. (2022). Autonomous electric vehicles can reduce carbon emissions and air pollution in cities. *Transportation Research Part D, Transport and Environment, 112*(September), 103472. doi:10.1016/j.trd.2022.103472

Erdener, B. C., Sergi, B., Guerra, O. J., Chueca, A. L., Pambour, K., Brancucci, C., & Hodge, B. M. (2023). A review of technical and regulatory limits for hydrogen blending in natural gas pipelines. *International Journal of Hydrogen Energy, 48*(14), 5595–5617. doi:10.1016/j.ijhydene.2022.10.254

Eroğlu, H. (2020). Effects of COVID-19 outbreak on the environment and renewable energy sector. *Environment, Development and Sustainability.* Advance online publication. doi:10.1007/s10668-020-00837-4 PMID:32837274

ET EnergyWorld. (2023). *India's first Hydrogen train: GreenH Electrolysis, Medha Servo to develop production facility in Haryana.* https://energy.economictimes.indiatimes.com/news/renewable/indias-first-hydrogen-train-greenh-electrolysis-medha-servo-to-develop-production-facility-in-haryana/105897700

Evans, D. L. (1980). *Simplified design guide for estimating photovoltaic flat array and system performanc.* Sandia: SAND 80-7185.

Evans, D. L. (1978.). *Combined photovoltaic/ thermal system studies.* Arizona State University – Tempe.

Evans, D. L., Rule, T. T., & Wood, B. D. (1982). A new look at long term collector performance and utilizability. *Solar Energy, 28*(1), 13–23. doi:10.1016/0038-092X(82)90219-5

Eyring, V., Corbett, J. J., Lee, D. S., &Winebrake, J. J. (2007). Brief summary of the impact of ship emissions on atmospheric composition, climate, and human health. *Document Submitted to the Health and Environment Sub-Group of the International Maritime Organization, 6*.

Farhad, S., Saffar-Avval, M., & Younessi-Sinaki, M. (2008). Efficient design of feedwater heaters network in steam power plants using pinch technology and exergy analysis. *International Journal of Energy Research, 32*(1), 1–11. doi:10.1002/er.1319

Farrell, N. (2023). Policy design for green hydrogen. *Renewable & Sustainable Energy Reviews, 178*, 113216. doi:10.1016/j.rser.2023.113216

Feng, G., Xia, G., Chen, B., Xiao, Y., & Zhou, R. (2019). A method for rockburst prediction in the deep tunnels of hydropower stations based on the monitored microseismicity and an optimized probabilistic neural network model. *Sustainability (Basel), 11*(11), 3212. https://www.irena.org/hydropower. doi:10.3390/su11113212

Ferraro, M., Gallo, P., Ippolito, M. G., Massaro, F., Sanseverino, E. R., Ruffino, S., Santamaria, M., & Sciumè, G. (2023). A test bench for the production of green hydrogen and its traceability and certification using blockchain technology. *Proceedings - 2023 IEEE International Conference on Environment and Electrical Engineering and 2023 IEEE Industrial and Commercial Power Systems Europe, EEEIC / I and CPS Europe 2023*. Scopus. 10.1109/EEEIC/ICPSEurope57605.2023.10194789

Fichtner, M., & Idrissova, F. (2009). Fundamental properties of hydrogen. In M. Ball & M. Wietschel (Eds.), *The Hydrogen Economy Opportunities and Challenges* (pp. 271–276). Cambridge University Press. doi:10.1017/CBO9780511635359.010

Financial Express. (2024). *Budget 2024 1 Indian Railways' green revolution: As India moves towards 100% railway electrification, let's take a look at progress so far.* https://www.financialexpress.com/budget/budget-2024-1-indian-railways-green-revolution-as-india-moves-towards-100-railway-electrification-lets-take-a-look-at-progress-so-far-3372124/

Font, A., Tremper, A. H., Lin, C., Priestman, M., Marsh, D., Woods, M., Heal, M. R., & Green, D. C. (2020). Air quality in enclosed railway stations: Quantifying the impact of diesel trains through deployment of multi-site measurement and random forest modelling. *Environmental Pollution, 262*, 114284. doi:10.1016/j.envpol.2020.114284 PMID:32443191

Foroutan, S. L. F., & Bamdad, S. (2023). Efficiency measurement of railway passenger stations through network data envelopment analysis. *Research in Transportation Business & Management, 466*, 100767. doi:10.1016/j.rtbm.2021.100767

Fottrell, Q. (2020). Will COVID-19 push more employees to work remotely after the pandemic? This economist says yes. *Market Watch*. https://www.marketwatch.com/story/does-COVID-19-have-the-power-to-revolutionize-work-this-economist-says-yes-2020-12-11

Fox-Penner, P. (2020). Will the COVID-19 Pandemic Slow The Global Shift to Renewable Energy? *The Brink, Boston University*. https://www.bu.edu/articles/2020/will-the-COVID-19-pandemic-slow-the-global-shift-to-renewable-energy/

Fraas, L. M. (2014). *Low-cost solar electric power*. Springer. doi:10.1007/978-3-319-07530-3

Francesco, R., Gabriele, M., & Stefano, R. (2016). Complex railway systems: Capacity and utilisation of interconnected networks. *European Transport Research Review, 8*(4), 1–21. doi:10.1007/s12544-016-0216-6

Fruhwirt, D., Sturm, P., Nöst, T., Leonhardt, P., Bode, G., Michael, S., & Rodler, J. (2023). PM emissions from railways Results of tests on a wheel-rail test bench. *Transportation Research Part D, Transport and Environment, 122*, 103858. doi:10.1016/j.trd.2023.103858

Fthenakis, V., & Raugei, M. (2017). Environmental life-cycle assessment of photovoltaic systems. In *The Performance of PV Systems* (pp. 209–232). Modelling, Measurement and Assessment. doi:10.1016/B978-1-78242-336-2.00007-0

Gabbar, H. A., Othman, A. M., & Singh, K. (2017). Control architecture of resilient interconnected microgrids (RIMGs) for rauway infrastructures. *Energy Procedia*, *141*, 594–603. doi:10.1016/j.egypro.2017.11.079

Gangwar, M., & Sharma, S. M. (2014). Evaluating choice of traction option for a sustainable Indian Railways. *Transportation Research Part D, Transport and Environment*, *33*, 135–145. doi:10.1016/j.trd.2014.08.025

Gao, Y., Zhang, T., Wang, Q., Yang, S., Zhuo, J., Tang, J., Zhu, Y., Wang, P., & He, Q. (2023). Low- construction-emissions Optimization for Railway Alignment Fine- grained Designs within Densely Populated Urban Areas. *Construction & Building Materials*, *366*, 130179. doi:10.1016/j.conbuildmat.2022.130179

García Martín, J. (2022). The future of hydrogen: Seizing today's opportunities. *Economía Industrial*, *424*, 183–184.

Garcia, M., & Oliva, S. (2023). Technical, economic, and CO2 emissions assessment of green hydrogen production from solar/wind energy: The case of Chile. *Energy*, *278*, 127981. doi:10.1016/j.energy.2023.127981

Germscheidt, R. L., Moreira, D. E. B., Yoshimura, R. G., Gasbarro, N. P., Datti, E., dos Santos, P. L., & Bonacin, J. A. (2021). Hydrogen Environmental Benefits Depend on the Way of Production: An Overview of the Main Processes Production and Challenges by 2050. *Advanced Energy and Sustainability Research*, *2*(10), 2100093. doi:10.1002/aesr.202100093

Ghimire, R., Niroula, S., Pandey, B., Subedi, A., & Thapa, B. S. (2024). Techno-economic assessment of fuel cell-based power backup system as an alternative to diesel generators in Nepal: A case study for hospital applications. *International Journal of Hydrogen Energy*, *56*, 289–301. doi:10.1016/j.ijhydene.2023.12.174

Global Energy Perspectives. (2023). *Outlook for Green and Blue Hydrogen Market*. https://www.gep.com/blog/strategy/Green-and-blue-hydrogen-current-levelized-cost-of-production-and-outlook

Głuchy, D., Kurz, D., & Trzmiel, G. (2014). The use of fuel cells to store energy from renewable sources. *Przegląd Elektrotechniczny, R. 90*(4), 81-84. doi:10.12915/pe.2014.04.18

Goltsov, V. A., Veziroglu, T. N., & Goltsova, L. F. (2006). Hydrogen civilization of the future—A new conception of the IAHE. *International Journal of Hydrogen Energy*, *31*(2), 153–159. doi:10.1016/j.ijhydene.2005.04.045

Goosen, M., Mahmoudi, H., & Ghaffour, N. (2010). Water Desalination using geothermal energy. *Energies*, *3*(8), 1423–1442. doi:10.3390/en3081423

Gorji, S. A. (2023). Challenges and opportunities in green hydrogen supply chain through metaheuristic optimization. *Journal of Computational Design and Engineering*, *10*(3), 1143–1157. doi:10.1093/jcde/qwad043

Gowri, N. V., Dwivedi, J. N., Krishnaveni, K., Boopathi, S., Palaniappan, M., & Medikondu, N. R. (2023). Experimental investigation and multi-objective optimization of eco-friendly near-dry electrical discharge machining of shape memory alloy using Cu/SiC/Gr composite electrode. *Environmental Science and Pollution Research International*, *0123456789*(49), 1–19. doi:10.1007/s11356-023-26983-6 PMID:37126160

GRAAF. (2022, December). *Hydrogen's decade*. IMF. https://www.imf.org/en/Publications/fandd/issues/2022/12/hydrogen-decade-van-de-graaf

Green hydrogen opportunities for emerging and developing economies: Identifying success factors for market development and building enabling conditions (OECD Environment Working Papers 205; OECD Environment Working Papers, Vol. 205). (2022). doi:10.1787/19970900

Green Hydrogen: A Guide to Policy Making. (2020). https://typeset.io/papers/green-hydrogen-a-guide-to-policy-making-yd42vr1uos

Green Hydrogen: Challenges for Commercialization - IEEE Smart Grid. (2021). https://smartgrid.ieee.org/bulletins/february-2021/green-hydrogen-challenges-for-commercialization

Griffin, P. W., & Hammond, G. P. (2021). The prospects for 'green steel'making in a net-zero economy: A UK perspective. *Global Transitions, 3*, 72–86. doi:10.1016/j.glt.2021.03.001

Grubbström, R. W. (2022). On the net present value distribution affected by a random walk process. *International Journal of Production Economics, 250*, 108686. doi:10.1016/j.ijpe.2022.108686

Gude, V. G. (2018). Geothermal Source for Water Desalination-Challenges and Opportunities. *Renewable Energy Powered Desalination Handbook: Application and Thermodynamics*, 141–176. doi:10.1016/B978-0-12-815244-7.00004-0

Gude, V. G. (2016). Geothermal source potential for water desalination - Current status and future perspective. *Renewable & Sustainable Energy Reviews, 57*, 1038–1065. doi:10.1016/j.rser.2015.12.186

Guidi, F. (2022). An economic evaluation of the technologies to power a hydrogen steel production process under a CO2 emission limit. *Energy, 2004*, 2965.

Gujer, W., & Zehnder, A. J. B. (1983). Conversion processes in anaerobic digestion. *Water Science and Technology, 15*(8–9), 127–167. doi:10.2166/wst.1983.0164

Guo, J., He, H., & Sun, C. (2019). ARIMA-based road gradient and vehicle velocity prediction for hybrid electric vehicle energy management. *IEEE Transactions on Vehicular Technology, 68*(6), 5309–5320. doi:10.1109/TVT.2019.2912893

Guo, Z., Sun, Y., Pan, S. Y., & Chiang, P. C. (2019). Integration of green energy and advanced energy-efficient technologies for municipal wastewater treatment plants. *International Journal of Environmental Research and Public Health, 16*(7), 1282. doi:10.3390/ijerph16071282 PMID:30974807

Gupta, S., Sinha, S., & Bhushan, B. (2020). Emergence of Blockchain Technology: Fundamentals, Working and its Various Implementations. SSRN *Electronic Journal*. doi:10.2139/ssrn.3569577

Gupta, N., & Agrawal, R. (2023). Applications of Blockchain in Cyber Security Industry. In R. Agrawal & N. Gupta (Eds.), *Blockchain Applications in Cybersecurity Solutions* (pp. 48–61). Bentham Science Publishers. doi:10.2174/97898150805991230100006

Gurri, S., Santacaterina, E., Guarrera, M., & Dalla Chiara, B. (2023). Driving modal shift on low- traffic railway lines through technological innovation: A case study in Piedmont (Italy) including hydrogen fuel-cells as an alternative. *Transportation Research Procedia, 69*, 99–106. doi:10.1016/j.trpro.2023.02.150

Gusev, A. L., Jabbarov, T. G., Mamedov, S. G., Malikov, R., Hajibalaev, N. M., Abdullaeva, S. D., & Abbasov, N. M. (2023). Production of hydrogen and carbon in the petrochemical industry by cracking of hydrocarbons in the process of heat utilization in steel production. *International Journal of Hydrogen Energy, 48*(40), 14954–14963. doi:10.1016/j.ijhydene.2022.12.341

Habash, D., Abdelsalam, E., Alnawafah, H., Almomani, F., Salameh, T., & Al Makky, A. (2023). Green Hydrogen: A Novel Hybrid Solar Chimney Power Plant Integrated with Electrolysis Station. *2023 Advances in Science and Engineering Technology International Conferences (ASET)*, 1–6. 10.1109/ASET56582.2023.10180725

Haberl, J. S. (2004). *Literature Review of Uncertainty of Analysis Methods*. Energy Systems Laboratory.

Hall, D. O., Mynick, H. E., & Williams, R. H. (1991). Cooling the greenhouse with bioenergy. *Nature, 353*(6339), 11–12. doi:10.1038/353011a0

Hammar, L., Ehnberg, J., Mavume, A., Cuamba, B. C., & Molander, S. (2012). Renewable ocean energy in the Western Indian Ocean. *Renewable & Sustainable Energy Reviews*, *16*(7), 4938–4950. doi:10.1016/j.rser.2012.04.026

Hamukoshi, S. S., Mama, N., Shimanda, P. P., & Shafudah, N. H. (2022). An overview of the socio-economic impacts of the green hydrogen value chain in Southern Africa. *Journal of Energy in Southern Africa*, *33*(3), 3. Advance online publication. doi:10.17159/2413-3051/2022/v33i3a12543

Hanif, M., & Song, H. (2019). Blocks' Network: Redesign Architecture Based on Blockchain Technology. *2019 IEEE International Conference on Industrial Internet (ICII)*, 34–39. 10.1109/ICII.2019.00017

Haribalaji, V., Boopathi, S., & Asif, M. M. (2021). Optimization of friction stir welding process to join dissimilar AA2014 and AA7075 aluminum alloys. *Materials Today: Proceedings*, *50*, 2227–2234. doi:10.1016/j.matpr.2021.09.499

Harikaran, M., Boopathi, S., Gokulakannan, S., & Poonguzhali, M. (2023). Study on the Source of E-Waste Management and Disposal Methods. In *Sustainable Approaches and Strategies for E-Waste Management and Utilization* (pp. 39–60). IGI Global. doi:10.4018/978-1-6684-7573-7.ch003

Hart, M. (2015). *Hubris: The troubling science, economics, and politics of climate change*. Lulu.com.

Hassan, Q., Algburi, S., Sameen, A. Z., Tariq, J., Al-Jiboory, A. K., Salman, H. M., ... & Jaszczur, M. (2024). *A Comprehensive Review of International Renewable Energy Growth. Energy and Built Environment*. Academic Press.

Hassan. (2022). *Toward a Green and Safe Energy: An Implication for Hospitality, Tourism and Family Usage*. Paper Presented at the 10th International Istanbul Scientific Research Congress.

Hassan, M., El-Sharkawy, I. I., & Harby, K. (2023). Study of an innovative combined absorption-adsorption cooling system employing the same evaporator and condenser. *Case Studies in Thermal Engineering*, *42*, 102690. doi:10.1016/j.csite.2022.102690

Hassan, Q., Algburi, S., Sameen, A. Z., Salman, H. M., & Jaszczur, M. (2024). Green hydrogen: A pathway to a sustainable energy future. *International Journal of Hydrogen Energy*, *50*, 310–333. doi:10.1016/j.ijhydene.2023.08.321

Hauwanga, A. S., & Musti, K. S. S. (2022). Evaluation of Energy Losses and Costs Based on the Location of PCC in Smart Grids. *IEEE 11th Int. Conf. on Communication Systems and Network Technologies (CSNT)*, 76-82. 10.1109/CSNT54456.2022.9787642

Hawkes, A., Staffell, I., Brett, D., & Brandon, N. (2009). Fuel cells for micro-combined heat and power generation. *Energy & Environmental Science*, *2*(7), 729–744. doi:10.1039/b902222h

Healing, D. (2020). Big Oil's interest in renewable energy investments is expected to waver, the report says. *The Globe and Mail*. https://www.theglobeandmail.com/business/article-big-oils-interest-in-renewable-energy-investments-expected-to-waver/

He, G., Mallapragada, D. S., Bose, A., Heuberger, C. F., & Gençer, E. (2021). Hydrogen Supply Chain Planning With Flexible Transmission and Storage Scheduling. *IEEE Transactions on Sustainable Energy*, *12*(3), 1730–1740. doi:10.1109/TSTE.2021.3064015

Heiman, M. K., & Solomon, B. D. (2007). Fueling US transportation: The hydrogen economy and its alternatives. *Environment*, *49*(8), 10–25. doi:10.3200/ENVT.49.8.10-25

Hermesmann, M., & Müller, T. E. (2022). Green, Turquoise, Blue, or Grey? Environmentally friendly Hydrogen Production in Transforming Energy Systems. *Progress in Energy and Combustion Science*, *90*, 100996. doi:10.1016/j.pecs.2022.100996

He, W., King, M., Luo, X., Dooner, M., Li, D., & Wang, J. (2021). Technologies and economics of electric energy storages in power systems: Review and perspective. *Advances in Applied Energy*, *4*(May), 100060. doi:10.1016/j.adapen.2021.100060

Hočevar, S., & Summers, W. (2008). Hydrogen Production. In A. Leon (Ed.), Hydrogen Technology Mobile and Portable Applications (pp. 15-79). Springer. doi:10.1007/978-3-540-69925-5_2

Hoffmann, C., Ziemann, N., Penske, F., & Müsgens, F. (2023). The value of secure electricity supply for increasing acceptance of green hydrogen First experimental evidence from the virtual reality lab. *2023 19th International Conference on the European Energy Market (EEM)*, 1–9. 10.1109/EEM58374.2023.10161844

Holappa, L. (2020). A general vision for reduction of energy consumption and CO2 emissions from the steel industry. *Metals*, *10*(9), 1117. doi:10.3390/met10091117

Holechek, J. L., Geli, H. M. E., Sawalhah, M. N., & Valdez, R. (2022). A Global Assessment: Can Renewable Energy Replace Fossil Fuels by 2050? *Sustainability (Basel)*, *14*(8), 1–22. doi:10.3390/su14084792

Holm-Nielsen, J. B., Al Seadi, T., & Oleskowicz-Popiel, P. (2009). The future of anaerobic digestion and biogas utilization. *Bioresource Technology*, *100*(22), 5478–5484. doi:10.1016/j.biortech.2008.12.046 PMID:19217772

Hoogwijk, M., & Graus, W. (2008). *Global potential of renewable energy sources: a literature assessment*. Background report prepared by order of REN21. Ecofys, PECSNL072975.

Hordeski, M. F. (2008). *Alternative Fuels-The Future of Hydrogen*. Fairmont Press.

Horst, G. H., & Hovorka, A. J. (2009). Fuelwood: The "other" renewable energy source for Africa? *Biomass and Bioenergy*, *33*(11), 1605–1616. doi:10.1016/j.biombioe.2009.08.007

Hosseini, S. E. (2023). Hydrogen fuel, a game changer for the world's energy scenario. *International Journal of Green Energy*, 1–17.

Hosseini, S. E., & Wahid, M. A. (2016). Hydrogen production from renewable and sustainable energy resources: Promising green energy carrier for clean development. *Renewable & Sustainable Energy Reviews*, *57*, 850–866. doi:10.1016/j.rser.2015.12.112

Huang, D., & Yu, T. (2017). Study on Energy Payback Time of Building Integrated Photovoltaic System. *Procedia Engineering*, *205*, 1087–1092. doi:10.1016/j.proeng.2017.10.174

Huang, H., Tian, J., Min, G., & Miao, W. (2020). Introduction to blockchains. In *Blockchains for Network Security: Principles, technologies and applications* (pp. 1–21). IET Digital Library. doi:10.1049/PBPC029E_ch1

Huang, J., Balcombe, P., & Feng, Z. (2023). Technical and economic analysis of different colours of producing hydrogen in China. *Fuel*, *337*, 127227. doi:10.1016/j.fuel.2022.127227

Huang, S., Wang, J., Fu, Y., Zuo, W., Hinkelman, K., Kaiser, R. M., He, D., & Vrabie, D. (2021). An open-source virtual testbed for a real Net-Zero Energy Community. *Sustainable Cities and Society*, *75*, 103255. doi:10.1016/j.scs.2021.103255

Huppmann, D., & Egging, R. (2014). Market power, fuel substitution and infrastructure–A large-scale equilibrium model of global energy markets. *Energy*, *75*, 483–500. doi:10.1016/j.energy.2014.08.004

Huseynli, B. (2022). The Role of Renewable Energy in the Development of the Tourism Sector. *African Journal of Hospitality, Tourism and Leisure*, *11*(6), 1924–1936.

Hwang, J., Maharjan, K., & Cho, H. J. (2023). A review of hydrogen utilization in power generation and transportation sectors: Achievements and future challenges. *International Journal of Hydrogen Energy*, *48*(74), 28629–28648. doi:10.1016/j.ijhydene.2023.04.024

Hydrogen Science Coalition. (2022, December 21). Clean Hydrogen Definition—Hydrogen Science Coalition %. *Hydrogen Science Coalition*. https://h2sciencecoalition.com/briefings/clean-hydrogen-definition/

Ibrahim, N. A.-e., Wan Alwi, S. R., Abd Manan, Z., Mustaffa, A. A., & Kidam, K. (2024). Climate change impact on solar system in Malaysia: Techno-economic analysis. *Renewable & Sustainable Energy Reviews*, *189*, 113901. doi:10.1016/j.rser.2023.113901

IDTechEx. (2023). *Battery Electric & Hydrogen Fuel Cell Trains 2023-2043*. https://www.idtechex.com/en/research-report/battery-electric-and-hydrogen-fuel-cell-trains- 2023-2043/930

IEA. (2020). *Exploring the impacts of the COVID-19 pandemic on global energy markets, energy resilience, and climate change*. IEA.

IEA. (2022). *Renewable Energy Market Update*. Paris: IEA. https://www.iea.org/reports/renewable-energy-market-update-may-2022

IEA. (2022). *World energy outlook 2022*. IEA.

Ilari, A., Toscano, G., Foppa Pedretti, E., Fabrizi, S., & Duca, D. (2020). Environmental sustainability of heating systems based on pellets produced in mobile and stationary plants from vineyard pruning residues. *Resources*, *9*(8), 94. doi:10.3390/resources9080094

Inamdar, S., & Singh, S. (2018). Performance Evaluation of 5kWp grid connected polycrystalline Si-based PV Plant. *National Power Systems Conf*, 1-6. 10.1109/NPSC.2018.8771846

Incer-Valverde, J., Korayem, A., Tsatsaronis, G., & Morosuk, T. (2023). "Colors" of hydrogen: Definitions and carbon intensity. *Energy Conversion and Management*, *291*, 117294. doi:10.1016/j.enconman.2023.117294

India Briefing. (2024). *How India Plans to Achieve Cost-Effective Green Hydrogen Production*. https://www.india-briefing.com/news/india-strives-for-cost-effective-green-hydrogen-production-30876.html/

International Energy Agency. (2020). *Renewables 2020 – Analysis and Forecast to 2025*. https://www.iea.org/reports/renewables-2020

International Energy Agency. (2022). *International Energy Agency (IEA) World Energy Outlook 2022*. https://www.iea.org/reports/world-energy-outlook-2022

International Energy Agency. (2022). *International Energy Agency (IEA) World Energy Outlook 2022*. Https://Www.Iea.Org/Reports/World-Energy-Outlook-2022/Executive-Summary

International Energy Agency. (2023) *Rail* https://www.iea.org/energy-system/transport/rail

International Renewable Energy Agency (IRENA). (2020). *Renewable Capacity Statistics 2020*. https://www.irena.org/publications/2020/Mar/Renewable-Capacity-Statistics-2020

International Renewable Energy Agency (IRENA). (2022a). *Renewables 2022: Global Status Report*. IRENA.

International Renewable Energy Agency (IRENA). (2022b). *World Energy Transitions Outlook 2022*. IRENA.

IRENA. (2020). *Renewables accounted for Almost Three-Quarters of New Capacity in 2019*. IRENA.

IRENA. (2023). *Creating a global hydrogen market: Certification to enable trade*. IRENA.

Ishaq, H., Dincer, I., & Crawford, C. (2022). A review on hydrogen production and utilization: Challenges and opportunities. *International Journal of Hydrogen Energy*, *47*(62), 26238–26264. doi:10.1016/j.ijhydene.2021.11.149

Ismoyo, K., Muhajirin, Iskendar, Ajidarmo, K., Triyanto, & Muis, A. (2023). Analysis of technology readiness level of dual fuel harbor tug boat research achievement. *IOP Conference Series. Earth and Environmental Science*, *1166*(1), 012004. doi:10.1088/1755-1315/1166/1/012004

Janardhana, K., Anushkannan, N. K., Dinakaran, K. P., Puse, R. K., & Boopathi, S. (2023). Experimental Investigation on Microhardness, Surface Roughness, and White Layer Thickness of Dry EDM. *Engineering Research Express*, *5*(2), 025022. doi:10.1088/2631-8695/acce8f

Jang, D., Kim, J., Kim, D., Han, W. B., & Kang, S. (2022). Techno-economic analysis and Monte Carlo simulation of green hydrogen production technology through various water electrolysis technologies. *Energy Conversion and Management*, *258*, 115499. doi:10.1016/j.enconman.2022.115499

Japan Times. (2023). *Just how green are Japan's trains?* https://www.japantimes.co.jp/news/2023/07/02/national/japan-trains-cars-carbon-emissions/

Jeevanantham, Y. A., A, S., V, V., J, S. I., Boopathi, S., & Kumar, D. P. (2023). Implementation of Internet-of Things (IoT) in Soil Irrigation System. *IEEE Explore*, 1–5. doi:10.1109/ICPECTS56089.2022.10047185

Jiang, T., Shen, Z., Liu, Y., & Hou, Y. (2018). Carbon footprint assessment of four normal size hydropower stations in China. *Sustainability (Basel)*, *10*(6), 2018. doi:10.3390/su10062018

Jin, Y. H., Sun, L., & Shen, J. (2019). Thermal economic analysis of hybrid open-cathode hydrogen fuel cell and heat pump cogeneration. *International Journal of Hydrogen Energy*, *44*(56), 29692–29699. doi:10.1016/j.ijhydene.2019.03.098

Jonas Moberg. (2023, June 22). *Finance for a just renewable energy and green hydrogen economy roundtable—OECD.* https://www.oecd.org/environment/cc/cefim/green-hydrogen/financeforajustrenewableenergyandgreenhydrogeneconomyroundtable.htm

Juangsa, F. B., Cezeliano, A. S., Darmanto, P. S., & Aziz, M. (2022). Thermodynamic analysis of hydrogen utilization as alternative fuel in cement production. *South African Journal of Chemical Engineering*, *42*, 23–31. doi:10.1016/j.sajce.2022.07.003

Junwen, C., Xiangfu, Q., Ga, G., Wenqiang, Z., & Bo, Y. (2021). Current status and prospects of hydrogen storage and transportation technology. *Acta Petrolei Sinica*, *37*(6), 1461.

Kabeyi, M. J., & Olanrewaju, O. A. (2023). The levelized cost of energy and modifications for use in electricity generation planning. *Energy Reports*, *9*, 495–534. doi:10.1016/j.egyr.2023.06.036

Kacker, R., & Srivastava, N. (2023). Elements and Possible Constraints for a Smooth Energy Transition in India. *Ecology. Economy and Society–the INSEE Journal*, *6*(1), 5–11.

Kaczmarczyk, M., Tomaszewska, B., & Bujakowski, W. (2022). Innovative desalination of geothermal wastewater supported by electricity generated from low-enthalpy geothermal resources. *Desalination*, *524*, 115450. doi:10.1016/j.desal.2021.115450

Kaiser, N., & Barstow, C. K. (2022). Rural Transportation Infrastructure in Low-and Middle-Income Countries: A Review of Impacts, Implications, and Interventions. *Sustainability (Basel)*, *14*(4), 2149. Advance online publication. doi:10.3390/su14042149

Kaleybar, H. J., Brenna, M., Foiadelli, F., & Zaninelli, D. (2023). Electric Railway System Integrated with EV Charging Station Realizing Train-to-Vehicle Technology. *IEEE Transactions on Vehicular Technology*.

Kamanja, L., Komarala, V., Dutta, V., Waita, S., & Wachira, K. (2022). *Techno-Economic Analysis of a Rooftop Grid-connected PV Solar System: A case study of JK University of Agriculture and Technology (SAJOREC).* IEEE PES/IAS PowerAfrica. doi:10.1109/PowerAfrica53997.2022.9905282

Kamel, T., Amor, P., Polater, N., Zhang, Y., Tian, Z., Hillmansen, S., & Tricoli, P. (2023). Development of a smart hybrid drive system with advanced logistics for railway applications. *International Journal of Hydrogen Energy.*

Kamel, T., Tian, Z., Zangiabadi, M., Wade, R., Pickert, V., & Tricoli, P. (2022). Smart soft open point to synergically improve the energy efficiencies of the interconnected electrical railways with the Cow voltage grids. *International Journal of Electrical Power & Energy Systems, 142*, 1082880. doi:10.1016/j.ijepes.2022.108288

Kang, J. B., Lee, J. Y., & Lee, J. H. (2022). Guidelines for Determining the Initial Shape and Specifications of High-Speed AFPM Based on Recent Research. *Energies, 15*(16), 6099. Advance online publication. doi:10.3390/en15166099

Karayel, G. K., Javani, N., & Dincer, I. (2023). Hydropower for green hydrogen production in Turkey. *International Journal of Hydrogen Energy, 48*(60), 22806–22817. doi:10.1016/j.ijhydene.2022.04.084

Karayel, K., Javani, N., & Dincer, I. (2022). Green hydrogen production potential for Turkey with solar energy. *International Journal of Hydrogen Energy, 47*(45), 19354–19364. doi:10.1016/j.ijhydene.2021.10.240

Karlin, M. (2020). Slow Job Growth in November Leaves 446, 000 Clean Energy Workers Unemployed as COVID Cases, Shutdowns Surge. *American Council on Renewable Energy.* https://goodmenproject.com/featured-content/slow-job-growth-in-november-leaves-446000-clean-energy-workers-unemployed-as-covid-case

Karn, A. L., Pandya, S., Mehbodniya, A., Arslan, F., Sharma, D. K., Phasinam, K., Aftab, M. N., Rajan, R., Bommisetti, R. K., & Sengan, S. (2021). An integrated approach for sustainable development of wastewater treatment and management system using IoT in smart cities. *Soft Computing*, 1–17. doi:10.1007/s00500-021-06244-9

Katterbauer, K., Shehri, A. A., Qasim, A., & Yousif, A. (2023). Sustainable hydrogen production utilizing artificial intelligence – a McKee reservoir study. *84th EAGE Annual Conference & Exhibition*, 1–5. 10.3997/2214-4609.202310062

Kellenberger, A., Duca, D., Vaszilcsin, N., & Craciunescu, C. M. (2020). Electrochemical Evaluation of Niobium Corrosion Resistance in Simulated Anodic PEM Electrolyzer Environment. *International Journal of Electrochemical Science, 15*(11), 10664–10673. doi:10.20964/2020.11.47

Khamharnphol, R., Kamdar, I., Waewsak, J., Chiwamongkhonkarn, S., Khunpetch, S., Kongruang, C., & Gagnon, Y. (2023). Techno-Economic Assessment of a 100 kWp Solar Rooftop PV System for Five Hospitals in Central Southern Thailand. *International Journal of Renewable Energy Development, 12*(1), 77–86. doi:10.14710/ijred.2023.46864

Khan, S. U.-D., Wazeer, I., & Almutairi, Z. (2023). Comparative Analysis of SAM and RETScreen Tools for the Case Study of 600 kW Solar PV System Installation in Riyadh, Saudi Arabia. *Sustainability (Basel), 15*(6), 5381. Advance online publication. doi:10.3390/su15065381

Khatib, T., Mohamed, A., Sopian, K., & Mahmoud, M. (2012). A new approach for optimal sizing of standalone photovoltaic systems. *International Journal of Photoenergy, 2012*, 2012. doi:10.1155/2012/391213

Khezami, N., Gharbi, N., Neji, B., & Braiek, N. B. (2022). Blockchain Technology Implementation in the Energy Sector: Comprehensive Literature Review and Mapping. *Sustainability (Basel), 14*(23), 15826. doi:10.3390/su142315826

Kiaghadi, A., Sobel, R. S., & Rifai, H. S. (2017). Modeling geothermal energy efficiency from abandoned oil and gas wells to desalinate produced water. *Desalination, 414*, 51–62. doi:10.1016/j.desal.2017.03.024

Kim, J. H., Hansora, D., Sharma, P., Jang, J.-W., & Lee, J. S. (2019). Toward practical solar hydrogen production – an artificial photosynthetic leaf-to-farm challenge. *Chemical Society Reviews*, *48*(7), 1908–1971. doi:10.1039/C8CS00699G PMID:30855624

Kirli, D., Couraud, B., Robu, V., Salgado-Bravo, M., Norbu, S., Andoni, M., Antonopoulos, I., Negrete-Pincetic, M., Flynn, D., & Kiprakis, A. (2022). Smart contracts in energy systems: A systematic review of fundamental approaches and implementations. *Renewable & Sustainable Energy Reviews*, *158*, 112013. doi:10.1016/j.rser.2021.112013

Kirmani, F., Pal, A., Mudgal, A., Shrestha, A., & Siddiqui, A. (2021). Advantages and disadvantages of hydroelectric power plant. *International Journal of Innovative Science and Research Technology*.

Kirmani, S., Jamil, M., & Akhtar, I. (2018). Economic feasibility of hybrid energy generation with reduced carbon emission. *IET Renewable Power Generation*, *12*(8), 934–942. doi:10.1049/iet-rpg.2017.0288

Klein, S. A. (1981). *TRNSYS - A transient system simulation program*. Medison: Solar Energy Laboratory.

Klein, S. A. (1985). *PV F-Chart User's Manual: DOS Version*. www.fchart.com

Klein, S. A. (1978). Calculation of flat-plate collector utilizability. *Solar Energy*, *21*(5), 393–402. doi:10.1016/0038-092X(78)90171-8

Koppers, F. (2020). *Lock for e.g. Chain, bracket or cable, forms closed circuit sensor device*. Academic Press.

Ko, S., Lautala, P., & Handler, R. M. (2018). Securing the feedstock procurement for bioenergy products: A literature review on the biomass transportation and logistics. *Journal of Cleaner Production*, *200*, 205–218. doi:10.1016/j.jclepro.2018.07.241

Koshariya, A. K., Kalaiyarasi, D., Jovith, A. A., Sivakami, T., Hasan, D. S., & Boopathi, S. (2023a). AI-Enabled IoT and WSN-Integrated Smart Agriculture System. In *Artificial Intelligence Tools and Technologies for Smart Farming and Agriculture Practices* (pp. 200–218). IGI Global. doi:10.4018/978-1-6684-8516-3.ch011

Kothari, R., Buddhi, D., & Sawhney, R. (2008). Comparison of environmental and economic aspects of various hydrogen production methods. *Renewable & Sustainable Energy Reviews*, *12*(2), 553–563. doi:10.1016/j.rser.2006.07.012

Kouchaki-Penchah, H., Bahn, O., Bashiri, H., Bedard, S., Bernier, E., Elliot, T., Hammache, A., Vaillancourt, K., & Levasseur, A. (2024). The role of hydrogen in a net-zero emission economy under alternative policy scenarios. *International Journal of Hydrogen Energy*, *49*, 173–187. doi:10.1016/j.ijhydene.2023.07.196

Kramarchuk, R., Klein, D., Brunetti, B., Joseph, I., Schiavo, M., Georges, P., Redmond, S., Anankina, E., Prabhu, A., Lu, G., Huang, D., Amiot, M., & Roache, S. (2020). *How is COVID-19 Impacting the Energy Transition?* Available at https://www.spglobal.com/en/research-insights/featured/how-is-COVID-19-impacting-the-energy-transition

Kuhnert, E., Mayer, K., Heidinger, M., Rienessel, C., Hacker, V., & Bodner, M. (2023, September). Impact of Intermittent Operation on Photovoltaic-PEM Electrolyzer Systems: A Degradation Study Based on Accelerated Stress Testing. *International Journal of Hydrogen Energy*, *55*, 683–695. doi:10.1016/j.ijhydene.2023.11.249

Kumar, M., Panda, K. P., Naayagi, R. T., Thakur, R., & Panda, G. (2023). *Applied sciences Comprehensive Review of Electric Vehicle Technology and Its Impacts : Detailed Investigation of Charging Infrastructure, Power Management, and Control Techniques*. Academic Press.

Kumara, V., Mohanaprakash, T. A., Fairooz, S., Jamal, K., Babu, T., & B., S. (2023). Experimental Study on a Reliable Smart Hydroponics System. In *Human Agro-Energy Optimization for Business and Industry* (pp. 27–45). IGI Global. doi:10.4018/978-1-6684-4118-3.ch002

Kumar, A. (2023). *Overview of Green Hydrogen Technology*. Presentation Slides for Lectures.

Kumar, G. M. S., & Cao, S. (2023). Leveraging energy flexibilities for enhancing the cost- effectiveness and grid-responsiveness of net-zero-energy metro railway and station systems. *Applied Energy*, *333*, 120632. doi:10.1016/j.apenergy.2022.120632

Kumar, S. S., & Lim, H. (2022). An overview of water electrolysis technologies for green hydrogen production. *Energy Reports*, *8*, 13793–13813. doi:10.1016/j.egyr.2022.10.127

Kumar, S., Arzaghi, E., Baalisampang, T., Garaniya, V., & Abbassi, R. (2023). Insights into decision-making for offshore green hydrogen infrastructure developments. *Process Safety and Environmental Protection*, *174*, 805–817. doi:10.1016/j.psep.2023.04.042

Kumar, V., Jethani, J. K., & Bohra, L. (2023). Combating elimate change through renewable sources of electricity-A review of rooftop solar projects in India. *Sustainable Energy Technologies and Assessments*, *60*, 103526. doi:10.1016/j.seta.2023.103526

Kuriqi, A., Pinheiro, A. N., Sordo-Ward, A., Bejarano, M. D., & Garrote, L. (2021). Ecological impacts of run-of-river hydropower plants—Current status and future prospects on the brink of energy transition. *Renewable & Sustainable Energy Reviews*, *142*, 110833. doi:10.1016/j.rser.2021.110833

Kuriqi, A., Pinheiro, A. N., Sordo-Ward, A., & Garrote, L. (2019). Influence of hydrologically based environmental flow methods on flow alteration and energy production in a run-of-river hydropower plant. *Journal of Cleaner Production*, *232*, 1028–1042. doi:10.1016/j.jclepro.2019.05.358

Kusakana, K., & Vermaak, H. J. (2013). Hybrid renewable power systems for mobile telephony base stations in developing countries. *Renewable Energy*, *51*, 419–425. doi:10.1016/j.renene.2012.09.045

Kwan, T. H., Wu, X. F., & Yao, Q. H. (2019). Performance comparison of several heat pump technologies for fuel cell micro-CHP integration using a multi-objective optimisation approach. *Applied Thermal Engineering*, *160*, 114002. doi:10.1016/j.applthermaleng.2019.114002

Lacko, R., Drobnic, B., Mori, M., Sekavcnik, M., & Vidmar, M. (2014). Stand-alone renewable combined heat and power system with hydrogen technologies for household application. *Energy*, *77*, 164–170. doi:10.1016/j.energy.2014.05.110

Lai, C. S., Locatelli, G., Pimm, A., Tao, Y., Li, X., & Lai, L. L. (2019). A financial model for lithium-ion storage in a photovoltaic and biogas energy system. *Applied Energy*, *113179*, 113179. Advance online publication. doi:10.1016/j.apenergy.2019.04.175

Lai, C. S., & McCulloch, M. D. (2017). Levelized cost of electricity for solar photovoltaic and electrical energy storage. *Applied Energy*, *190*, 191–203. doi:10.1016/j.apenergy.2016.12.153

Laurent, T., & Bazarkina, E. (2019). *Natural hydrogen the fuel of the 21st century*. Academic Press.

Ledari, M. B., Khajehpour, H., Akbarnavasi, H., & Edalati, S. (2023). Greening steel industry by hydrogen: Lessons learned for the developing world. *International Journal of Hydrogen Energy*.

Lee, K. C. (2001). Classification of geothermal resources by exergy. *Geothermics*, *30*(4), 431–442. doi:10.1016/S0375-6505(00)00056-0

Legasov, V. (1979). *Nuclear hydrogen energy and technology. Part 2*. Atomno-vodorodnaya energetika i tekhnologiya.

Lehner, B., Czisch, G., & Vassolo, S. (2005). The impact of global change on the hydropower potential of Europe: A model-based analysis. *Energy Policy*, *33*(7), 839–855. doi:10.1016/j.enpol.2003.10.018

Leon, A. (2008). Introduction. In A. Leon (Ed.), Hydrogen Technology Mobile and Portable Applications (pp. 11-13). Springer. doi:10.1007/978-3-540-69925-5_1

Li, W., Qiu, W., Yin, K., & Hu, M. (2021). *Blockchain-based multi-stage signing method, device, computer apparatus, and storage medium.* Academic Press.

Li, X., Zhu, C. & Liu, Y. (2023). Traction power supply system of China high-speed railway under Jow-carbon target: Form evolution and operation control. *Electric Power Systems Researc*h, 223.

Li, C. (2018). Comparative performance analysis of grid-connected PV power systems with different PV technologies in the hot summer and cold winter zone. *International Journal of Photoenergy*, *2018*, 1–9. Advance online publication. doi:10.1155/2018/8307563

Li, C., Xu, C., & Li, X. (2020). A multi-criteria decision-making framework for site selection of distributed PV power stations along high-speed railway. *Journal of Cleaner Production*, *277*, 124068. doi:10.1016/j.jclepro.2020.124086

Lin, R. H., Zhao, Y. Y., & Wu, B. D. (2020). Toward a hydrogen society: Hydrogen and smart grid integration. *International Journal of Hydrogen Energy*, *45*(39), 20164–20175. doi:10.1016/j.ijhydene.2020.01.047

Li, T., Roskilly, A. P., & Wang, Y. (2018). Life cycle sustainability assessment of grid-connected photovoltaic power generation: A case study of Northeast England. *Applied Energy*, *227*, 465–479. doi:10.1016/j.apenergy.2017.07.021

Liu, B. Y. (1963). A rational procedure for predicting the long term performance of flat plate solar energy collectors. *Solar Energy*, *7*, 53. doi:10.1016/0038-092X(63)90006-9

Liu, B. Y. H., & Jordan, R. C. (1960). The Interrelationship and Characteristic Distribution of Direct, Diffuse and Total Solar Radiation. *Solar Energy*, *4*(3), 1–19. doi:10.1016/0038-092X(60)90062-1

Liu, H. C., Wu, S. M., Wang, Z. L., & Li, X. Y. (2021). A New Method for Quality Function Deployment with Extended Prospect Theory under Hesitant Linguistic Environment. *IEEE Transactions on Engineering Management*, *68*(2), 442–451. doi:10.1109/TEM.2018.2864103

Liu, J., Zhou, Y., Yang, H., & Wu, H. (2022). Net-zero energy management and optimization of commercial building sectors with hybrid renewable energy systems integrated with energy storage of pumped hydro and hydrogen taxis. *Applied Energy*, *321*, 119312. doi:10.1016/j.apenergy.2022.119312

Liu, Y., Liu, H., Wang, C., Hou, S. X., & Yang, N. (2013). Sustainable energy recovery in wastewater treatment by microbial fuel cells: Stable power generation with nitrogen-doped graphene cathode. *Environmental Science & Technology*, *47*(23), 13889–13895. doi:10.1021/es4032216 PMID:24219223

Liu, Z. J., Xu, W., Zhai, X., Qian, C., & Chen, X. (2017). Feasibility and performance study of the hybrid ground-source heat pump system for one office building in Chinese heating dominated areas. *Renewable Energy*, *101*, 1131–1140. doi:10.1016/j.renene.2016.10.006

Liu, Z., Chen, Y., Yang, X., & Yan, J. (2023). Power to Heat: Opportunity of Flexibility Services Provided by Building Energy Systems. *Advances in Applied Energy*, *11*, 100149. doi:10.1016/j.adapen.2023.100149

Liu, Z., Deng, Z., Davis, S., & Ciais, P. (2023). Monitoring global carbon emissions in 2022. *Nature Reviews. Earth & Environment*, *4*(4), 205–206. doi:10.1038/s43017-023-00406-z PMID:37065615

Li, Z., Du, H., Xiao, Y., & Guo, J. (2017). Carbon footprints of two large hydro-projects in China: Life-cycle assessment according to ISO/TS 14067. *Renewable Energy*, *114*, 534–546. doi:10.1016/j.renene.2017.07.073

Li, Z., Shao, T., Zheng, T. Q., Li, H., & Huang, B. (2019). Interference source analysis and EMC design for All-SiC power module in EV charger. *Microelectronics and Reliability*, *100–101*(September), 113458. Advance online publication. doi:10.1016/j.microrel.2019.113458

Li, Z., Wang, Q., Fang, M., & Luo, Z. (2021). Thermodynamic and economic analysis of a new 600 MWe coal-fired power plant integrated with CaO-based carbon capture system. *International Journal of Greenhouse Gas Control*, *109*, 103386. doi:10.1016/j.ijggc.2021.103386

Llácer-Iglesias, R. M., López-Jiménez, P. A., & Pérez-Sánchez, M. (2021). Hydropower technology for sustainable energy generation in wastewater systems: Learning from the experience. *Water (Basel)*, *13*(22), 3259. doi:10.3390/w13223259

Longoria, G., Lynch, M., & Curtis, J. (2021). Green hydrogen for heating and its impact on the power system. *International Journal of Hydrogen Energy*, *46*(53), 26725–26740. doi:10.1016/j.ijhydene.2021.05.171

Loy-Benitez, J., Safder, U., Nguyen, H. T., Li, Q., Woo, T., & Yoo, C. (2021). Techno-economic assessment and smart management of an integrated fuel cell-based energy system with absorption chiller for power, hydrogen, heating, and cooling in an electrified railway network. *Energy*, *233*, 121099. doi:10.1016/j.energy.2021.121099

Lubosny, Z., & Bialek, J. W. (2007). Supervisory control of a wind farm. *IEEE Transactions on Power Systems*, *22*(3), 985–994. doi:10.1109/TPWRS.2007.901101

Lucia, U. (2014). Overview on fuel cells. *Renewable & Sustainable Energy Reviews*, *30*, 164–169. doi:10.1016/j.rser.2013.09.025

Luqman, M., Ahmad, S. R., Khan, S., Ahmad, U., Raza, A., & Akmal, F. (2015). Estimation of solar energy potential from rooftop of Punjab government servants cooperative housing society Lahore using GIS. *Smart Grid and Renewable Energy*, *6*(05), 128–139. doi:10.4236/sgre.2015.65012

Maammeur, H. H., Hamidat, A., Loukarfi, L., Missoum, M., Abdeladim, K., & Nacer, T. (2017). Performance investigation of grid-connected PV systems for family farms: Case study of North-West of Algeria. *Renewable & Sustainable Energy Reviews*, *78*, 1208–1220. doi:10.1016/j.rser.2017.05.004

Mahmoudi, A., Rahim, N. A., & Hew, W. P. (2011). Axial-flux permanent-magnet machine modeling, design, simulation and analysis. *Scientific Research and Essays*, *6*(12), 2525–2549.

Mai-Moulin, T., Hoefnagels, R., Grundmann, P., & Junginger, M. (2021). Effective sustainability criteria for bioenergy: Towards the implementation of the european renewable directive II. *Renewable & Sustainable Energy Reviews*, *138*, 110645. doi:10.1016/j.rser.2020.110645

Majorro, N. (2020). COVID-19 is a game-changer for renewable energy. Here's why. *World Economic Forum*. https://www.weforum.org/agenda/2020/06/COVID-19-is-a-game-changer-for-renewable-energy/

Majumdar, A., Haas, M., Elliot, I., & Nazari, S. (2023). Control and Control-Oriented Modeling of PEM Water Electrolyzers: A Review. *International Journal of Hydrogen Energy*, *48*(79), 1–21. doi:10.1016/j.ijhydene.2023.04.204

Malafeh, S., & Sharp, B. (2015). Role of royalties in sustainable geothermal energy development. *Energy Policy*, *85*, 235–242. doi:10.1016/j.enpol.2015.06.023

Mann, S. A., de Wild-Scholten, M. J., Fthenakis, V. M., van Sark, W. G. J. H. M., & Sinke, W. C. (2014). The energy payback time of advanced crystalline silicon PV modules in 2020: A prospective study. *Progress in Photovoltaics: Research and Applications*, *22*(11), 1180–1194. doi:10.1002/pip.2363

Manoharan, Y., Hosseini, S. E., Butler, B., Alzhahrani, H., Senior, B. T. F., Ashuri, T., & Krohn, J. (2019). Hydrogen fuel cell vehicles; current status and future prospect. *Applied Sciences (Basel, Switzerland), 9*(11), 2296. doi:10.3390/app9112296

Manwell, J. F., McGowan, J. G., & Rogers, A. L. (2010). *Wind energy explained: theory, design and application.* John Wiley & Sons.

Manzardo, A., Ren, J., Toniolo, S., & Scipioni, A. (2023). Critical factors and cause-effect analysis for enhancing the sustainability of hydrogen supply chain. In A. Scipioni, A. Manzardo, & J. Ren (Eds.), *Hydrogen Economy* (2nd ed., pp. 67–111). Academic Press. doi:10.1016/B978-0-323-99514-6.00010-8

Mariano-Hernández, D., Hernández-Callejo, L., Zorita-Lamadrid, A., Duque-Pérez, O., & García, F. S. (2021). A review of strategies for building energy management system: Model predictive control, demand side management, optimization, and fault detect & diagnosis. *Journal of Building Engineering, 33*, 101692. doi:10.1016/j.jobe.2020.101692

Marketsandmarket. (2023). *Green Hydrogen Market, Industry Size Forecast Report.* MarketsandMarkets. https://www.marketsandmarkets.com/Market-Reports/green-hydrogen-market-92444177.html

Marocco, P., Gandiglio, M., Audisio, D., & Santarelli, M. (2023). Assessment of the role of hydrogen to produce high-temperature heat in the steel industry. *Journal of Cleaner Production, 388*, 135969. doi:10.1016/j.jclepro.2023.135969

Matheus, M. N., & Musti, K. S. (2023). Design and Simulation of a Floating Solar Power Plant for Goreagab Dam, Namibia. In Human Agro-Energy Optimization for Business and Industry (pp. 1-26). IGI Global. doi:10.4018/978-1-6684-4118-3.ch001

Matter, U. (2023). *Big Data Analytics: A Guide to Data Science Practitioners Making the Transition to Big Data* (1st ed.). Chapman and Hall/CRC. doi:10.1201/9781003378822

Ma, Z., Tian, T., Cui, Q., Shu, J., Zhao, J., & Wang, H. (2023). Rapid sizing of a hydrogen-battery storage for an offshore wind farm using convex programming. *International Journal of Hydrogen Energy, 48*(58), 21946–21958. doi:10.1016/j.ijhydene.2023.03.037

Mbah, J., Weaver, E., Srinivasan, S., Krakow, B., Wolan, J., Goswami, Y., & Stefanakos, E. (2010). Low voltage H2O electrolysis for enhanced hydrogen production. *Energy, 35*(12), 5008–5012. doi:10.1016/j.energy.2010.08.021

McCalmont, J. P., Hastings, A., McNamara, N. P., Richter, G. M., Robson, P., Donnison, I. S., & Clifton-Brown, J. (2017). Environmental costs and benefits of growing Miscanthus for bioenergy in the UK. *Global Change Biology. Bioenergy, 9*(3), 489–507. doi:10.1111/gcbb.12294 PMID:28331551

McElroy, B. (2020). *How COVID-19 is impacting renewable energy.* Available at https://www.renewableenergyworld.com/2020/04/29/how-COVID-19-is-impacting-renewable-energy/

McKibbin, W. J., & Fernando, R. (2020). The Global Macroeconomic Impacts of COVID-19: Seven Scenarios. SSRN *Electronic Journal,* 1–43. doi:10.2139/ssrn.3547729

McKinsey. (2022). Global Energy Perspective 2022 McKinsey's Global Energy Perspective Is a Collaboration between Energy Insights and Adjacent Practices. *Executive Summary.*

Mehmeti, A., Angelis-Dimakis, A., Arampatzis, G., McPhail, S. J., & Ulgiati, S. (2018). Life Cycle Assessment and Water Footprint of Hydrogen Production Methods: From Conventional to Emerging Technologies. *Environments (Basel, Switzerland), 5*(2), 2. Advance online publication. doi:10.3390/environments5020024

Meisen, P., & Loiseau, A. (2009). *Ocean energy technologies for renewable energy generation.* Global Energy Network Institute.

Melikoglu, M. (2017). Geothermal energy in Turkey and around the World: A review of the literature and an analysis based on Turkey's Vision 2023 energy targets. *Renewable & Sustainable Energy Reviews*, *76*, 485–492. doi:10.1016/j.rser.2017.03.082

Meneses-Jácome, A., Diaz-Chavez, R., Velásquez-Arredondo, H. I., Cárdenas-Chávez, D. L., Parra, R., & Ruiz-Colorado, A. A. (2016). Sustainable Energy from agro-industrial wastewaters in Latin-America. *Renewable & Sustainable Energy Reviews*, *56*, 1249–1262. doi:10.1016/j.rser.2015.12.036

Meng, X., Gu, A., Wu, X., Zhou, L., Zhou, J., Liu, B., & Mao, Z. (2021). Status quo of China hydrogen strategy in the field of transportation and international comparisons. *International Journal of Hydrogen Energy*, *46*(57), 28887–28899. doi:10.1016/j.ijhydene.2020.11.049

Meng, X., Zhang, Q., Hu, G., Zhang, G., Zhang, Y., Liu, F., & Liu, Z. (2023). Multi-vehicle accessed railway vehicle-grid system stability analysis and optimization based on OLTC. *Control Engineering Practice*, *141*, 105693. doi:10.1016/j.conengprac.2023.105693

Messalti, S., Harrag, A., & Loukriz, A. (2017). A new variable step size neural networks MPPT controller: Review, simulation and hardware implementation. *Renewable and Sustainable Energy Reviews, 68*(August), 221–233. doi:10.1016/j.rser.2016.09.131

Metro Rail News. (2023). *Sustainable & Green Mobility: Green Initiatives of Metro & Railways in India* https://www.metrorailnews.in/sustainable-green-mobility-green-initiatives-of-metro-railways-in-india/

Meziane, S., & Bentebbiche, A. (2019). Numerical study of blended fuel natural gas-hydrogen combustion in rich/quench/lean combustor of a micro gas turbine. *International Journal of Hydrogen Energy*, *44*(29), 15610–15621. doi:10.1016/j.ijhydene.2019.04.128

Millet, P., &Grigoriev, S. (2013). Water electrolysis technologies. *Renewable Hydrogen Technologies: Production, Purification, Storage, Applications and Safety, 19.*

Milosavljević, D., Kevkić, T., & Jovanović, S. (2022). Review and validation of photovoltaic solar simulation tools/software based on case study. *Open Physics*, *20*(1), 431–451. doi:10.1515/phys-2022-0042

Mio, A., Barbera, E., Pavan, A. M., Bertucco, A., & Fermeglia, M. (2024). Sustainability analysis of hydrogen production processes. *International Journal of Hydrogen Energy*, *54*, 540–553. doi:10.1016/j.ijhydene.2023.06.122

Möckl, M., Ernst, M. F., Kornherr, M., Allebrod, F., Bernt, M., Byrknes, J., Eickes, C., Gebauer, C., Moskovtseva, A., & Gasteiger, H. A. (2022). Durability Testing of Low-Iridium PEM Water Electrolysis Membrane Electrode Assemblies. *Journal of the Electrochemical Society*, *169*(6), 064505. doi:10.1149/1945-7111/ac6d14

Mohamed, N., Aymen, F., Alqarni, M., Turky, R. A., Alamri, B., Ali, Z. M., & Abdel Aleem, S. H. E. (2022). A new wireless charging system for electric vehicles using two receiver coils. *Ain Shams Engineering Journal*, *13*(2), 101569. Advance online publication. doi:10.1016/j.asej.2021.08.012

Mohammed, M., Salmiaton, A., Azlina, W. W., Amran, M. M., Fakhru'l-Razi, A., & Taufiq-Yap, Y. (2011). Hydrogen rich gas from oil palm biomass as a potential source of renewable energy in Malaysia. *Renewable & Sustainable Energy Reviews*, *15*(2), 1258–1270. doi:10.1016/j.rser.2010.10.003

Mohanakrishna, G., Srikanth, S., & Pant, D. (2016). Bioprocesses for waste and wastewater remediation for sustainable energy. In *Bioremediation and Bioeconomy* (pp. 537–565). Elsevier. doi:10.1016/B978-0-12-802830-8.00021-6

Mohanty, P., Muneer, T., Gago, E. J., & Kotak, Y. (2015). Solar radiation fundamentals and PV system components. In *Solar Photovoltaic System Applications: A Guidebook for Off-Grid Electrification* (pp. 7–47). Springer International Publishing.

Mojarro, N. (2020). *COVID-19 is a game-changer for renewable energy. Here's why.* Available at https://www.weforum.org/agenda/2020/06/COVID-19-is-a-game-changer-for-renewable-energy

Monrat, A. A., Schelen, O., & Andersson, K. (2019). A Survey of Blockchain From the Perspectives of Applications, Challenges, and Opportunities. *IEEE Access : Practical Innovations, Open Solutions*, *7*, 117134–117151. doi:10.1109/ACCESS.2019.2936094

Monteiro, E., & Brito, P. S. D. (2023). Hydrogen supply chain: Current status and prospects. *Energy Storage*, *5*(7), e466. doi:10.1002/est2.466

Moriarty, P., & Honnery, D. (2019). Prospects for hydrogen as a transport fuel. *International Journal of Hydrogen Energy*, *44*(31), 16029–16037. doi:10.1016/j.ijhydene.2019.04.278

Mori, D., & Hirose, K. (2009). Recent challenges of hydrogen storage technologies for fuel cell vehicles. *International Journal of Hydrogen Energy*, *34*(10), 4569–4574. doi:10.1016/j.ijhydene.2008.07.115

Morvaj, B., Evins, R., & Carmeliet, J. (2016). Optimization framework for distributed energy systems with integrated electrical grid constraints. *Applied Energy*, *171*, 296–313. doi:10.1016/j.apenergy.2016.03.090

Moudgil, V., Hewage, K., Hussain, S. A., & Sadiq, R. (2023). Integration of IoT in building energy infrastructure: A critical review on challenges and solutions. *Renewable & Sustainable Energy Reviews*, *174*, 113121. doi:10.1016/j.rser.2022.113121

Moulebe, L. P., Touati, A., Akpoviroro, E. O., & Rabbah, N. (2022). Study on performance of a green hydrogen production system integrated with the thermally activated cooling. *Acta Innovations*, *47*(47), 5–19. doi:10.32933/ActaInnovations.47.1

Mousavi, S., Marroquín, M. G. V., Hajiaghaei-Keshteli, M., & Smith, N. R. (2023). Data-driven prediction and optimization toward net-zero and positive-energy buildings: A systematic review. *Building and Environment*, *242*, 110578. doi:10.1016/j.buildenv.2023.110578

Müller, V. P., & Eichhammer, W. (2023). Economic complexity of green hydrogen production technologies—A trade data-based analysis of country-specific industrial preconditions. *Renewable & Sustainable Energy Reviews*, *182*, 113304. doi:10.1016/j.rser.2023.113304

Muradov, N. (2001). Hydrogen via methane decomposition: An application for decarbonization of fossil fuels. In International Journal of Hydrogen Energy (Vol. 26, Issue 11, pp. 1165–1175). doi:10.1016/S0360-3199(01)00073-8

Musti, K. S. S. (2020). Quantification of Demand Response in Smart Grids. *IEEE Int. Conf. INDISCON*, 278-282. 10.1109/INDISCON50162.2020.00063

Musti, K. S., & Kapali, D. (2021). Digital Transformation of SMEs in the Energy Sector to Survive in a Post-COVID-19 Era. In *Handbook of Research on Strategies and Interventions to Mitigate COVID-19 Impact on SMEs* (pp. 186–207). IGI Global. doi:10.4018/978-1-7998-7436-2.ch009

N, V. T. (1974). Dynamics of a Universal Hydrogen Fuel System. *Hydrogen Energy, 0*, 1309–1326.

Nagendra, G. R., Chen, L., Covic, G. A., & Boys, J. T. (2014). Detection of EVs on IPT highways. *Conference Proceedings - IEEE Applied Power Electronics Conference and Exposition - APEC*, 1604–1611. 10.1109/APEC.2014.6803521

Naik, O., Jaiswal, S., Bhuyar, K., & Kodape, S. (2023). Green Hydrogen Production as a Sustainable Initiative for Alternative Energy Source: A Review. *International Journal of Basic Sciences and Applied Computing*, *9*(10), 1–5. doi:10.35940/ijbsac.I0503.0691023

Nakomcic-Smaragdakis, B., Stajic, T., Cepic, Z., & Djuric, S. (2012). Geothermal energy potentials in the province of Vojvodina from the aspect of the direct energy utilization. *Renewable & Sustainable Energy Reviews*, *16*(8), 5696–5706. doi:10.1016/j.rser.2012.05.038

Nami, H., Rizvandi, O. B., Chatzichristodoulou, C., Hendriksen, P. V., & Frandsen, H. L. (2022). Techno-economic analysis of current and emerging electrolysis technologies for green hydrogen production. *Energy Conversion and Management*, *269*, 116162. doi:10.1016/j.enconman.2022.116162

Nature Research Custom Media & Kawasaki. (2024). *Hydrogen gas turbine offers promise of clean electricity*. Nature Portfolio. Retrieved 14.01.2024 from https://www.nature.com/articles/d42473-022-00211-0

Nazari, M. A., Alavi, M. F., Salem, M., & Assad, M. E. (2022). Utilization of hydrogen in gas turbines: A comprehensive review. *The International Journal of Low Carbon Technologies*, *17*, 513–519. doi:10.1093/ijlct/ctac025

Nematollahi, O., & Kim, K. C. (2017). A feasibility study of solar energy in South Korea. *Renewable & Sustainable Energy Reviews*, *77*, 566–579. doi:10.1016/j.rser.2017.03.132

Neuwirth, M., Fleiter, T., Manz, P., & Hofmann, R. (2022). The future potential hydrogen demand in energy-intensive industries-a site-specific approach applied to Germany. *Energy Conversion and Management*, *252*, 115052. doi:10.1016/j.enconman.2021.115052

News, E. T. N. (2023) *No looking back: Energy transition in Brazil and Latin America* https://etn.news/buzz/brazil-latin-america-energy-transition-2023

Ngo, N. T., Truong, N. S., Truong, T. T. H., Pham, A. D., & Huynh, N. T. (2023). Implementing a Web-based Optimized Artificial Intelligence (AI)System with Metaheuristic Optimization for Improving Building Energy Performance. *Journal of Asian Architecture and Building Engineering*.

Nguyen, B. H., Trovão, J. P. F., German, R., & Bouscayrol, A. (2020). Real-time energy management of parallel hybrid electric vehicles using linear quadratic regulation. *Energies*, *13*(21), 1–19. doi:10.3390/en13215538

Nicolin, B. A., & Nicolin, I. (2023). Green hydrogen as an environmentally-friendly power source. *INCAS BULLETIN*, *15*(2), 141–147. doi:10.13111/2066-8201.2023.15.2.13

Ning, F., Ji, L., Ma, J., Jia, L., & Yu, Z. (2021). Research and analysis of a flexible integrated developmentmodel of railway system and photovoltaic in China. *Renewable Energy*, *175*, 853–867. doi:10.1016/j.renene.2021.04.119

Njong, V. A., Musti, K. S., & van der Walt, J. P. (2022), Modeling and Simulation of Solar District Heating Systems. *IEEE 11th Int. Conf. on Communication Systems and Network Technologies (CSNT)*, 125-131. 10.1109/CSNT54456.2022.9787573

Norazahar, N., Khan, F., Rahmani, N., & Ahmad, A. (2023). Degradation Modelling and Reliability Analysis of PEM Electrolyzer. *International Journal of Hydrogen Energy*.

Nosratabadi, S. M., Moshizi, H. N., & Guerrero, J. M. (2022). Strategy for demand side management effectiveness assessment via a stochastic risk-based bidding approach in a multi-energy microgrid containing combined cooling, heat and power and photovoltaic units. *IET Renewable Power Generation*, *16*(10), 2036–2058. doi:10.1049/rpg2.12482

Nouicer, I., Menia, S., Kabouche, N., Ghenai, C., & Meziane, F. (2023). Production : Case Study in Constantine, Algeria. *2023 14th International Renewable Energy Congress (IREC)*, 1–4.

Nouicer, Yaiche, Menia, Meziane, & Kabouche. (2021). *Estimation of Hydrogen Production in Three Cities in the North of Algeria*. Academic Press.

Nouicer, I., Khellaf, A., Menia, S., Yaiche, M. R., Kabouche, N., & Meziane, F. (2019). Solar Hydrogen Production Using Direct Coupling of SO2 Depolarized Electrolyser to a Solar Photovoltaic System. *International Journal of Hydrogen Energy*, *44*(39), 22408–22418. doi:10.1016/j.ijhydene.2018.11.106

Nowak, D. J. (1994). Atmospheric carbon dioxide reduction by Chicago's urban forest. *Chicago's Urban Forest Ecosystem: Results of the Chicago Urban Forest Climate Project*, 83–94.

Ogden, J., Jaffe, A. M., Scheitrum, D., McDonald, Z., & Miller, M. (2018). Natural gas as a bridge to hydrogen transportation fuel: Insights from the literature. *Energy Policy*, *115*, 317–329. doi:10.1016/j.enpol.2017.12.049

Oh, S. (2013). *Origins of analysis methods in energy simulation programs used for high performance commercial buildings*. PhD Diss.

Oh, S., & Haberl, J. S. (2016). Origins of analysis methods used to design high-performance commercial buildings: Solar energy analysis. *Science and Technology for the Built Environment*, *22*(1), 87–106. doi:10.1080/23744731.2015.1090277

Ohunakin, O. S., Adaramola, M. S., Oyewola, O. M., & Fagbenle, R. O. (2014). Solar energy applications and development in Nigeria: Drivers and barriers. *Renewable & Sustainable Energy Reviews*, *32*, 294–301. doi:10.1016/j.rser.2014.01.014

Oh, Y. K., Hwang, K. R., Kim, C., Kim, J. R., & Lee, J. S. (2018). Recent developments and key barriers to advanced biofuels: A short review. *Bioresource Technology*, *257*, 320–333. doi:10.1016/j.biortech.2018.02.089 PMID:29523378

Olabi, A. G., Abdelkareem, M. A., Mahmoud, M. S., Elsaid, K., Obaideen, K., Rezk, H., Wilberforce, T., Eisa, T., Chae, K.-J., & Sayed, E. T. (2023). Green hydrogen: Pathways, roadmap, and role in achieving sustainable development goals. *Process Safety and Environmental Protection*, *177*, 664–687. doi:10.1016/j.psep.2023.06.069

Olabi, A. G., Wilberforce, T., Sayed, E. T., Elsaid, K., & Abdelkareem, M. A. (2020). Prospects of Fuel Cell Combined Heat and Power Systems. *Energies*, *13*(16), 4104. doi:10.3390/en13164104

Oliveira, A. M., Beswick, R. R., & Yan, Y. S. (2021). A green hydrogen economy for a renewable energy society. *Current Opinion in Chemical Engineering*, *33*, 100701. doi:10.1016/j.coche.2021.100701

Omer, A. M. (2008). Energy, environment, and sustainable development. *Renewable & Sustainable Energy Reviews*, *12*(9), 2265–2300. doi:10.1016/j.rser.2007.05.001

Omid, M. A., Şahin, M. E., & Cora, Ö. N. (2024). Challenges and Future Perspectives on Production, Storage Technologies, and Transportation of Hydrogen: A Review. *Energy Technology (Weinheim)*. Advance online publication. doi:10.1002/ente.202300997

Omkar, K., Srikanth, M. v., Swaroop, K. P., & Rama Rao, P. V. V. (2015). Performance evaluation of 50KWp rooftop solar PV plant. *2015 International Conference on Industrial Instrumentation and Control, ICIC 2015*, 761–765. 10.1109/IIC.2015.7150844

Ozcelep, Y., Bekdas, G., & Apak, S. (2023). Meeting the electricity demand for the heating of greenhouses with hydrogen: Solar photovoltaic-hydrogen-heat pump system application in Turkey. *International Journal of Hydrogen Energy*, *48*(7), 2510–2517. doi:10.1016/j.ijhydene.2022.10.125

Öztürk Aydın, A., & Bayrakçeken Yurtcan, A. (2023). Hydrogen Safety Issues: The Inherent Challenges of Hydrogen and Possible Precautions. In R. A. Felseghi (Ed.), *Hydrogen Fuel Cell Technology for Mobile Applications* (pp. 207–237). IGI Global. doi:10.4018/978-1-6684-6721-3.ch009

Paksoy, T., & Deveci, M. (Eds.). (2023). *Smart and sustainable operations and supply chain management in industry 4.0* (1st ed.). CRC Press. doi:10.1201/9781003180302

Palaniappan, M., Tirlangi, S., Mohamed, M. J. S., Moorthy, R. M. S., Valeti, S. V., & Boopathi, S. (2023). Fused Deposition Modelling of Polylactic Acid (PLA)-Based Polymer Composites. In Development, Properties, and Industrial Applications of 3D Printed Polymer Composites (pp. 66–85). IGI Global. doi:10.4018/978-1-6684-6009-2.ch005

Palani, G., Sengamalai, U., Vishnuram, P., & Nastasi, B. (2023). Challenges and Barriers of Wireless Charging Technologies for Electric Vehicles. *Energies*, *16*(5), 2138. Advance online publication. doi:10.3390/en16052138

Palit, T., Bari, A. B. M. M., & Karmaker, C. L. (2022). An integrated Principal Component Analysis and Interpretive Structural Modeling approach for electric vehicle adoption decisions in sustainable transportation systems. *Decision Analytics Journal*, *4*(August), 100119. doi:10.1016/j.dajour.2022.100119

Panda, D. K., & Das, S. (2021). Smart grid architecture model (SGAM) for control, optimization and data analytics of future power networks with more renewable energy. *Journal of Cleaner Production*, *301*, 126877. doi:10.1016/j.jclepro.2021.126877

Panic, I., Cuculic, A., & Celic, J. (2022). Color-Coded Hydrogen: Production and Storage in Maritime Sector. *Journal of Marine Science and Engineering*, *10*(12), 1995. doi:10.3390/jmse10121995

Pan, R., Liu, D., Yang, Y., & Yang, J. (2024). Network based impedance analysis of grid forming based MMC-HVDC with wind farm integration. *Electric Power Systems Research*, *229*, 110120. doi:10.1016/j.epsr.2024.110120

Pant, D., Singh, A., Van Bogaert, G., Irving Olsen, S., Singh Nigam, P., Diels, L., & Vanbroekhoven, K. (2012). Bioelectrochemical systems (BES) for sustainable energy production and product recovery from organic wastes and industrial wastewaters. *RSC Advances*, *2*(4), 1248–1263. doi:10.1039/C1RA00839K

Pant, D., Van Bogaert, G., Diels, L., & Vanbroekhoven, K. (2010). A review of the substrates used in microbial fuel cells (MFCs) for sustainable energy production. *Bioresource Technology*, *101*(6), 1533–1543. doi:10.1016/j.biortech.2009.10.017 PMID:19892549

Panwar, N. L., Kaushik, S. C., & Kothari, S. (2011). Role of renewable energy sources in environmental protection: A review. *Renewable & Sustainable Energy Reviews*, *15*(3), 1513–1524. doi:10.1016/j.rser.2010.11.037

Pan, Y. (2019). Theoretical discovery of high capacity hydrogen storage metal tetrahydrides. *International Journal of Hydrogen Energy*, *44*(33), 18153–18158. doi:10.1016/j.ijhydene.2019.05.145

Pan, Y., Zhu, M., Lv, Y., Yang, Y., Liang, Y., Yin, R., Yang, Y., Jia, X., Wang, X., Zeng, F., Huang, S., Hou, D., Xu, L., Yin, R., & Yuan, X. (2023). Building energy simulation and its application for building performance optimization: A review of methods, tools, and case studies. *Advances in Applied Energy*, *10*, 100135. doi:10.1016/j.adapen.2023.100135

Pasupathi, S., Gomez, J. C. C., Su, H., Reddy, H., Bujlo, P., & Sita, C. (2016). *Recent Advances in High Temperature PEM Fuel Cells*. Academic Press. doi:10.1016/C2015-0-06917-9

Patnaik, D., Pattanaik, A. K., Bagal, D. K., & Rath, A. (2023). Reducing CO2 emissions in the iron industry with green hydrogen. *International Journal of Hydrogen Energy*, *48*(61), 23449–23458. doi:10.1016/j.ijhydene.2023.03.099

Pechlivanis, C. (2023). Implementing IoT Technology in Practice: Monitoring the Supply Chain for Sustainable Operation. *WSEAS Transactions on Systems*, *22*, 349–359. doi:10.37394/23202.2023.22.38

Pimm, A. J., Cockerill, T. T., & Gale, W. F. (2023). Reducing industrial hydrogen demand through preheating with very high temperature heat pumps. *Applied Energy*, *347*, 121464. doi:10.1016/j.apenergy.2023.121464

Piraino, F., Genovese, M., Pagnotta, L., Caposciutti, M., Dei, L.F.N. & Fragiacomo, P., (2023). Integrated hydrogen and battery energy systems as emergency backup in electric trains. *Energy Conversion ana Management: X, 18*, 100382.

Pitra, G. M., & Musti, K. S. S. (2021). Duck Curve with Renewable Energies and Storage Technologies. *IEEE 13th Int. Conf. (CICN)*, 66-71. 10.1109/CICN51697.2021.9574671

Pivetta, D., Dall'Armi, C., Sandrin, P., Bogar, M., & Taccani, R. (2024). The role of hydrogen as enabler of industrial port area decarbonization. *Renewable & Sustainable Energy Reviews*, *189*, 113912. doi:10.1016/j.rser.2023.113912

Prakasam, C., & Saravanan, R. (2020). Environmental flow—a mitigation measure for impact of hydropower projects. In *Sustainable Civil Engineering Practices: Select Proceedings of ICSCEP 2019* (pp. 207-213). Springer Singapore.

Prakash, R., & Bhat, I. K. (2012). Life cycle greenhouse gas emissions estimation for small hydropower schemes in India. *Energy*, *44*(1), 498–508. doi:10.1016/j.energy.2012.05.052

Prasad, S., Singh, A., Korres, N. E., Rathore, D., Sevda, S., & Pant, D. (2020). Sustainable utilization of crop residues for energy generation: A life cycle assessment (LCA) perspective. *Bioresource Technology*, *303*, 122964. doi:10.1016/j.biortech.2020.122964 PMID:32061494

Prol, J. L., & Steininger, K. W. (2020). Photovoltaic self-consumption is now profitable in Spain: Effects of the new regulation on prosumers' internal rate of return. *Energy Policy*, *146*, 111793. doi:10.1016/j.enpol.2020.111793

Qamar, A., Iqbal, J., Saher, S., Shah, A. A., & Basit, A. (2022). Design of optimized energy system based on active energy-saving technologies in very low-energy smart buildings. *Transactions on Emerging Telecommunications Technologies*, *33*(2), e3691. doi:10.1002/ett.3691

Qi, H., Huang, X., & Sheeraz, M. (2023). Green financing for renewable energy development: Driving the attainment of zero-emission targets. *Renewable Energy*, *213*, 30–37. doi:10.1016/j.renene.2023.05.111

Rachoutis, E., & Koubogiannis, D. (2016). Energy Payback Time of a Rooftop Photovoltaic System in Greece. *IOP Conference Series. Materials Science and Engineering*, *161*(1), 012092. Advance online publication. doi:10.1088/1757-899X/161/1/012092

Radeef, A. Y., & Ismail, Z. Z. (2021). Bioelectrochemical treatment of actual carwash wastewater associated with sustainable energy generation in three-dimensional microbial fuel cell. *Bioelectrochemistry (Amsterdam, Netherlands)*, *142*, 107925. doi:10.1016/j.bioelechem.2021.107925 PMID:34392137

Rahman, M., Liu, H., Masri, M., Durazo-Cardenas, 1. & Starr, A. (2023). A railway track reconstruction method using robotic vision on a mobile manipulator: A proposed strategy. Computers in Industry, 148, 103900.

Rail Tech. (2023). *Will battery or hydrogen trains be the future? The vision of Siemens Mobility*. https://www.railtech.com/rolling-stock/2023/05/04/will-battery-or-hydrogen-trains- be-the-future-the-vision-of-siemens-mobility/?gdpr-accept

Raj, K., Lakhina, P., & Stranger, C. (2022). *Harnessing green hydrogen opportunities for deep decarbonisation in India*. NITI Aayog, RMI. Available: https://www. niti. gov. in/sites/default/files/2022-06/Harnessing_Green_Hydrogen_V21_DIGITAL_29062022. pdf

Rajput, P., Singh, Y. K., Tiwari, G. N., Sastry, O. S., Dubey, S., & Pandey, K. (2018). Life cycle assessment of the 3.2 kW cadmium telluride (CdTe) photovoltaic system in composite climate of India. *Solar Energy*, *159*, 415–422. doi:10.1016/j.solener.2017.10.087

Rajput, S. K., & Dheer, D. K. (2022). Integration of 100-kWp PV with low-voltage distribution power system in composite climate: Performance and energy metrics analysis. *International Journal of Ambient Energy*, *43*(1), 8176–8192. doi:10.1080/01430750.2022.2092775

Ratanasopitkul, P. (2018). Blockchain – Revolutionize Green Energy Management. *2018 International Conference and Utility Exhibition on Green Energy for Sustainable Development (ICUE)*, 1–6. 10.23919/ICUE-GESD.2018.8635666

Rathore, N. S., & Panwar, N. L. (2007). *Renewable energy sources for sustainable development.* New India Publishing. doi:10.59317/9789390083862

Ravindranath, N. H., & Hall, D. O. (1995). *Biomass, energy and environment: a developing country perspective from India.* Oxford University Press. doi:10.1093/oso/9780198564362.001.0001

Ray, B. e., Shah, R., Islam, M. R., & Islam, S. (2020). A new data driven long-term solar yield analysis model of photovoltaic power plants. *IEEE Access : Practical Innovations, Open Solutions*, *8*, 136223–136233. doi:10.1109/ACCESS.2020.3011982

Razeghi, M., Hajinezhad, A., Naseri, A., Noorollahi, Y., & Moosavian, S. F. (2023). An overview of renewable energy technologies for the simultaneous production of high-performance power and heat. *Future Energy*, *2*(2), 1–11. doi:10.55670/fpll.fuen.2.2.1

Rechberger, K., Spanlang, A., Sasiain Conde, A., Wolfmeir, H., & Harris, C. (2020). Green hydrogen-based direct reduction for low-carbon steelmaking. *Steel Research International, 91*(11), 2000110.

Reddy, M. A., Reddy, B. M., Mukund, C. S., Venneti, K., Preethi, D. M. D., & Boopathi, S. (2023). Social Health Protection During the COVID-Pandemic Using IoT. In *The COVID-19 Pandemic and the Digitalization of Diplomacy* (pp. 204–235). IGI Global. doi:10.4018/978-1-7998-8394-4.ch009

Reil, Baumann, Althaus, & Gebhard. (2013). SunPower Maxeon 3 Datasheet. *Conference Record of the IEEE Photovoltaic Specialists Conference,* 1537–41.

Renewable Energy Agency. (2021). *International* (Vol. 56). Estadísticas De Energía Renovable.

Ren, X., Dong, L., Xu, D., & Hu, B. (2020). Challenges towards hydrogen economy in China. *International Journal of Hydrogen Energy*, *45*(59), 34326–34345. doi:10.1016/j.ijhydene.2020.01.163

Richter, A. (2019). *The Top 10 Geothermal Countries 2018.* Academic Press.

Riego Martinez, J., Perez-Alonso, M., & Duque-Perez, O. (2020). Influence of the rai electrification system topology on the energy consumption of train trajectories. *IET Renewable Power Generation*, *14*(18), 3589–3598. doi:10.1049/iet-rpg.2020.0119

Risbud, M., Kaur, G., Dhawale, D. S., Zhu, H., Haque, N., & Giddey, S. (2023). Electrolyzer technologies for hydrogen economy. In *Hydrogen Economy* (pp. 459–485). Elsevier. doi:10.1016/B978-0-323-99514-6.00003-0

Rituraj, R., & Kadar, P. (2021). Performance Analysis of 50 KWp Rooftop Grid-Connected PV System. *SISY 2021 - IEEE 19th International Symposium on Intelligent Systems and Informatics, Proceedings*, 177–182, 10.1109/SISY52375.2021.9582513

Rosen, M. A., & Koohi-Fayegh, S. (2016). The prospects for hydrogen as an energy carrier: An overview of hydrogen energy and hydrogen energy systems. *Energy, Ecology & Environment*, *1*(1), 10–29. doi:10.1007/s40974-016-0005-z

Roy, S. K., Rajput, S. K., Prakash, R., & Dheer, D. K. (2022). Performance and Energy Metrics Analysis of Stand Alone Rooftop PV plant. *IEEE International Conf. on Emerging Frontiers in Electrical & Electronic Tech, ICEFEET.* 10.1109/ICEFEET51821.2022.9848352

S&P Global Market Intelligence. (2020). *As the market expands, renewable energy unemployment barely budges*. https://www.spglobal.com/marketintelligence/en/news-insights/latest-news-headlines/as-market-expands-renewable-energy-unemployment-barely-budges-60666681

Sadhukhan, J., Sen, S., Randriamahefasoa, T. M. S., & Gadkari, S. (2022). Energy system optimization for net-zero electricity. *Digital Chemical Engineering*, *3*, 100026. doi:10.1016/j.dche.2022.100026

Saha1, B. C., R, D., A, A., Thrinath, B. V. S., Boopathi, S., J. R., & Sudhakar, M. (2022). *IoT based smart energy meter for smart grid.* Academic Press.

Saheb-Koussa, D., Haddadi, M., & Belhamel, M. (2009). Economic and technical study of a hybrid system (wind–photovoltaic–diesel) for rural electrification in Algeria. *Applied Energy*, *86*(7-8), 1024–1030. doi:10.1016/j.apenergy.2008.10.015

Sahouane, N. e., Dabou, R., Ziane, A., Neçaibia, A., Bouraiou, A., Rouabhia, A., & Mohammed, B. (2019). Energy and economic efficiency performance assessment of a 28 kWp photovoltaic grid-connected system under desertic weather conditions in Algerian Sahara. *Renewable Energy*, *143*, 1318–1330. doi:10.1016/j.renene.2019.05.086

Salvi, B. L., & Subramanian, K. (2015). Sustainable development of road transportation sector using hydrogen energy system. *Renewable & Sustainable Energy Reviews*, *51*, 1132–1155. doi:10.1016/j.rser.2015.07.030

Samikannu, R., Koshariya, A. K., Poornima, E., Ramesh, S., Kumar, A., & Boopathi, S. (2023). Sustainable Development in Modern Aquaponics Cultivation Systems Using IoT Technologies. In *Human Agro-Energy Optimization for Business and Industry* (pp. 105–127). IGI Global. doi:10.4018/978-1-6684-4118-3.ch006

Sampath, B., Pandian, M., Deepa, D., & Subbiah, R. (2022). Operating parameters prediction of liquefied petroleum gas refrigerator using simulated annealing algorithm. *AIP Conference Proceedings*, *2460*(1), 70003. doi:10.1063/5.0095601

Sansaniwal, S., Rosen, M., & Tyagi, S. (2017). Global challenges in the sustainable development of biomass gasification: An overview. *Renewable & Sustainable Energy Reviews*, *80*, 23–43. doi:10.1016/j.rser.2017.05.215

Sarad, N. J., Alexander, M. J., & Donovan, L. P. (2020). COVID-19 Impact on US Renewable Energy Projects Article By. *The National Law Review*, *10*(363), 1–5.

Saray, M., Saray, M., Kazan, C., & Guner, S. (2024). Optimization of renewable energy usage in public transportation: Mathematical model for energy management of plug-in PV-based electric metrobuses. *Journal of Energy Storage*, *78*, 109946. doi:10.1016/j.est.2023.109946

Sasaki, K., Li, H.-W., Hayashi, A., Yamabe, J., Ogura, T., & Lyth, S. M. (2016). *Hydrogen Energy Engineering A Japanese Perspective*. Springer. doi:10.1007/978-4-431-56042-5

Sastry Musti, K. S. (2023a). Energy Sector Growth and Sustainability: Role of Government. In Leadership and Governance for Sustainability (pp. 186-203). IGI Global. doi:10.4018/978-1-6684-9711-1.ch010

Sastry Musti, K. S. (2023b). Multicriteria Decision Analysis for Sustainable Green Financing in Energy Sector. In *Green Finance Instruments, FinTech, and Investment Strategies. Sustainable Finance*. Springer. doi:10.1007/978-3-031-29031-2_1

Sastry Musti, K. S., & Van der Merwe, M. (2022). A Novel MS Excel Tool for Multi-Criteria Decision Analysis in Energy Systems. In *Optimal Planning of Smart Grid With Renewable Energy Resources* (pp. 83–109). IGI Global. doi:10.4018/978-1-6684-4012-4.ch003

Sastry, M. K. S., & Sahadeo, A. (2015). Distributed Cloud Computing Based GIS Solution for Electrical Power Utility Asset Management. *IEEE International Conf. on Computational Intelligence and Communication Networks (CICN)*, 832-838. 10.1109/CICN.2015.169

Sathish, T., Sunagar, P., Singh, V., Boopathi, S., Sathyamurthy, R., Al-Enizi, A. M., Pandit, B., Gupta, M., & Sehgal, S. S. (2023). Characteristics estimation of natural fibre reinforced plastic composites using deep multi-layer perceptron (MLP) technique. *Chemosphere*, *337*(June), 139346. doi:10.1016/j.chemosphere.2023.139346 PMID:37379988

Savian, F., Tomasin Da Silva, L., Sellitto, M., Sauer Oliveira, H. L., Neri, A., & Butturi, M. A. (2023). Green Hydrogen, Industrial Symbiosis, And Blockchain: Enhancing Sustainability And Resilience In Supply Chains. ECMS 2023 Proceedings Edited by Enrico Vicario, Romeo Bandinelli, Virginia Fani, Michele Mastroianni, 350–356. doi:10.7148/2023-0350

Savu, L. M., Amalia, M. A., Remus, S. E., Vasile, A. M., & Tabita, N. D. (2021). Possibilities to combine the hydrogen with heat pumps in heating systems of buildings. *Annals of 'Constantin Brancusi' University of Targu-Jiu. Engineering Series*, (4), 36–41.

Schiel, K., Baume, O., Caruso, G., & Leopold, U. (2016). GIS-based modelling of shallow geothermal energy potential for CO2 emission mitigation in urban areas. *Renewable Energy*, *86*, 1023–1036. doi:10.1016/j.renene.2015.09.017

Schmidhalter, I., Mussati, M. C., Mussati, S. F., Oliva, D. G., Fuentes, M., & Aguirre, P. A. (2024). Green hydrogen levelized cost assessment from wind energy in Argentina with dispatch constraints. *International Journal of Hydrogen Energy*, *53*, 1083–1096. doi:10.1016/j.ijhydene.2023.12.052

Schneider, S., Bajohr, S., Graf, F., & Kolb, T. (2020). State of the Art of Hydrogen Production via Pyrolysis of Natural Gas. In ChemBioEng Reviews (Vol. 7, Issue 5, pp. 150–158). doi:10.1002/cben.202000014

Segura, F., Vivas, F. J., Andujar, J. M., & Martinez, M. (2023). Hydrogen-powered refrigeration system for environmentally friendly transport and delivery in the food supply chain. *Applied Energy*, *338*, 120945. doi:10.1016/j.apenergy.2023.120945

Selvakumar, S., Adithe, S., Isaac, J. S., Pradhan, R., Venkatesh, V., & Sampath, B. (2023). A Study of the Printed Circuit Board (PCB) E-Waste Recycling Process. In Sustainable Approaches and Strategies for E-Waste Management and Utilization (pp. 159–184). IGI Global.

Shandiz, S. C., Rismanchi, B., & Foliente, G. (2021). Energy master planning for net-zero emission communities: State of the art and research challenges. *Renewable & Sustainable Energy Reviews*, *137*, 110600. doi:10.1016/j.rser.2020.110600

Sharifi, S., Ferencz, I. F., Kamel, T., Petreus, D., & Tricoli, P. (2022). Medium-voltage DC electric railway systems: A review on feeding arrangements and power converter topologies. *IET Electrical Systems in Transportation*, *12*(4), 223–237. doi:10.1049/els2.12054

Sharma, R., & Tiwari, G. N. (2013). Life cycle assessment of stand-alone photovoltaic (SAPV) system under on-field conditions of New Delhi, India. *Energy Policy*, *63*, 272–282. doi:10.1016/j.enpol.2013.08.081

Sharma, S., & Ghoshal, S. K. (2015). Hydrogen the future transportation fuel: From production to applications. *Renewable & Sustainable Energy Reviews*, *43*, 1151–1158. doi:10.1016/j.rser.2014.11.093

Shen, R., Han, Z., Liu, S., & Gao, S. (2021), October. Design and Application of Dynamic Test Platform for Railway System Protective Relays. *International Conference on Electrical and Information Technologies for Rail Transportation*, 66-73.

Sherif, S. A., Goswami, Y., Stefanakos, E., & Steinfeld, A. (2014). Handbook of Hydrogen Energy (S. A. Sherif, Y. Goswami, E. Stefanakos, & A. Steinfeld, Eds. 1 ed.). CRC Press.

Shinde, A. M., Dikshit, A. K., Singh, R. K., & Campana, P. E. (2018). Life cycle analysis based comprehensive environmental performance evaluation of Mumbai Suburban Railway, India. *Journal of Cleaner Production*, *188*, 989–1003. doi:10.1016/j.jclepro.2018.04.022

Shin, J., Hwang, W.-S., & Choi, H. (2019). Can hydrogen fuel vehicles be a sustainable alternative on vehicle market?: Comparison of electric and hydrogen fuel cell vehicles. *Technological Forecasting and Social Change, 143*, 239–248. doi:10.1016/j.techfore.2019.02.001

Siegel, M. D., Klein, S. A., & Beckman, W. A. (1981). A simplified method for estimating the monthly-average performance of photovoltaic systems. *Solar Energy, 26*(5), 413–418. doi:10.1016/0038-092X(81)90220-6

Sims, R. E. H. (2003). Bioenergy to mitigate for climate change and meet the needs of society, the economy and the environment. *Mitigation and Adaptation Strategies for Global Change, 8*(4), 349–370. doi:10.1023/B:MITI.0000005614.51405.ce

Sky. (2022). *The future of energy may lie with hydrogen, but the journey to get there won't be easy*. https://news.sky.com/story/the-future-of-energy-may-lie-with-hydrogen-but-the-journey-to-get-there-wont-be-easy-12732319

Sorenson, B. (2020). Renewable Power Generation Costs in 2019. *IRENA*. https://irena.org/publications/2020/Jun/Renewable-Power-Costs-in-2019

Souza Filho, I. R., Springer, H., Ma, Y., Mahajan, A., da Silva, C. C., Kulse, M., & Raabe, D. (2022). Green steel at its crossroads: Hybrid hydrogen-based reduction of iron ores. *Journal of Cleaner Production, 340*, 130805. doi:10.1016/j.jclepro.2022.130805

Staffell, I. (2015). Zero carbon infinite COP heat from fuel cell CHP. *Applied Energy, 147*, 373–385. doi:10.1016/j.apenergy.2015.02.089

Staffell, I., Scamman, D., Abad, A. V., Balcombe, P., Dodds, P. E., Ekins, P., Shah, N., & Ward, K. R. (2019). The role of hydrogen and fuel cells in the global energy system. *Energy & Environmental Science, 12*(2), 463–491. doi:10.1039/C8EE01157E

Streets blog. (2023). *California Needs Leadership on Electric Rail*. https://cal.streetsblog.org/2023/07/06/california-needs-leadership-on-electric-mil

Subha, S., Inbamalar, T. M., Komala, C. R., Suresh, L. R., Boopathi, S., & Alaskar, K. (2023, February). A Remote Health Care Monitoring system using internet of medical things (IoMT). *Proceedings of 2023 3rd International Conference on Innovative Practices in Technology and Management, ICIPTM 2023*. 10.1109/ICIPTM57143.2023.10118103

Su, S. B., He, C., Shu, Y., Chen, Q. L., & Zhang, B. J. (2023). Total site modeling and optimization for petrochemical low-carbon retrofits using multiple CO2 emission reduction methods. *Journal of Cleaner Production, 383*, 135450. doi:10.1016/j.jclepro.2022.135450

Swennenhuis, F., de Gooyert, V., & de Coninck, H. (2022). Towards a CO2-neutral steel industry: Justice aspects of CO2 capture and storage, biomass-and green hydrogen-based emission reductions. *Energy Research & Social Science, 88*, 102598. doi:10.1016/j.erss.2022.102598

Tariq, R., Cetina-Quiñones, A. J., Cardoso-Fernández, V., Daniela-Abigail, H. L., Soberanis, M. E., Bassam, A., & De Lille, M. V. (2021). Artificial intelligence assisted technoeconomic optimization scenarios of hybrid energy systems for water management of an isolated community. *Sustainable Energy Technologies and Assessments, 48*, 101561. doi:10.1016/j.seta.2021.101561

Team LightBox. (2020). *The Effects of COVID-19 on the Renewable Energy Industry, and How this Industry is Important to Economic Recovery*. https://www.digmap.com/blog/the-effects-of-COVID-19-on-the-renewable-energy-industry-and-how-this-industry-is-important-to-economic-recovery/

Tejashwini, P. S., Bharath Gowda, B., Shukla, A., Chiranth, J. S., & Harish, V. (2023). Trusted Crowdfunding Platform using Smart Contracts. *2023 World Conference on Communication & Computing (WCONF)*, 1–7. 10.1109/WCONF58270.2023.10234998

Temiz, M., & Dincer, I. (2022). Development of Solar and Wind Based Hydrogen Energy Systems for Sustainable Communities. *Energy Conversion and Management*, *269*(June), 116090. doi:10.1016/j.enconman.2022.116090

The Economist. (2020). *COVID-19 to send almost all G20 countries into a recession - Economist Intelligence Unit.* https://www.eiu.com/n/COVID-19-to-send-almost-all-g20-countries-into-a-recession/

Thirunavukkarasu, M., Sawle, Y., & Lala, H. (2023). A comprehensive review on optimization of hybrid renewable energy systems using various optimization techniques. *Renewable & Sustainable Energy Reviews*, *176*, 113192. doi:10.1016/j.rser.2023.113192

Toscano, G., Alfano, V., Scarfone, A., & Pari, L. (2018). Pelleting vineyard pruning at low cost with a mobile technology. *Energies*, *11*(9), 2477. doi:10.3390/en11092477

Toscano, G., Leoni, E., Gasperini, T., & Picchi, G. (2022). Performance of a portable NIR spectrometer for the determination of moisture content of industrial wood chips fuel. *Fuel*, *320*, 123948. doi:10.1016/j.fuel.2022.123948

Toscano, G., Pizzi, A., Pedretti, E. F., Rossini, G., Ciceri, G., Martignon, G., & Duca, D. (2015). Torrefaction of tomato industry residues. *Fuel*, *143*, 89–97. doi:10.1016/j.fuel.2014.11.039

Toscano, G., Riva, G., Pedretti, E. F., & Duca, D. (2013). Effect of the carbon oxidation state of biomass compounds on the relationship between GCV and carbon content. *Biomass and Bioenergy*, *48*, 231–238. doi:10.1016/j.biombioe.2012.11.002

Toure, I., Payman, A., Camara, M. B., & Dakyo, B. (2023). A Review on Electrolyser and Hydrogen production from wind Energy. *2023 11th International Conference on Smart Grid (icSmartGrid)*, 1–6. 10.1109/icSmartGrid58556.2023.10171040

Tripathy, M., Joshi, H., & Panda, S. K. (2017). Energy payback time and life-cycle cost analysis of building integrated photovoltaic thermal system influenced by adverse effect of shadow. *Applied Energy*, *208*, 376–389. doi:10.1016/j.apenergy.2017.10.025

U.S. Energy Information Administration. (2019). *Renewable energy explained: Portfolio standards.* https://www.eia.gov/energyexplained/renewable-sources/portfolio-standards.php

Ullah, K., Hafeez, G., Khan, I., Jan, S., & Javaid, N. (2021). A multi-objective energy optimization in smart grid with high penetration of renewable energy sources. *Applied Energy*, *299*, 117104. doi:10.1016/j.apenergy.2021.117104

Vallejos-Romero, A., Cordoves-Sánchez, M., Cisternas, C., Sáez-Ardura, F., Rodríguez, I., Aledo, A., Boso, Á., Prades, J., & Álvarez, B. (2022). Green Hydrogen and Social Sciences: Issues, Problems, and Future Challenges. *Sustainability (Basel)*, *15*(1), 303. doi:10.3390/su15010303

Van Lier, J. B., Tilche, A., Ahring, B. K., Macarie, H., Moletta, R., Dohanyos, M., Hulshoff Pol, L. W., Lens, P., & Verstraete, W. (2001). New perspectives in anaerobic digestion. *Water Science and Technology*, *43*(1), 1–18. doi:10.2166/wst.2001.0001 PMID:11379079

Vanitha, S. K. R., & Boopathi, S. (2023). Artificial Intelligence Techniques in Water Purification and Utilization. In *Human Agro-Energy Optimization for Business and Industry* (pp. 202–218). IGI Global. doi:10.4018/978-1-6684-4118-3.ch010

Vasudev, K. P., Mathew, M., Anand, A., & Hossain, J. (2018). Performance Analysis of a 48 kWp Grid connected Rooftop Photovoltaic System. *2018 4th International Conference for Convergence in Technology (I2CT)*. 10.1109/I2CT42659.2018.9058318

Velazquez Abad, A., & Dodds, P. E. (2020). Green hydrogen characterisation initiatives: Definitions, standards, guarantees of origin, and challenges. *Energy Policy*, *138*, 111300. doi:10.1016/j.enpol.2020.111300

Veneri, O. (2011). Hydrogen as Future Energy Carrier. In P. Corbo, F. Migliardini, & O. Veneri (Eds.), *Hydrogen Fuel Cells for Road Vehicles* (pp. 33–70). Springer. doi:10.1007/978-0-85729-136-3_2

Veziroglu, T. (1975c). Solar–hydrogen energy system and hydrogen production. Academic Press.

Veziroglu, T. (1975a). A universal hydrogen energy system and world parameters. *Proceedings of a Symposium on Introduction to Hydrogen Energy*.

Veziroglu, T. (1975b). Hydrogen Energy. *Hydrogen Energy*. Advance online publication. doi:10.1007/978-1-4684-2607-6

Villegas-Mier, C. G., Rodriguez-Resendiz, J., Álvarez-Alvarado, J. M., Rodriguez-Resendiz, H., Herrera-Navarro, A. M., & Rodríguez-Abreo, O. (2021). Artificial neural networks in mppt algorithms for optimization of photovoltaic power systems: A review. *Micromachines*, *12*(10), 1–19. doi:10.3390/mi12101260 PMID:34683311

Vögele, S. P.-J., Poganietz, W.-R., Kleinebrahm, M., Weimer-Jehle, W., Bernhard, J., Kuckshinrichs, W., & Weiss, A. (2022). Dissemination of PV-Battery systems in the German residential sector up to 2050: Technological diffusion from multidisciplinary perspectives. *Energy*, *248*, 123477. doi:10.1016/j.energy.2022.123477

Vyas, S., Ahmed, S., & Parashar, A. (2014). BEE (Bureau of energy efficiency) and Green Buildings. *International Journal of Research*, *1*(3), 23–32.

Waewsak, J., Khamharnphol, R., Khunpetch, S., Kamdar, I., Chiwamongkhonkarn, S., Kongruang, C., & Gagnon, Y. (2023). Techno-Economic Assessment of a 1 MW Solar PV Rooftop System at Thaksin University (Phatthalung Campus), Thailand. ASEAN Journal of Scientific and Technological Reports, 26(1), 59–75. doi:10.55164/ajstr.v26i1.248159

Wang, Y., Liu, Z., Wang, J., Du, B., Qin, Y., Liu, X. & Liu, L. (2023). A Stackelberg game-based approach to transaction optimization for distributed integrated energy system. Energy, 128475.

Wang, D., Almojil, S. F., Ahmed, A. N., Chaturvedi, R., & Almohana, A. I. (2023). An intelligent design and environmental consideration of a green-building system utilizing biomass and solar having a bidirectional interaction with the grid to achieve a sustainable future. *Sustainable Energy Technologies and Assessments*, *57*, 103287. doi:10.1016/j.seta.2023.103287

Wang, L., Jiao, S., Xie, Y., Xia, S., Zhang, D., Zhang, Y., & Li, M. (2022). Two-way dynamic pricing mechanism of hydrogen filling stations in electric-hydrogen coupling system enhanced by blockchain. *Energy*, *239*, 122194. doi:10.1016/j.energy.2021.122194

Wang, R., Lam, C. M., Hsu, S. C., & Chen, J. H. (2019). Life cycle assessment and energy payback time of a standalone hybrid renewable energy commercial microgrid: A case study of Town Island in Hong Kong. *Applied Energy*, *250*, 760–775. doi:10.1016/j.apenergy.2019.04.183

Wang, T., Cao, X., & Jiao, L. (2022). PEM Water Electrolysis for Hydrogen Production: Fundamentals, Advances, and Prospects. *Carbon Neutrality*, *1*(1), 1–19. doi:10.1007/s43979-022-00022-8

Wani, F., Dong, J., & Polinder, H. (2020). Tidal turbine generators. In *Advances in Modelling and Control of Wind and Hydrogenerators*. IntechOpen. doi:10.5772/intechopen.90433

Wannack, V. (2023). Blockchain Technology as a Catalyst for a Sustainable and Efficient Green Hydrogen Supply Chain. In K. Arai (Ed.), *Proceedings of the Future Technologies Conference (FTC) 2023, Volume 3* (pp. 367–372). Springer Nature Switzerland. 10.1007/978-3-031-47457-6_23

Wannack, V. (2022). Blockchain Based Hydrogen Market (BBH2)—A Paradigm-Shifting, Innovative Solution for a Climate-Friendly and Sustainable Structural Change. *Open Conference Proceedings, 2*, 59–63. 10.52825/ocp.v2i.147

Ward, A. J., Hobbs, P. J., Holliman, P. J., & Jones, D. L. (2008). Optimisation of the anaerobic digestion of agricultural resources. *Bioresource Technology*, *99*(17), 7928–7940. doi:10.1016/j.biortech.2008.02.044 PMID:18406612

Weger, L., Abánades, A., & Butler, T. (2017). Methane cracking as a bridge technology to the hydrogen economy. *International Journal of Hydrogen Energy, 42*(1), 720–731. Advance online publication. doi:10.1016/j.ijhydene.2016.11.029

Weiß, A., Siebel, A., Bernt, M., Shen, T. H., Tileli, V., & Gasteiger, H. A. (2019). Impact of Intermittent Operation on Lifetime and Performance of a PEM Water Electrolyzer. *Journal of the Electrochemical Society*, *166*(8), F487–F497. doi:10.1149/2.0421908jes

Whillier, A. (1953a). The Utilization of Solar Energy in South Africa. *Journal of South Africa Institute of Mechanical Engineers*, *2*(9), 260–266.

Whitaker, J., Field, J. L., Bernacchi, C. J., Cerri, C. E., Ceulemans, R., Davies, C. A., DeLucia, E. H., Donnison, I. S., McCalmont, J. P., Paustian, K., Rowe, R. L., Smith, P., Thornley, P., & McNamara, N. P. (2018). Consensus, uncertainties and challenges for perennial bioenergy crops and land use. *Global Change Biology. Bioenergy*, *10*(3), 150–164. doi:10.1111/gcbb.12488 PMID:29497458

White, T. A., Whitaker, B. N., & Brewer, K. J. (2011). Discovering the balance of steric and electronic factors needed to provide a new structural motif for photocatalytic hydrogen production from water. *Journal of the American Chemical Society*, *133*(39), 15332–15334. doi:10.1021/ja206782k PMID:21875106

Wind, G. (2017). *Sun onto the Grid. A Manual for Policy Makers*. Insight Series.

Winter, C. J. (2005). Into the hydrogen energy economy - milestones. *International Journal of Hydrogen Energy*, *30*(7), 681–685. doi:10.1016/j.ijhydene.2004.12.011

Wire, F. (2020). Oil and Coronavirus Shocks Add Pressure for MEA Sovereigns. *FitchRatings*. https://www.fitchratings.com/research/sovereigns/oil-coronavirus-shocks-add-pressure-for-mea-sovereigns-10-03-2020

World Economic Forum. (2021, July 27). *Grey, blue, green – why are there so many colours of hydrogen?* World Economic Forum. https://www.weforum.org/agenda/2021/07/clean-energy-green-hydrogen/

World Economy. (2020). *5 charts show what the global economy looks like heading into 2021*. https://www.cnbc.com/2020/12/28/5-charts-show-covid-impact-on-the-global-economy-in-2020.html

Worldometer. (2020). *COVID-19 coronavirus pandemic*. https://www.worldometers.info/coronavirus/

Wu, Y., Wu, Y., Guerrero, J. M., & Vasquez, J. C. (2022). Decentralized transactive energy community in edge grid with positive buildings and interactive electric vehicles. *International Journal of Electrical Power & Energy Systems*, *135*, 107510. doi:10.1016/j.ijepes.2021.107510

Xiao, Z. e., Huang, X., Liu, J., Li, C., & Tai, Y. (2023). A novel method based on time series ensemble model for hourly photovoltaic power prediction. *Energy*, *276*, 127542. doi:10.1016/j.energy.2023.127542

Xie, L., Shi, Y., Thomas Hou, Y., & Lou, W. (2013). W Ireless P Ower T Ransfer and a Pplications To S Ensor N Etworks. *IEEE Wireless Communications*, *20*(4), 140–145.

Xinyi, Y., Yi, Z., & He, Y. (2018). Technical Characteristics and Model of Blockchain. *2018 10th International Conference on Communication Software and Networks (ICCSN)*, 562–566. 10.1109/ICCSN.2018.8488289

Xu, X., Weber, I., & Staples, M. (2019). Varieties of Blockchains. In X. Xu, I. Weber, & M. Staples, Architecture for Blockchain Applications (pp. 45–59). Springer International Publishing. doi:10.1007/978-3-030-03035-3_3

Xu, B., Ge, Z., & He, Z. (2015). Sediment microbial fuel cells for wastewater treatment: Challenges and opportunities. *Environmental Science. Water Research & Technology*, *1*(3), 279–284. doi:10.1039/C5EW00020C

Xu, S., & Liu, J. (2019). Metal-based direct hydrogen generation as unconventional high density energy. *Frontiers in Energy*, *13*(1), 27–53. doi:10.1007/s11708-018-0603-x

Xu, Y., Yan, C., Liu, H., Wang, J., Yang, Z., & Jiang, Y. (2020). Smart energy systems: A critical review on design and operation optimization. *Sustainable Cities and Society*, *62*, 102369. doi:10.1016/j.scs.2020.102369

Yale Environment 360. (2020). https://e360.yale.edu/digest/despite-covid-driven-economic-slowdown-renewable-energy-to-hit-new-record-in-2020

Yang, J., Sudik, A., Wolverton, C., & Siegel, D. J. (2010). High capacity hydrogen storage materials: Attributes for automotive applications and techniques for materials discovery. *Chemical Society Reviews*, *39*(2), 656–675. doi:10.1039/B802882F PMID:20111786

Yang, M., Hunger, R., Berrettoni, S., Sprecher, B., & Wang, B. (2023). A review of hydrogen storage and transport technologies. *Clean Energy*, *7*(1), 190–216. doi:10.1093/ce/zkad021

Yang, N., Pouget, J., Letrouvé, T., Jecu, C., & Joseph-Auguste, L. (2019). Techno-economic design methodology of hybrid energy systems connected to electrical grid: An application of hybrid railway power substation. *Mathematics and Computers in Simulation*, *158*, 107–119. doi:10.1016/j.matcom.2018.06.012

Yao, P., & Zhu, Z. (2020). Industrial design and application of a railway electric special power supply. *CPSS Transactions on Power Electronics and Applications*, *5*(4), 317–328. doi:10.24295/CPSSTPEA.2020.00026

Yaqoob, A. A., Ibrahim, M. N. M., Umar, K., Parveen, T., Ahmad, A., Lokhat, D., & Setapar, S. H. M. (2021). A glimpse into the microbial fuel cells for wastewater treatment with energy generation. *Desalination and Water Treatment*, *214*, 379–389. doi:10.5004/dwt.2021.26737

Yardeni, E., Johnson, D., & Quintana, M. (2020). Global Economic Briefing: Industrial Production. *Global Economic Briefing: Industrial Production*. www.yardeni.com

Yenugula, M., Sahoo, S. K., & Goswami, S. S. (2023). Cloud computing in supply chain management: Exploring the relationship. *Management Science Letters*, *13*(3), 193–210. doi:10.5267/j.msl.2023.4.003

Yilmaz, M., & Krein, P. T. (2013). Review of battery charger topologies, charging power levels, and infrastructure for plug-in electric and hybrid vehicles. *IEEE Transactions on Power Electronics*, *28*(5), 2151–2169. doi:10.1109/TPEL.2012.2212917

Yingcheng, X., & Nengling, T. (2012). System frequency regulation in doubly fed induction generators. *International Journal of Electrical Power & Energy Systems*, *43*(1), 977–983. doi:10.1016/j.ijepes.2012.05.039

Yoshimura, T., McDonell, V., & Samuelsen, S. (2005). Evaluation of hydrogen addition to natural gas on the stability and emissions behavior of a model gas turbine combustor. *Proceedings of the ASME Turbo Expo 2005*, 2, 573-581. 10.1115/GT2005-68785

Youm, I., Sarr, J., Sall, M., & Kane, M. M. (2000). Renewable energy activities in Senegal: A review. *Renewable & Sustainable Energy Reviews*, *4*(1), 75–89. doi:10.1016/S1364-0321(99)00009-X

Yu, M. L., Wang, K., & Vredenburg, H. (2021). Insights into low-carbon hydrogen production methods: Green, blue and aqua hydrogen. *International Journal of Hydrogen Energy*, *46*(41), 21261–21273. doi:10.1016/j.ijhydene.2021.04.016

Yupapin, P., Trabelsi, Y., Nattappan, A., & Boopathi, S. (2023). Performance Improvement of Wire-Cut Electrical Discharge Machining Process Using Cryogenically Treated Super-Conductive State of Monel-K500 Alloy. *Iranian Journal of Science and Technology. Transaction of Mechanical Engineering*, *47*(1), 267–283. doi:10.1007/s40997-022-00513-0

Yu, S., Fan, Y., Shi, Z., Li, J., Zhao, X., Zhang, T., & Chang, Z. (2023). Hydrogen-based combined heat and power systems: A review of technologies and challenges. *International Journal of Hydrogen Energy*, *48*(89), 34906–34929. Advance online publication. doi:10.1016/j.ijhydene.2023.05.187

Yusuf, A., & Ballikaya, S. (2023). Performance analysis of concentrated photovoltaic syste using thermoelectric module with phase change material. *Journal of Energy Storage*, *59*, 106544. doi:10.1016/j.est.2022.106544

Zainal, B. S., Ker, P. J., Mohamed, H., Ong, H. C., Fattah, I. M. R., Rahman, S. A., Nghiem, L. D., & Mahlia, T. I. (2024). Recent advancement and assessment of green hydrogen production technologies. *Renewable & Sustainable Energy Reviews*, *189*, 113941. doi:10.1016/j.rser.2023.113941

Zakhidov, R. A. (2008). Central Asian countries energy system and role of renewable energy sources. *Applied Solar Energy*, *44*(3), 218–223. doi:10.3103/S0003701X08030201

Zgonnik, V. (2020). The occurrence and geoscience of natural hydrogen: A comprehensive review. *Earth-Science Reviews*, *203*, 103140. doi:10.1016/j.earscirev.2020.103140

Zhang, J. Z., Li, J., Li, Y., & Zhao, Y. (2014). Introduction to Basic Properties of Hydrogen. In *Hydrogen Generation, Storage and Utilization* (pp. 1–3). Wiley. doi:10.1002/9781118875193.ch1

Zhang, J., Wu, J., Zou, X., Hackenberg, K., Zhou, W., Chen, W., Yuan, J., Keyshar, K., Gupta, G., Mohite, A., Ajayan, P. M., & Lou, J. (2019). Discovering superior basal plane active two-dimensional catalysts for hydrogen evolution. *Materials Today*, *25*, 28–34. doi:10.1016/j.mattod.2019.02.014

Zhang, W., Yan, C., Xu, Y., Fang, J., & Pan, Y. (2022). A critical review of the performance evaluation and optimization of grid interactions between zero-energy buildings and power grids. *Sustainable Cities and Society*, *86*, 104123. doi:10.1016/j.scs.2022.104123

Zhang, Y., Zhu, N., Zhao, X., Luo, Z., Hu, P., & Lei, F. (2023). Energy performance and enviroeconomic analysis of a novel PV-MCHP-TEG system. *Energy*, *274*, 127342. Advance online publication. doi:10.1016/j.energy.2023.127342

Zhao, C., Zu, B., Xu, Y., Wang, Z., Zhou, J., & Liu, L. (2020). Design and analysis of an engine-start control strategy for a single-shaft parallel hybrid electric vehicle. *Energy*, *202*, 117621. doi:10.1016/j.energy.2020.117621

Zhao, G.-J., & Han, K.-L. (2012). Hydrogen Bonding in the Electronic Excited State. *Accounts of Chemical Research*, *45*(3), 404–413. doi:10.1021/ar200135h PMID:22070387

Zhao, W., Yan, P., Li, B., Bahri, M., Liu, L., Zhou, X., Clowes, R., Browning, N. D., Wu, Y., Ward, J. W., & Cooper, A. I. (2022). Accelerated Synthesis and Discovery of Covalent Organic Framework Photocatalysts for Hydrogen Peroxide Production. *Journal of the American Chemical Society*, *144*(22), 9902–9909. doi:10.1021/jacs.2c02666 PMID:35635501

Zheng, Z., LixiongWang, & HienWong, N. (2016). Intelligent control system integration and optimization for zero energy buildings (ZEBs)to mitigate urban heat island (UHI). *Procedia Engineering*, *169*, 100–107. doi:10.1016/j.proeng.2016.10.012

Zheng, Z., Xie, S., Dai, H. N., Chen, X., & Wang, H. (2018). Blockchain challenges and opportunities: A survey. *International Journal of Web and Grid Services*, *14*(4), 352. doi:10.1504/IJWGS.2018.095647

Zheng, Z., Xie, S., Dai, H., Chen, X., & Wang, H. (2017). An Overview of Blockchain Technology: Architecture, Consensus, and Future Trends. *2017 IEEE International Congress on Big Data (BigData Congress)*, 557–564. 10.1109/BigDataCongress.2017.85

Zhou, L. (2005). Progress and problems in hydrogen storage methods. *Renewable & Sustainable Energy Reviews*, *9*(4), 395–408. doi:10.1016/j.rser.2004.05.005

Zhu, C., Lu, J., & Li, X. (2023). Review of Studies on Energy-Efficient Train Operation in High- Speed Railways. *IEEJ Transactions on Electrical and Electronic Engineering*, *18*(3), 451–462. doi:10.1002/tee.23741

Zhu, F., Yang, Z., Lin, F., & Xin, Y. (2020). Decentralized cooperative control of multiple energy storage systems in urban railway based on multiagent deep reinforcement learning. *IEEE Transactions on Power Electronics*, *35*(9), 9368–9379. doi:10.1109/TPEL.2020.2971637

Zhu, Q., Johal, J., Widdowson, D. E., Pang, Z., Li, B., Kane, C. M., Kurlin, V., Day, G. M., Little, M. A., & Cooper, A. I. (2022). Analogy Powered by Prediction and Structural Invariants: Computationally Led Discovery of a Mesoporous Hydrogen-Bonded Organic Cage Crystal. *Journal of the American Chemical Society*, *144*(22), 9893–9901. doi:10.1021/jacs.2c02653 PMID:35634799

Zhu, Q., Leibowicz, B. D., Busby, J. W., Shidore, S., Adelman, D. E., & Olmstead, S. M. (2022). Enhancing policy realism in energy system optimization models: Politically feasible decarbonization pathways for the United States. *Energy Policy*, *161*, 112754. doi:10.1016/j.enpol.2021.112754

Zhu, T., Qi, M., Yin, M., Dang, J., Zhang, X., & Zhao, D. (2023). On the Hazards of Proton Exchange Membrane (PEM) Water Electrolysis Systems Using System Theoretic Process Analysis (STPA). *Process Safety and Environmental Protection*, *180*(November), 1118–1131. doi:10.1016/j.psep.2023.10.062

Zier, M., Stenzel, P., Kotzur, L., & Stolten, D. (2021). A review of decarbonization options for the glass industry. *Energy Conversion and Management-X, 10*. https://doi.org/ARTN 10008310.1016/j.ecmx.2021.100083

Zimmer, M., & Hoffmann, P. (2023). *The Green Industrial Revolution–Investment Pathways to Decarbonize the Industrial Sector in Europe*. Academic Press.

Zun, M. T., Ahmad, M. S., Fayaz, H., Selvaraj, J., Ahmed, W., Wang, Y. R., Ben Khedher, N., Silitonga, A. S., Elfasakhany, A., Kalam, M. A., & Rashid, B. (2023). Towards techno-economics of green hydrogen as a primary combustion fuel for recreational vehicle vapor absorption refrigeration system. *Sustainable Energy Technologies and Assessments*, *56*. https://doi.org/ARTN 10300710.1016/j.seta.2022.103007

Zuo, J., Dong, L., Yang, F., Guo, Z., Wang, T., & Zuo, L. (2022). Energy harvesting solutions for railway transportation: A comprehensive review. *Renewable Energy*.

Züttel, A., Borgschulte, A., Schlapbach, L., Chorkendorff, I., & Suda, S. (2008). Properties of Hydrogen. In A. Züttel, A. Borgschulte, & L. Schlapbach (Eds.), *Hydrogen as a Future Energy Carrier* (pp. 71–147). Wiley. doi:10.1002/9783527622894.ch4

About the Contributors

Maria Simona Raboaca is working as a Researcher at to National Research and Development Institute for Cryogenics and Isotopic Technologies ICSI Rm. Valcea, Hydrogen and Fuel Cell Department. Her Ph.D. is "Theoretical and practical Contribution regarding to sustain with hybrid energy a Passive House" in Faculty of Building Services Engineering in Technical University of Cluj-Napoca, Romania. Now, she is a project manager at ICSI to project "Smart conductive charging station, fixed and mobile, for electric propulsion transportation (SMiLE-EV)" proposes the deployment of fixed and mobile EV & PHEV charging stations to meet the mobility needs of tomorrow's society and to prepare active/potential industrial partners for knowledge/technology transfer at the component or system level in prepare launching new products. She has been contributing to the field of Renewable Energy, Green Buildings, Passive House Concept, Hydrogen Energy and Stationary and Mobile applications. She is the author and co-author of more technical papers in scientific conference proceedings and ISI journals.

Omar Moussaoui received his PhD in Computer Science at the University of Cergy-Pontoise France in 2006. He is an Associate Professor at the Higher School of Technology (ESTO) of the University Mohammed Premier, Oujda – Morocco. He has been a member of the Computer Science Department of ESTO since 2013. He is currently director of the MATSI research laboratory. His research interests lie in the fields of IoT, AI, Wireless Networks and Cybersecurity. He has actively collaborated with researchers in several other computer science disciplines. He participated in several scientific & organizing committees of national and international conferences. He served as reviewer for numerous international journals. He has more than 40 publications in international journals and conferences and several co-authored book chapters, and he has h-index 13 (Scopus). Omar is an instructor for CISCO Networking Academy on CCNA Routing & Switching and CCNA Security.

* * *

Jtn Anand is currently pursuing Bachelor of Technology from Asansol Engineering College in Electrical Engineering. My area of interest is the renewable computing soft computing.

Ashok Kumar Asthana is an alumni and a gold medallist from Jiwaji University, Gwalior. He is a highly motivated, organized and a versatile person with a passion for inculcating innovative methods in teaching. He has done his PGDBM (Approved by AICTE and Eqv. to MBA by AIU) in Marketing and International Business from NDIM and MBA in HRM form Alagappa University. He has qualified UGC-NET in management. He has over 13 years of rich teaching experience. He has done his doctoral research in the area of OB and HR from ICFAI University, Ranchi. His topic of research was related to Attitude and Performance of teachers in higher educational institutions. He is currently working as Assistant professor in Faculty of Commerce and Management, Sarala Birla University, Ranchi, Jharkhand and also holding an additional responsibility of "Dean Student`s Welfare". Earlier he was associated with New Delhi Institute of Management (Affiliated to GGSIP University), New Delhi. To his credit there are several national, international publications in journal of repute. He has also presented many papers in National and international seminar and conferences. His research areas of interest are Organization Behaviour and HR. He has received many awards and recognition including "Best Teacher Award", "Chairman`s Appreciation Award for Outstanding Performance", "Researcher of the Year Award" etc. He has also participated in FDP`S and MDP`s. He is a life time member of TERA. He has been actively mentoring Undergraduates and as well as Post Graduates students so that they can build their career.

Bessem Kaghouche obtained his engineering degree, magister degree and doctorate in electronics from the Constantine University of Mentouri Brothers Constantine 1, Algeria in 2006, 2010 and 2014 respectively.

Gmze Bozkurt received her BS degree in Chemical Engineering from Atatürk University Turkey in 2010. First MSc degree in Nanoscience and Nanoengineering from Atatürk University Turkey in 2013. She completed her second master's degree in Chemical Engineering (Atatürk University) and received her degree in 2017. In addition, she started her doctorate education in the field of Nanoscience and Nanoengineering in 2013 and completed it in 2018. She started working as a lecturer at Erzurum Technical University in Turkey 2018. She was at Gazi University in Turkey between August 2022 and August 2023 to do postdoctoral studies. She currently continues as lecturer at Erzurum Technical University.

Eif Daş is an Assistant Professor in the Physics Department at Atatürk University. She received her MS and PhD degrees in Nanoscience and Nanoengineering at the same university in 2014 and 2018, respectively. Her research interests include the synthesis and characterization of nanostructured materials for various applications such as PEM fuel cells, supercapacitors, and optoelectronic devices.

Pasad G. is currently working as faculty in the department of Aerospace Engineering Chandigarh University, India. He has 9 years of experience in Teaching and research. He received his Bachelor of Engineering and Master of Engineering degrees with first class and distinction and Doctor of Philosophy in the field of Aeronautical Engineering from the Anna University, Chennai, India. Prof. Prasad G is the author of over 30 technical publications, 1 Patent Published and 1 Book published. His research interests include Unmanned Aerial Vehicles, Aircraft Structures, Computational Fluid Dynamics and Interdisciplinary research. He has completed three funded project sponsored by The Institution of Engineers (India) and Tamilnadu State Council for Science and Technology. Awarded Indian National Science Academy (INSA) Visiting Scientist Programme 2019 and Awarded Science Academies' Summer Research Fellowship Programme (SRPF) 2019. Further he is SPOC of NPTEL- IIT Madras, SPOC of ISRO IIRS outreach coordinator and SPOC for NASA International SpaceApps Challenge. Further key positions such as coordinator of Board of Studies, Outcome Based Education, Project Coordinator, Research Coordinator NBA, NAAC, ABET. He is a Professional Membership of American Institute of Aeronautics and Astronautics, Institution of Engineers, Life Member in Indian Cryogenic Council and Life Member in Shock Wave Society. Reviewer in Aircraft Engineering and Aerospace Technology, International Journal of Engine Research and Journal of Institution of Engineers (India): Series C.

Chaouki Ghenai received the Ph.D. degree in mechanical engineering from the Orleans University, Orleans, France, in 1995. He is a Professor in the Department of Sustainable and Renewable Energy Engineering, and a Coordinator of the Sustainable Energy Development Research Group, Research Institute of Science and Engineering, University of Sharjah, Sharjah, UAE. His research interests include renewable energy, energy efficiency, sustainability, modeling, and simulation of microgrid power systems.

Suibu Hassan is a Principal Lecturer at Isa Kaita College of Education in the Department of Home Economics. An Alumni of the University of Bedfordshire, Usmanu Ɗan Fodio University, Kaduna Polytechnic and Katsina Polytechnic. Coordinator for Students' Industrial Work Experience Scheme (SIWES), Department Seminar Coordinator, Deputy Director Educational Supports Services, and Assistant Representative Katsina State Nutrition services. A Fellow at the Nigeria Institute of Management Consultants, member of the National Association of Vocational & Technical Educators, certified by Nigeria Teachers Registration Council.

Uomasa Kasenina is presently with the Department of Electrical and Computer Engineering, Namibia University of Science and Technology, Windhoek, Namibia. He specializes in Power Engineering. His research interests include renewable energy systems, solar power engineering and power systems.

Snjeev Kumar is an accomplished expert in Food and Beverage. He currently holds the positions of Professor at the Lovely Professional University, Punjab, India. With over a decade of experience in the field, food Service Industry, his research focuses on Alcoholic beverages, Event management and Sustainable Management Practices, Metaverse and Artificial Intelligence. He has published more than 35 research papers, articles and chapters in Scopus Indexed, UGC Approved and peer reviewed Journals and books. Dr. Sanjeev Kumar participated and acted as resource person in various National and International conferences, seminars, research workshops and industry talks and his work has been widely cited.

Fares Meziane, a skilled mechanical engineer specializing in advanced energy systems, holds a PhD degree in Mechanical Engineering from the University of Boumerdes, Algeria (2021). He achieved a Magister's degree in partnership with École des Mines de Nantes, France (2011), along with a specialized diploma in Industrial Engineering from the same institution. With a Bachelor's degree in Energy Engineering from the University of Blida, Algeria (2009), He has excelled in his academic journey. Since 2014, he has been a devoted researcher at Algiers' Center for Renewable Energy Development (CDER), contributing significantly to the field. His research interests encompass a wide spectrum of areas, including wind potential assessment, offshore energy, hydrogen production, water electrolysis, fuel cells, wind mapping, and wind atlas development.

K. S. Sastry Musti is presently with Namibia University of Science and Technology, Windhoek, Namibia. He has than 30 years of professional experience. He obtained his BTech in Electrical Engineering from JNTU, Kakinada (A.P.) in 1990, MTech and PhD from NIT, Warangal in the years 1996 and 2002 respectively. Some of his earlier employments include University of the West Indies, St Augustine, Trinidad, ESIGELEC, Rouen, France and Sophia University, Tokyo, Japan. His research interests include Power and Energy Systems, Information Systems and Engineering Education.

Aşnur Öztürk Aydın holds a bachelor's degree from Hacettepe University (Ankara, Turkey), Chemical Engineering in 2013 and Atatürk University (Erzurum, Turkey), Chemistry Departments in 2022. She also received her master's degree in 2016 and her doctorate degree in 2022 from the Department of Chemical Engineering at Ataturk University. She has worked as a research assistant in the same department since 2014. She continues her studies on developing different nanomaterials for the gas diffusion layer and catalyst layer of PEM fuel cells.

Sabyasachi Pramanik is a professional IEEE member. He obtained a PhD in Computer Science and Engineering from Sri Satya Sai University of Technology and Medical Sciences, Bhopal, India. Presently, he is an Associate Professor, Department of Computer Science and Engineering, Haldia Institute of Technology, India. He has many publications in various reputed international conferences, journals, and book chapters (Indexed by SCIE, Scopus, ESCI, etc). He is doing research in the fields of Artificial Intelligence, Data Privacy, Cybersecurity, Network Security, and Machine Learning. He also serves on the editorial boards of several international journals. He is a reviewer of journal articles from IEEE, Springer, Elsevier, Inderscience, IET and IGI Global. He has reviewed many conference papers, has been a keynote speaker, session chair, and technical program committee member at many international conferences. He has authored a book on Wireless Sensor Network. He has edited 8 books from IGI Global, CRC Press, Springer and Wiley Publications.

Dharmbir Prasad was born in 1986 at Nalanda, Bihar, India. He received his B.Tech. degree in electrical engineering from Hooghly Engineering & Technology College (under West Bengal University of Technology), Hooghly, India and M.Tech. in Power System from Dr. B.C. Roy Engineering College (under West Bengal University of Technology), Durgapur, India, respectively. Currently, he is working in the capacity of assistant professor in the department of electrical engineering, Asansol Engineering College, Asansol, West Bengal, India. He has completed Ph.D. degree from Indian School of Mines, Dhanbad, Jharkhand, India. His research interest includes economic operation of power system.

Rda Pratap Singh was born in 1983 at Chittaranjan, Burdwan, West Bengal, India. He received his B. Tech. and M. Tech degree in electrical engineering from Asansol Engineering (under the West Bengal University of Technology), Asansol, West Bengal, India and MBA in Power Management from the University of Petroleum and Energy Studies, Dehradun, Uttrakhand, India, respectively. He received his Ph.D. degree from the Indian School of Mines, Dhanbad, Jharkhand, India. Currently, he is working in the capacity of assistant professor in the department of electrical engineering, Asansol Engineering College, Asansol, West Bengal, India. His research interest includes the application of state estimation and optimization techniques in various fields of engineering.

Swati Srivastava is a Faculty in the Department of Management at K R Mangalam University. Her interest includes research in social sciences.

Index

S

T

U

V

W

Z

www.ingramcontent.com/pod-product-compliance
Lightning Source LLC
LaVergne TN
LVHW060304011026
849161LV00011B/2769